Handbook of Medical Engineering

Handbook of Medical Engineering

Edited by **Morris McMahon**

CLANRYE
INTERNATIONAL

New Jersey

Published by Clanrye International,
55 Van Reypen Street,
Jersey City, NJ 07306, USA
www.clanryeinternational.com

Handbook of Medical Engineering
Edited by Morris McMahon

International Standard Book Number: 978-1-63240-278-3 (Hardback)

Printed in the United States of America.

Contents

Preface

Medical engineering is a field which studies the basic and advanced applications of engineering principles and design concepts of biology and medicine science for healthcare issues and diagnostic purposes. This field brings together the two completely diverse branches of engineering and medicine science together. It combines design, problem detection and solving skills of different engineering fields with medical sciences to introduce advancements in healthcare treatment which includes diagnosis, therapy and monitoring. This subject can be viewed and studied from two completely different perspectives, the engineering and the medical applications angle. Ultimately, both these branches are combined together and used for revolutionizing medical sciences.

Medical engineering can be approached through different area of studies according to their applications and the purpose they serve. Some of the areas are tissue engineering, which is used for creating artificial organs and artificially producing tissues for organ transplant and healing damaged tissues; neural engineering, which utilises engineering skills to understand, enhance, repair, or replace neural systems; genetic engineering, a technique used to manipulate the organism's genes indirectly; pharmaceutical engineering, which is related to chemical engineering and mainly deals with the engineering of drugs, Medical devices, whose applications range from medical devices, implants, bionics, dental implants, pacemaker, etc. and clinical engineering, which is the implementation of medical engineering in clinics and hospital settings.

This book is a collection of some very interesting contributions from internationally renowned authors. I wish to thank all the contributors for their efforts, patience and timely submission. I must also acknowledge the team at the publishing house, who have done a tremendous job with the data collection. Last but not the least, I wish to thank my family and friends, who have supported me in my life through everything. I invite students, researchers, teachers and people interested in medical engineering issues, to read this book and enlighten themselves on the recent advancements and developments in this field.

Editor

Hybrid Discrete Wavelet Transform and Gabor Filter Banks Processing for Features Extraction from Biomedical Images

Salim Lahmiri and Mounir Boukadoum

Department of Computer Science, University of Quebec at Montreal, 201 President-Kennedy, Local PK-4150, Montreal, QC, Canada H2X 3Y7

Correspondence should be addressed to Salim Lahmiri; lahmiri.salim@courrier.uqam.ca

Academic Editor: Ying Zhuge

A new methodology for automatic feature extraction from biomedical images and subsequent classification is presented. The approach exploits the spatial orientation of high-frequency textural features of the processed image as determined by a two-step process. First, the two-dimensional discrete wavelet transform (DWT) is applied to obtain the HH high-frequency subband image. Then, a Gabor filter bank is applied to the latter at different frequencies and spatial orientations to obtain new Gabor-filtered image whose entropy and uniformity are computed. Finally, the obtained statistics are fed to a support vector machine (SVM) binary classifier. The approach was validated on mammograms, retina, and brain magnetic resonance (MR) images. The obtained classification accuracies show better performance in comparison to common approaches that use only the DWT or Gabor filter banks for feature extraction.

1. Introduction

Computer-aided diagnosis (CAD) has been the subject of a lot of research as a tool to help health professionals in medical decision making. As a result, many CAD systems integrate image processing, computer vision, and intelligent and statistical machine learning methods to aid radiologists in the interpretation of medical images and ultimately help improve diagnostic accuracy. These systems have been employed to analyze and classify various types of digitized biomedical images, including retina [1, 2], mammograms [3–5], brain magnetic resonance images [6–8], skin cancer images [9, 10], lung images [11, 12], and ulcer detection in endoscopy images [13, 14], just to name a few.

The typical CAD process starts with a segmentation stage to identify one or more regions of interest (ROI) in the image of interest. Then, the ROI(s) is processed for image enhancement and/or feature extraction before classification. Because the segmentation step requires prior knowledge of discriminant image features and its implementation typically calls for numerous parameter settings, recent works have attempted to eliminate it. These approaches realize feature space reduction by applying one or more transforms to the whole image and extracting the feature vector to classify from one or more of the obtained components [3, 5, 7–14].

Texture analysis has played an important role in the characterization of biomedical images. Texture analysis methods can be categorized as statistical, geometrical, and signal processing types [14]. Statistical methods are mainly based on the spatial distribution of pixel gray values, while geometrical approaches depend on the geometric properties of texture primitives. As for signal processing methods, they use texture filtering in the spatial or frequency domain to extract relevant features.

Multiresolution analysis is the most widely employed signal processing technique for characterizing biomedical images due to its capability to obtain high time-frequency resolutions. The wavelet-transform family methods are typical examples of multiresolution analysis techniques. The basic wavelet transform [15, 16] starts with a basis function, the mother wavelet, and decomposes a signal into components of different time and frequency scales; longer time intervals are used to obtain low-frequency information and shorter intervals are used to obtain high-frequency information.

The most commonly used wavelet transform in biomedical image processing is the discrete wavelet transform (DWT) [14] whose discrete time shifting and stretching variables lead to a sparse and efficient representation. The DWT takes an input image and decomposes into four subimage components that characterize it for different orientations in the horizontal and vertical frequency axes. The process can be repeated with one or more subimages if needed. More precisely, the DWT decomposition yields the approximation subband (LL), the horizontal detail subband (LH), the vertical detail subband (HL), and the diagonal detail subband (HH). These describe, respectively, the low-frequency components in the horizontal and vertical directions, the low-frequency components in the horizontal direction and high-frequency components in the vertical direction, the high-frequency components in the horizontal direction and low-frequency components in the vertical direction, and the high-frequency components in both directions. Thus, in essence, the standard DWT algorithm yields horizontal, vertical, and diagonal directional information about the frequency spectrum of an image. However, these three directions may not be sufficient to express all the directional information in digital images, particularly biomedical images [4, 14]. In an attempt to express the directional features more efficiently, several directional wavelet systems have been proposed. These include the Gabor wavelets [17], the dual-tree complex wavelet transform (DT-CWT) [18], the ridgelet [19], the curvelet [20], and the contourlet [21]. There exist also reports on biomedical applications of Gabor filter banks [22], DT-CWT [4], ridgelets [23], curvelets [24], and contourlets [5].

The two-dimensional (2D) Gabor filter decomposes an image into components corresponding to different scales and orientations. As a result, it captures visual properties such as spatial localization, orientation selectivity, and spatial frequency. The 2D Gabor filter has real and imaginary parts and is highly flexible in its representation as its parameters can be adapted to the structure of the patterns that one wants to analyze in the image. It is however difficult to find the optimal set of parameters to characterize a given image. In comparison, the DT-CWT transform provides directional selectivity, shift invariant features, and complex images. However, it suffers from limited orientation selectivity [25] and redundancy of information [26]. The ridgelet transform is appropriate to capture radial directional details in the frequency domain; in particular it is optimal for representing straight-line singularities. However, those structures are not dominant in medical images and are rarely observed in real world images. This limits the suitability of the ridgelet transform to characterize the texture of real images [27]. The curvelet transform is an extension of the ridgelet transform for detecting image edges and singularities along curves while analyzing images at multiple scales, locations, and desired orientations. It is particularly suitable for image features with discontinuities across straight lines. Unfortunately, the curvelet transform is highly redundant [28] and only few choices of mother functions are available for the curvelets as opposed to the many choices available for the standard

wavelet transform [29]. Finally, the contourlet transform can capture directional details and smooth contours in a given image. In particular, it is suitable in the analysis of images containing textures and oscillatory patterns. Its main drawback is the high degree of information redundancy and occurrence of artefacts [30, 31].

In past works, we proposed several transform-based approaches to account for directional features in classifying biomedical images. For instance, in the case of brain magnetic resonance images, we proposed a simple methodology in [32, 33] where features are extracted from the LH and HL components of the DWT instead of the more common LL, or image approximation, component. We found that the LH and HL coefficients are efficient at characterizing changes in the biological tissue and help distinguish normal and abnormal image textures. For mammograms, we investigated in [34] a hybrid processing system that sequentially uses the discrete cosine transform (DCT) to obtain the high-frequency component of the mammogram and then applies the Radon transform (RT) to the result in order to extract its directional features. The validation results showed that the RT helps improve the recognition rate of the detection system. In subsequent work, we combined the DWT and RT transforms [35]. The approach targeted the HH component of the DWT decomposition and improved classification accuracy when compared to using the DWT or RT alone or the DCT-RT used in [34]. Our previous works clearly showed that directional information helps improve classification accuracy. In addition, the DWT-RT detection system was more efficient for classifying normal and abnormal images than the DCT-RT, possibly because of the multiresolution capability of the DWT and the fact that it leads to a sparser signal representation than the DCT. Still, the RT cannot capture spatial frequency, a potential feature to improve further the classification accuracy.

In this paper, we describe a hybrid biomedical image processing and classification system that uses both the DWT and Gabor filter as directional transforms and statistical features derived from them for the classification task which is accomplished by support vector machines (SVMs) [36]. As stated before, the DWT is powerful at providing sparse and efficient image representations [14]. However, except for the LH and HL subbands whose coefficients depend on image row and column information, respectively (an effect of the subband coding used by the algorithm), the standard DWT is essentially an image compression tool and it cannot perform directional analysis at arbitrary directions. On the other hand, the Gabor filter can process images in terms of preferred orientations at arbitrary spatial frequencies. Moreover, it provides nonredundant information and can offer high directional selectivity. Thus, combining DWT and Gabor filter banks in sequence may lead to improved feature extraction from biomedical images and better classification of normal versus abnormal images in comparison to using DWT or Gabor filter banks alone. In this hybrid processing scheme, the DWT acts both as high-frequency filter to extract abrupt changes in image texture and image compression engine to reduce image dimensionality and a Gabor filter bank extracts the directional information.

In a preliminary work [37], the previously mentioned DWT-Gabor hybrid system was successfully applied to mammograms to extract features that allow discriminating normal and cancer images. More specifically, the goal was to detect the presence of malign microcalcifications (specs of calcium in the breast tissue that appear in the mammogram as small bright spots that are scattered or grouped in clusters), whose early detection is important for cancer screening [38, 39]. The results showed the superiority of the approach over simply using the DWT alone. In the present work, we widen our study to retina digital images and brain magnetic resonance images to investigate the effectiveness of the DWT-Gabor approach across application domains with similar image features. Indeed, the images of some pathologies related to brain, retina, and breast present similar contrast features characterized by abrupt changes in image texture with directional properties (see examples in Figures 1, 2, and 3). For instance, breast cancer is characterized by dense concentration of contrast cells in the biological tissue, cancer in brain magnetic resonance images is often characterized by large cells with high contrast, and many forms of retinopathy are characterized by the presence of spots on the retina or covering the macula. As a result, the DWT-Gabor hybrid system we have used in our previous work [37] to detect cancer in mammograms could potentially also be applied to brain magnetic resonance images and retina digital images with similar properties. Next is a brief description of the pathologies that were studied in this work.

Circinate retinopathy is a retinal degeneration characterized by a circle of white spots encircling the macula that causes complete foveal blindness [40]; retinal microaneurysms are due to a swelling of the capillaries caused by a weakening of the vessel wall [41] and are considered to be the earliest sign of diabetic retinopathy, among others. Magnetic resonance imaging (MRI) is a noninvasive imaging modality largely used for brain imaging to detect diseases such as Alzheimer's and multiple sclerosis [6, 8]. Alzheimer's disease is the most frequent cause of age-related dementia and multiple sclerosis is a progressive neurological disorder that can result in various dysfunctions [42]. Additional brain pathologies that can be detected from MR images and that are investigated in this work include glioma, herpes encephalitis, and metastatic bronchogenic carcinoma (Figure 1). All of these are characterized by large cells with high contrast, hence the interest in being able to detect them with the same algorithm.

The contribution of our work can be summarized as follows. First, we propose a relatively simple and fast approach to biomedical image characterization that relies on the directional properties of high-frequency components. The DWT is applied first to extract high-frequency components that characterize abrupt changes in the biological tissue and, then, the Gabor filter is applied to the obtained HH subimage to extract directional features. Second, the statistical features extracted from the hybrid DWT-Gabor transform are processed by an SVM for classification. This statistical binary classifier has proven its efficiency [4–6, 32–35, 37] and ease of tuning in comparison to alternatives such as artificial neural networks. Another desirable feature is its scalability and ability to avoid local minima [36]. Third, contrary to alternatives that focus on ROIs or specific image details, the proposed methodology is of more general reach as three different types of images used for validation show.

The paper is organized as follows. Section 2 reviews previous works related to the automatic classification of normal versus abnormal images in the context of brain magnetic (MR) resonance imaging, mammograms, and retina digital images. Section 3 describes our proposed approach for directional features extraction from biomedical images using discrete wavelet transform followed by Gabor filter banks and support vector machines classifier. Section 4 presents experimental results. Finally Section 5 draws the conclusions and gives future work to be done.

2. Related Works

Mammograms, retina, and MR images are the subject of many research efforts on feature extraction and subsequent classification. Next is a summary of some recent works related to DWTs and/or Gabor filters. In the problem of automatic classification of mammograms, the authors in [3] used Gabor filter banks to process images and k nearest neighbour (k-NN) algorithm as classifier. The obtained classification rate was 80%. In [4] the dual-tree complex wavelet transform (DT-CWT) and support vector machine (SVM) were employed to classify benign and malignant images. The experimental result achieved 88.64% classification accuracy. The authors in [5] employed the contourlet transform and successive enhancement learning (SEL) weighted SVM to obtain 96.6% correct classification rate. The previous studies all used images of size 1024×1024 pixels.

In the problem of retina digital image classification, the authors in [1] employed the Belkyns's shift-invariant DWT to classify normal against abnormal retina images of size 700×605 pixels. The pathologies of the abnormal images included exudates, large drusen, fine drusen, choroidal neovascularization, central vein and artery occlusion, histoplasmosis, arteriosclerotic retinopathy, hemicentral retinal vein occlusion and more. In order to capture texture directional features, they employed normalized gray level cooccurrence matrices (GLCMs). The obtained classification accuracy with linear discriminant analysis (LDA) was 82.2%. The authors in [2] employed the probabilistic boosting algorithm and morphological scale space analysis and GLCM to extract texture features. The purpose was to classify normal images versus drusen images with various texture complexities. The detection accuracy of normal images varied between 81.3% and 92.2%, and that of abnormal images varied between 71.7% and 85.2% depending on texture complexity (grade of pathology). The authors in [43] used four approaches to extract features from retina digital images of size 300×300 pixels to automatically classify glaucoma images. The first set of features is obtained by taking the pixel intensities as input to principal component analysis. The second features are obtained from Gabor texture filter responses. The third set of features is computed from the coefficients of the fast Fourier transform and the fourth set of features is obtained from

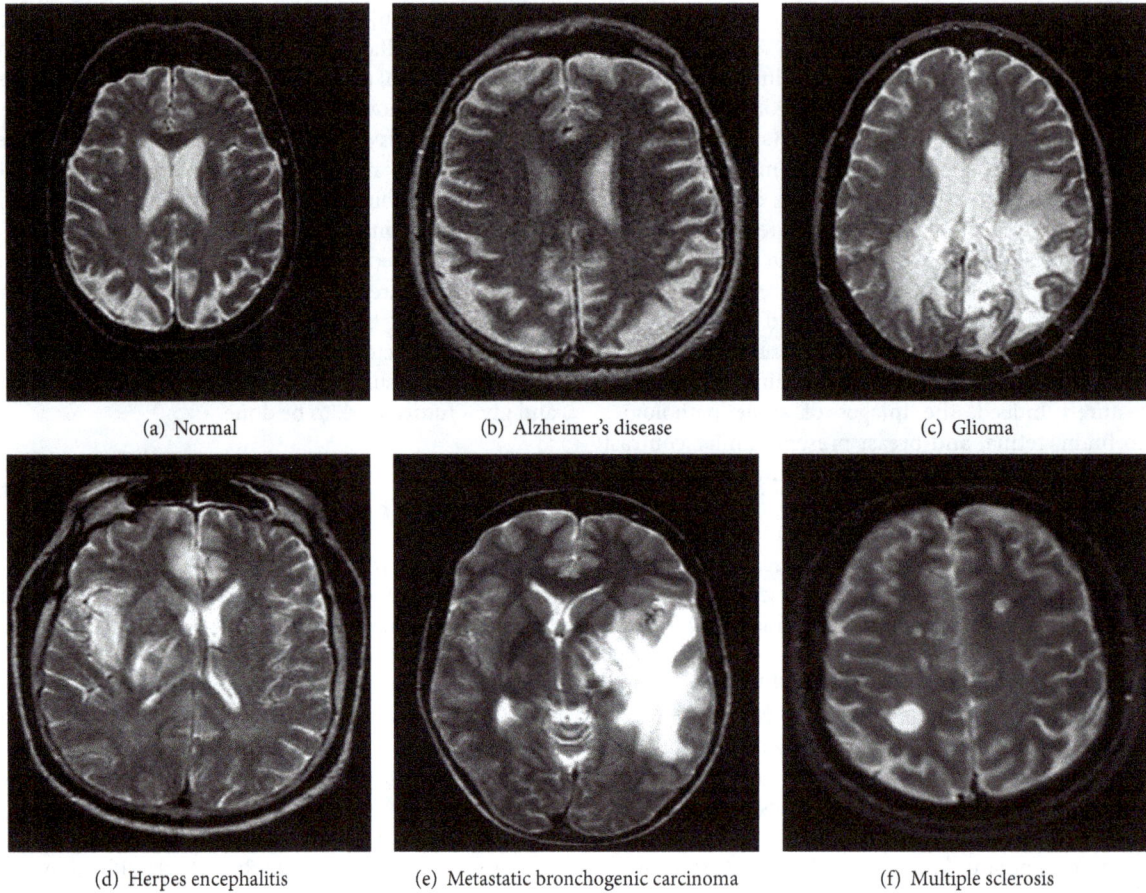

(a) Normal   (b) Alzheimer's disease   (c) Glioma

(d) Herpes encephalitis   (e) Metastatic bronchogenic carcinoma   (f) Multiple sclerosis

FIGURE 1: Examples of brain MR images.

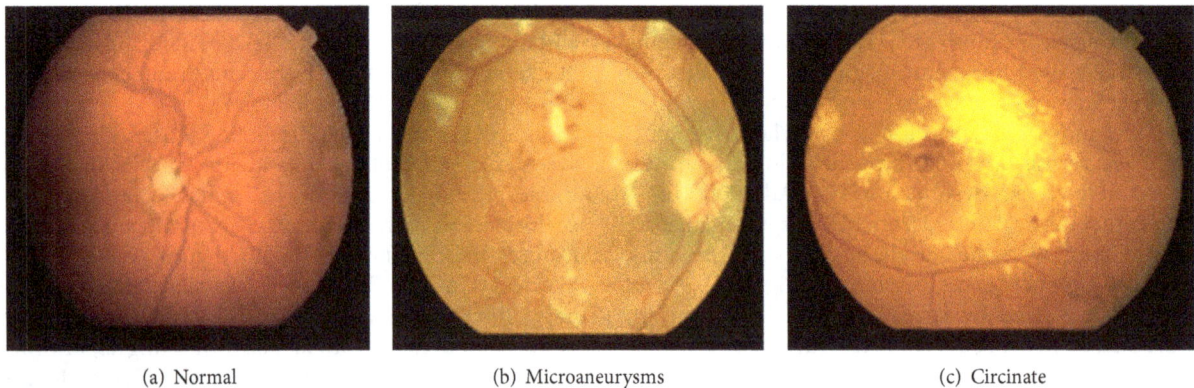

(a) Normal   (b) Microaneurysms   (c) Circinate

FIGURE 2: Examples of retina images.

the histogram of the intensity distribution of the image. Finally, support vector machines were employed for the classification task. The performance of the classifications using one feature set only was 73% with the histogram features, 76% with the fast Fourier transform coefficients, 80% with the Gabor textures, and 83% with the pixel intensities.

Finally, in the problem of brain MRI classification, the authors in [6] used the wavelet coefficients as input to a support vector machine to classify normal and abnormal

Alzheimer's disease images of size 256×256 pixels. The classification accuracy was 98% using SVM with a radial basis kernel. More recently, the authors in [42] used voxels to represent each brain MRI of size 512×512 pixels. Using cross-validated tests, the obtained correct classification rates of normal and Alzheimer images were 90%, 92%, and 78%, respectively, when using classification by SVM, naïve Bayes classifier, and voting feature intervals (VFIs). Still more recently, the authors in [8] employed the DWT to

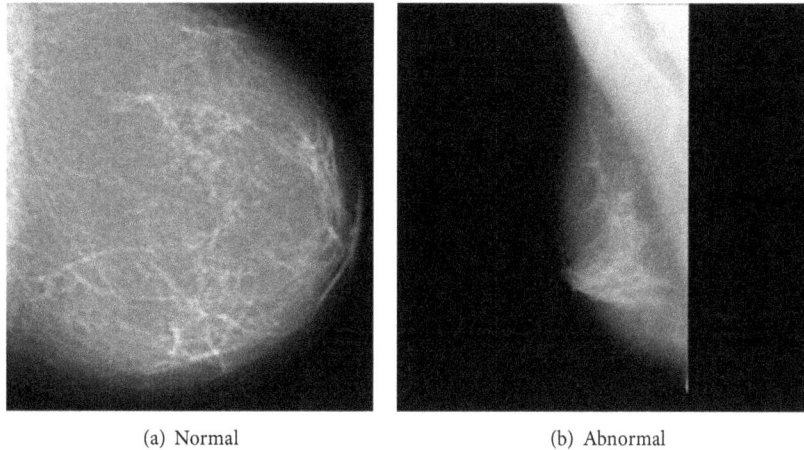

(a) Normal   (b) Abnormal

FIGURE 3: Examples of mammograms.

extract features from brain magnetic resonance images of size 256×256 pixels, and then principal component analysis was used to reduce the dimensions of the features space. The abnormal images included glioma, meningioma, Alzheimer's, Alzheimer's plus visual agnosia, Pick's disease, sarcoma, and Huntington's disease. The classification accuracies using backpropagation neural network (BPNN) were 100% using learning and testing sets of 33 images each.

In this work, we are interested in how a DWT-Gabor-based approach for feature extraction may provide better classification results than those reported in the previous works, particularly those based on the DWT alone. The next section provides the details of our methodology.

3. Methodology

The overall methodology proceeds as follows. First, the DWT is applied to the biomedical image to obtain its high-frequency image component since it often contains most of the desired information about the biological tissue [39]. Indeed, sudden changes in the texture of the image are typical indicators of the presence of abnormal biological tissue. Second, a bank of Gabor filters with different scales and orientations is applied to the high-frequency image to obtain Gabor-filtered images along different spatial orientations. Third, statistical features are extracted from the Gabor-filtered images. Finally, the SVM is used to classify the resulting feature vector for final diagnosis. The block diagram of the DWT-Gabor system is shown in Figure 4. Figure 5 summarizes the DWT approach in comparison.

3.1. Discrete Wavelet Transform. The two-dimensional discrete wavelet transform (2D-DWT) [14–16] performs a subband coding of an image in terms of spectral spatial/frequency components, using an iterative and recursive process. Figure 6 illustrates the case of two-level decomposition. The image is first represented by LH, HL, and HH subbands that encode the image details in three directions and an LL subband which provides an approximation of it.

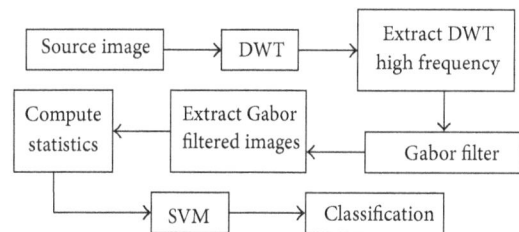

FIGURE 4: Schematic diagram of the DWT-Gabor approach.

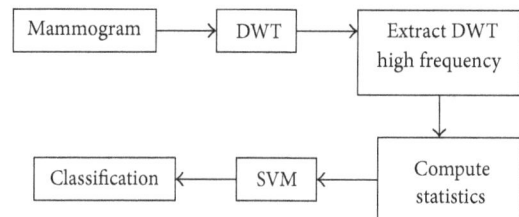

FIGURE 5: Schematic diagram of the DWT approach.

The obtained detail or approximation images can be decomposed again to obtain second-level detail and approximation images, and the process can be repeated for finer analysis as each iteration doubles the image scale.

The computation of the 2D-DWT proceeds from that of the 1D-DWT, the discrete version of the one-dimensional continuous wavelet transform. The one-dimensional continuous wavelet transform of a signal $x(t)$ is defined by [7, 8]

$$W_\psi(a, b) = \int_{-\infty}^{+\infty} x(t)\, \psi_{a,b}^*(t)\, dt, \tag{1}$$

$$\psi_{a,b}(t) = \frac{1}{\sqrt{|a|}} \psi\left(\frac{t - b}{a}\right), \tag{2}$$

where $\psi_{a,b}(\cdot)$ stands for a given wavelet function and a and b are the scale and translation parameters, respectively. The 1D-DWT is obtained by sampling a and b so that (1) becomes that

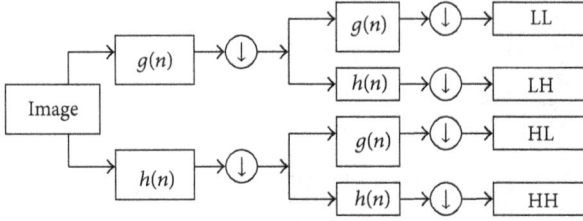

FIGURE 6: 2D-DWT decomposition of an image.

of a sequence. In dyadic sampling, a and b are, respectively, a power of 2 and multiples thereof, and the sequence elements (wavelet coefficients) are given by

$$c_{jk} = W_\psi \left(2^{-j}, 2^{-j}k \right), \tag{3a}$$

where j represents the discrete scale factor and k the discrete translation factor. In other words, a and b in (1) are replaced by 2^j and $2^j k$, respectively.

The one-dimensional wavelet decomposition is extended to an image by applying it to the row variable first and then to the column variable of the obtained result [44]. At each step, two subimages are created with half the number of pixels of the row or column that was processed. In the end, an $M \times N$ image is decomposed into 4 subimages, each with $M/2 \times N/2$ resolution and preserved scale. However, (1) has only theoretical merit due to the infinite ranges of a and b. For a practical implementation, the fact that (1) is essentially a measure of correlation between a signal and various wavelets derived from a mother is exploited, and the DWT decomposition is turned into a filtering operation with a sequence of high-pass and low-pass filters [45]. Following the notation in [7, 8], the discrete form of (1) can then be written as

$$ca_{j,k}\left[x\left(t\right)\right] = \mathrm{DS}\left[\sum x\left(t\right) g_j^*\left(t - 2^j k\right)\right],$$
$$cd_{j,k}\left[x\left(t\right)\right] = \mathrm{DS}\left[\sum x\left(t\right) h_j^*\left(t - 2^j k\right)\right], \tag{4}$$

where coefficients $ca_{j,k}$ and $cd_{j,k}$ specify approximation and details components provided by the $g(n)$ low-pass and $h(n)$ high-pass impulse responses, respectively, and the DS operator performs downsampling by a factor of 2. The one-dimensional wavelet decomposition is extended to two-dimensional objects by using row and column decompositions as shown in Figure 6. In our work, the most frequently used wavelet (Daubechies-4) [25] is considered to extract the HH image component.

3.2. Gabor Filter.

The two-dimensional (2D) Gabor filter decomposes an image into components corresponding to different scales and orientations [22], thus capturing visual properties such as spatial localization, orientation selectivity, and spatial frequency. The 2D Gabor filter consists of a complex exponential centered at a given frequency and modulated by a Gaussian envelope. Because of the complex

exponential, the filter has both real and imaginary parts. The general form of the real part is defined as follows:

$$G\left(x, y, \sigma_x, \sigma_y, f, \theta\right) = \exp\left[-\frac{1}{2}\left(\left(\frac{x'}{\sigma_x}\right)^2 + \left(\frac{y'}{\sigma_y}\right)^2\right)\right]$$
$$\times \cos\left(2\pi f x'\right), \tag{5}$$

where

$$x' = x\cos\left(\theta\right) + y\sin\left(\theta\right),$$
$$y' = y\cos\left(\theta\right) - x\sin\left(\theta\right) \tag{6}$$

and where σ_x and σ_y are the standard deviations of the Gaussian envelope along the x and y axes. The parameters f and θ are, respectively, the central frequency and the rotation of the Gabor filter. To obtain the Gabor-filtered image $f(x, y)$ of a given image $i(x, y)$ the 2D convolution operation ($*$) is performed:

$$f\left(x, y\right) = G\left(x, y, \sigma_x, \sigma_y, f, \theta\right) * i\left(x, y\right). \tag{7}$$

The selection of parameters σ_x, σ_y, f, and θ plays an important role in the filter's operation. However, no formal technique exists for choosing them and experience-guided intuition, trial and error, or heuristic search must be used. For retina digital images and brain MR images, σ_x and σ_y were arbitrarily set to unity. In the case of mammograms, σ_x and σ_y were set to the values used in [46], which were determined empirically. Consequently, we used $\sigma_x = \tau/2.35$, where τ is the full width at half-maximum of the Gaussian and $\sigma_y = 8\sigma_x$. No values of σ_x and σ_y other than the previous ones were tried since optimality was not the primary concern of this work and we obtained satisfactory results with these values.

Four orientations, $\theta = 0, \pi/4, \pi/2,$ and $3\pi/4$, were used as in [22, 33]. These values seemed reasonable as a first try since they covered both image axis directions and the forward and backward diagonals. Finally, the central frequency f was set to 2, 2.5, and 3. Given that the Gabor filter is modulated by the cosine of f, large values of f lead to a compressed cosine and, consequently, the filter output is more likely to show fast or frequent changes in biological tissue texture. This in turn would help verify our hypothesis that abnormal images are characterized by sudden and frequent variations in image texture. In the end, the application of the Gabor filter bank to the HH image component obtained with the 2D-DWT leads to twelve Gabor-filtered HH images components, for each choice of f and θ.

3.3. Feature Extraction.

Statistical measures are employed to extract features from both the DWT HH subband image and the real Gabor-filtered HH image components. More precisely, the entropy (E) and uniformity (U) of the coefficients of each one are computed. Entropy and uniformity were selected as features because previous works on mammograms have shown that uniformity is correlated with suspicious malignancy [47] and that entropy can successfully characterize breast biological tissue [48]. In this study, the entropy and uniformity statistics are hypothesized to also characterize retina and brain MR images with similar contrast information (i.e., abrupt and/or frequent variations in texture). Entropy (E) and uniformity (U) are defined by [49]

$$E = -\sum p(z) \times \log(p(z)),$$
$$U = \sum p^2(z), \tag{8}$$

where z is a random variable that represents a coefficient in the Gabor filtered image and $p(z)$ is its probability of occurrence as estimated by its relative frequency.

To investigate the performance of the previous approaches, the image features were extracted from HH at both level-one DWT decomposition (HH1) and level-two DWT decomposition (HH2), with and without filtering by a Gabor filter bank. We also applied the Gabor filter directly to the original image without the DWT for comparison purpose. For each DWT HH subband image, the feature vector is given by

$$V_{\text{DWT},a} = [E_a, U_a], \tag{9}$$

where a is the level of wavelet analysis (decomposition). Similarly, for each of the twelve outputs generated by the Gabor filter bank (4 angles × 3 central frequencies), the entropy and uniformity are computed and a twenty-four component feature vector is formed to represent the initial image. We thus have

$$V_{\text{Gabor},a} = [E_{1,a}, E_{2,a}, \ldots, E_{12,a}, U_{1,a}, U_{2,a}, \ldots, U_{12,a}]. \tag{10}$$

Either feature vector is subsequently fed to the SVM to classify normal versus pathological images.

3.4. The Support Vector Machine Classifier. Introduced by Vapnik [36], the support vector machine (SVM) classifier is based on statistical learning theory. It implements the principle of structural risk minimization and has excellent generalization ability as a result, even when the data sample is small. Moreover, SVM can tolerate high-dimensional and/or incomplete data [50]. It has been used with great success in various applications, including speech emotion recognition [51], card-sharing traffic detection [52], fault diagnosis [53], cardiac decision making [54], Parkinson's disease diagnosis [55], and Alzheimer's disease detection [56].

The support vector machine performs classification tasks by constructing an optimal separating hyperplane that maximizes the margin between the two nearest data points belonging to two separate classes. Given a training set $\{(x_i, y_i), i = 1, 2, \ldots, m\}$, where the input $x_i \in R^d$ and class labels $y_i \in \{+1, -1\}$, the separation hyperplane for a linearly separable binary classification problem is given by

$$f(x) = \langle w \cdot x \rangle + b, \tag{11}$$

where w is a weight vector and b is a bias. The optimal separation hyperplane is found by solving the following optimization problem:

$$\underset{w,b,\xi}{\text{minimize}} \quad \frac{1}{2} \langle w \cdot w \rangle + C \sum_{i=1}^{m} \xi_i \tag{12}$$

$$\text{Subject to:} \quad y_i(\langle w \cdot x_i \rangle + b) + \xi_i - 1 \geq 0, \quad \xi_i \geq 0,$$

where C is a penalty parameter that controls the tradeoff between the complexity of the decision function and the number of misclassified training examples and ξ is a positive slack variable. The previous optimization model can be solved by introducing Lagrange multipliers and using the Karush-Kuhn-Tucker theorem of optimization to obtain the solution as

$$w = \sum_{i=1}^{m} \alpha_i y_i x_i. \tag{13}$$

The x_i values corresponding to positive Lagrange multipliers α_i are called support vectors which define the decision boundary. The x_i values corresponding to zero α_i are irrelevant. Once the optimal solution α_i^* is found, the optimal hyperplane parameters w^* and b^* are determined. Then, the discriminant function of the SVM for a linearly separable binary classification problem is [32]

$$g(x) = \text{sign}\left(\sum_{i=1}^{m} y_i \alpha_i^* \langle x_i \cdot x \rangle + b^*\right). \tag{14}$$

In the nonlinearly separable case, the SVM classifier nonlinearly maps the training points to a high-dimensional feature space using a kernel function Φ, where linear separation can be possible. The scalar product $\langle \Phi(x_i) \cdot \Phi(x_j) \rangle$ is computed by Mercer kernel function K as $K(x_i, x_j) = \langle \Phi(x_i) \cdot \Phi(x_j) \rangle$. Then, the nonlinear SVM classifier has the following form:

$$g(x) = \text{sign}\left(\sum_{i=1}^{m} y_i \alpha_i^* K \langle x, x_i \rangle + b^*\right). \tag{15}$$

In this study, a polynomial kernel of degree 2 was used for the SVM. As a global kernel, it allows data points that are far away from each other to also have an influence on the kernel values. The general polynomial kernel is given by

$$K(x, x_i) = ((x_i \cdot x) + 1)^d, \tag{16}$$

where d is the degree of the polynomial to be used.

4. Experimental Results

As mentioned previously, mammograms and retina and brain MR images corresponding to given pathologies are considered in this work, and the aim is to classify normal versus abnormal images for each image category. To do so, one hundred digital mammograms (171 × 364 pixels) consisting of fifty normal images and fifty cancer images were taken from The Digital Database for Screening Mammography (DDSM)

TABLE 1: Average SVM classification accuracy as a function of feature extraction method and level of DWT decomposition[*].

	DWT	DWT-Gabor	DWT	DWT-Gabor
Decomposition level	One	One	Two	Two
Mammograms	95.98% (±0.04)	96.67% (±0.05)	89.13% (±0.01)	91.09% (±0.05)
Retina	74.69% (±0.05)	100%	90.98% (±0.03)	100%
Brain MRI	87.80% (±0.00)	97.36% (±0.02)	85.76% (±0.00)	91.18% (±0.04)

[*]Tenfold cross-validation used for mammograms and retina images, leave-one-out used for brain MRIs.

TABLE 2: SVM classification specificity as a function of feature extraction method and level of DWT decomposition[*].

	DWT	DWT-Gabor	DWT	DWT-Gabor
Decomposition level	One	One	Two	Two
Mammograms	97.81% (±0.02)	100%	88.85% (±0.05)	92.09% (±0.06)
Retina	6.81% (±0.04)	100%	2.32% (±0.10)	100%
Brain MRI	21.55% (±0.01)	99.58% (±0.01)	0%	97.24% (±0.01)

[*]Tenfold cross-validation used for mammograms and retina images, leave-one-out used for brain MRIs.

[57]. For retina, a set of 69 color images (150 × 130 pixels) from the STARE [58] database were employed including 23 normal images, 24 with microaneurysms, and 22 with circinate. Finally, a collection of 56 axial, T2-weighted, and MR brain images (256 × 256 pixels) were taken from the AANLIB database [59] of the Harvard Medical School. They consisted of 7 normal images, 9 with Alzheimer's disease, 13 with glioma, 8 with Herpes encephalitis, 8 with metastatic bronchogenic carcinoma, and 14 with multiple sclerosis. It is unfortunate that the number of images was not constant across pathologies, but we had no control over this and used what was available, with tenfold cross-validation or leave-one-out cross-validation of the results depending on sample size. All experiments were based on a binary classification approach of normal versus abnormal images. Many kinds of biomedical images could be considered for our experiments; we focused on mammograms, retina, and brain magnetic resonance images mainly because of public availability. An example of the processing of a normal retina and a retina with circinate is illustrated in Figures 7 and 8, respectively.

For each image type, the average and standard deviation of the correct classification rate (CCR), sensitivity, and specificity were computed to evaluate the performance feature extraction technique when used in conjunction with the SVM classifier. The three performance measures are defined by

$$\text{CCR} = \frac{\text{Classified Samples}}{\text{Total Number of Samples}},$$

$$\text{Sensitivity} = \frac{\text{Correctly Classified Positive Samples}}{\text{True Positive Samples}}, \quad (17)$$

$$\text{Specificity} = \frac{\text{Correctly Classified Negative Samples}}{\text{True Negative Samples}},$$

where positive samples and negative samples are, respectively, abnormal and normal images.

Finally, all experiments were performed with tenfold cross-validation, except those for MR images which used leave-one-out cross-validation due to the small sample size of each brain image category.

Table 1 shows the obtained average results for the three types of images that were investigated. The performance of the SVM classifier improved for all types of images and all levels of HH decomposition by the DWT. At level one, the average correct classification rate increased by, respectively, 0.69, 25.31, and 9.56 percentage points for mammograms, retina, and brain magnetic resonance images when using the DWT-Gabor approach. At level-two decomposition, the improvement was, respectively, 1.96, 9.02, and 5.42 percentage points.

Tables 2 and 3 provide the average results for classifier sensitivity and specificity. At level-one DWT decomposition, the DWT-Gabor approach improved classification specificity for mammograms and retina images to make it reach 100%, while it improved it by 78.03 percentage points for brain MR images. At level-two DWT decomposition, the improvement was 3.24 percentage points for mammograms, 97.68 percentage points for retina images (100% specificity), and 97.24 percentage points for MR images. Regarding sensitivity, the results were mixed.

At level-one DWT decomposition, the values were about the same for the DWT-Gabor and DWT-only approaches for mammograms and retina images, with, respectively, −0.81 and 0 percentage points differences, but there was a degradation of −11.33 and −46.16 percentage points for brain MR images at level one and level two of decomposition, respectively.

Following the same cross-validation protocol, we also conducted classification experiments with features extracted from a Gabor filtered image of the original biomedical image. The purpose was to check whether Gabor-based features alone help characterize images better than DWT or DWT-Gabor-based features. The results are given in Table 4.

(a) Original normal retina image   (b) Normal retina HH1 subimage   (c) Normal retina HH2 subimage

(d) Gabor filtered HH1 at $f = 2$ and $\theta = \pi/4$ (e) Gabor filtered HH2 $f = 2$ and $\theta = \pi/4$

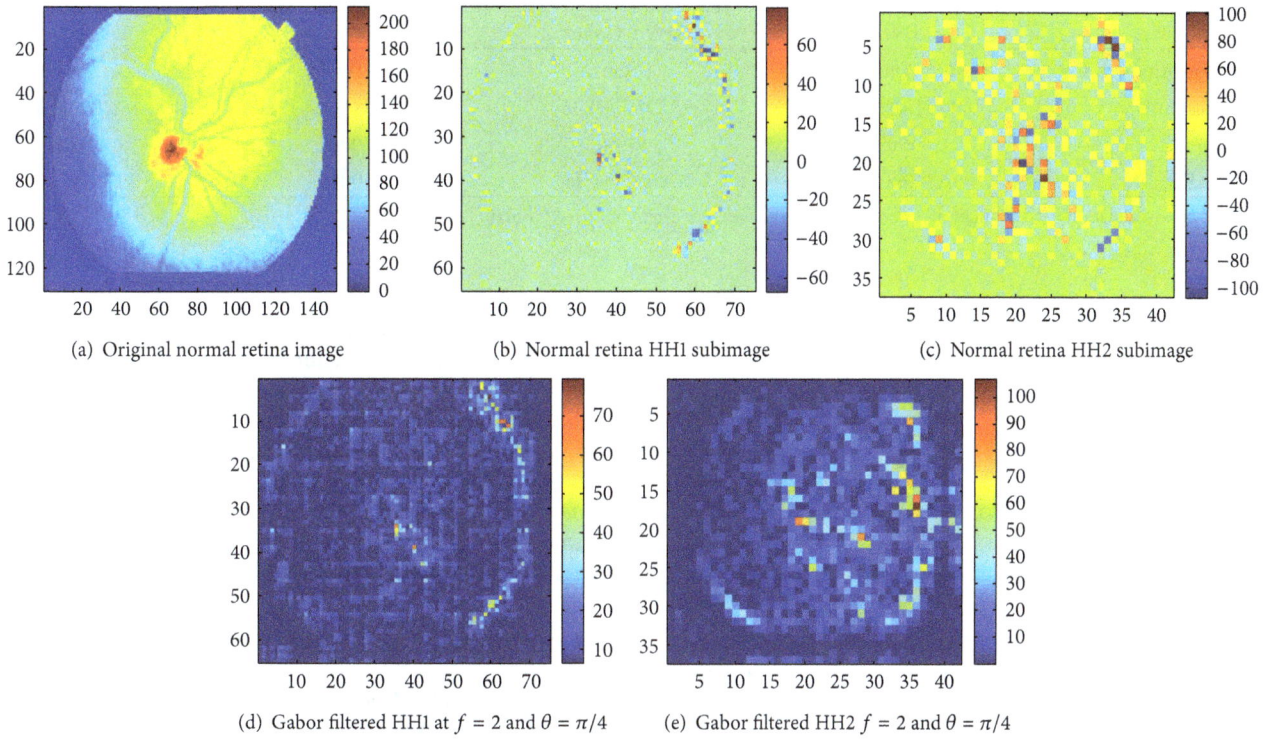

FIGURE 7: Analysis of a normal retina.

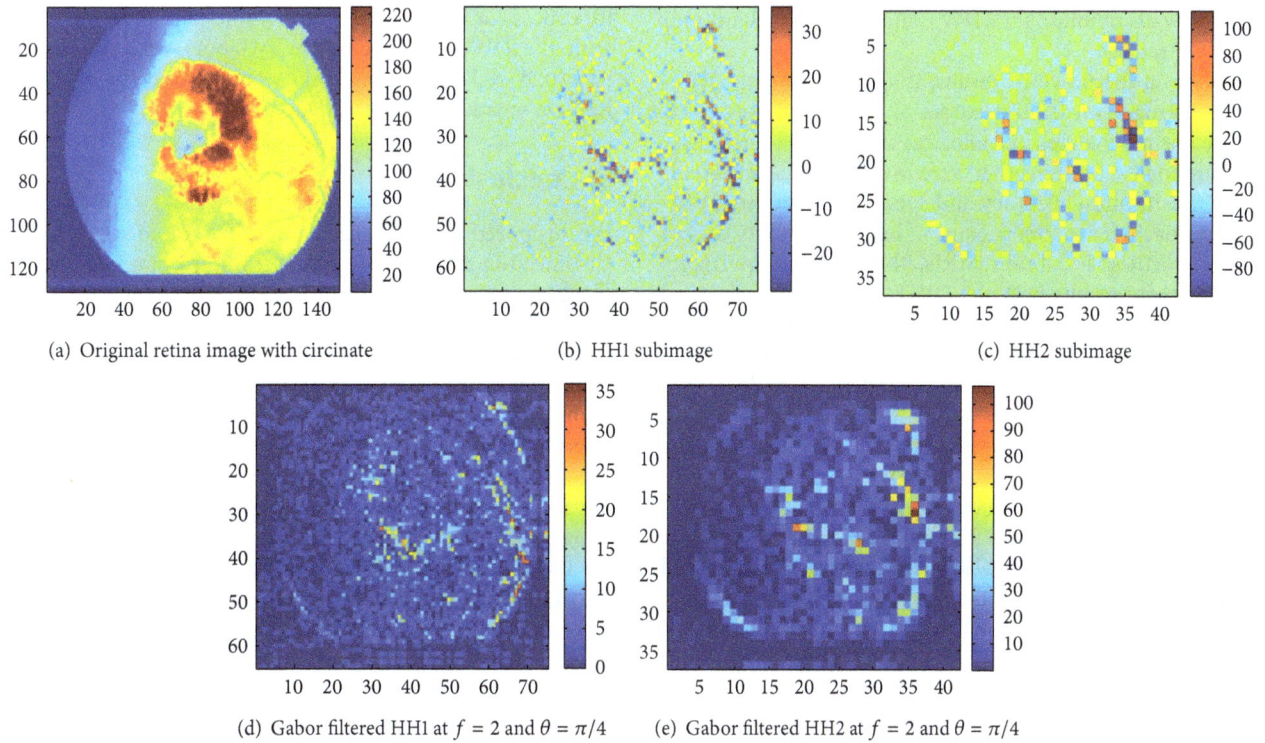

(a) Original retina image with circinate   (b) HH1 subimage   (c) HH2 subimage

(d) Gabor filtered HH1 at $f = 2$ and $\theta = \pi/4$ (e) Gabor filtered HH2 at $f = 2$ and $\theta = \pi/4$

FIGURE 8: Analysis of a retina with circinate.

TABLE 3: SVM classification sensitivity as a function of feature extraction method and level of DWT decomposition*.

	DWT	DWT-Gabor	DWT	DWT-Gabor
Decomposition level	One	One	Two	Two
Mammograms	94.14% (±0.06)	93.33% (±0.06)	90.29% (±0.039)	89.78% (±0.04)
Retina	100%	100%	100%	100%
Brain MRI	93.84% (±0.00)	82.51% (±0.16)	100%	53.84% (±0.23)

*Tenfold cross-validation used for mammograms and retina images, leave-one-out used for brain MRIs.

TABLE 4: SVM classification performance measures obtained with Gabor-based features.

	Accuracy	Specificity	Sensitivity
Mammograms	68.03% (±0.01)	100%	0%
Retina	50.00% (±0.00)	100%	0%
Brain MRI	86.61% (±0.03)	100%	0%

The obtained correct classification rate of mammograms, retina, and brain magnetic resonance images is, respectively, 68.03% (±0.01), 50.00% (±0.00), and 86.61% (±0.03). The average results for classifier specificity and sensitivity for all images are 100% and 0%. This finding indicates that Gabor-based features are suitable to detect pathological images, but fails to detect normal images. In sum, the results show that Gabor-based features do not perform better than DWT and DWT-Gabor-based features. These findings confirm the superiority of combining the DWT and Gabor filter banks for feature extraction.

Based on the previous results, it appears that the DWT-Gabor approach for feature extraction is effective for detecting the abrupt changes in biological tissue that characterize the pathological patterns that were investigated and it yields better classification accuracy and specificity than the DWT-only approach. It also offers equal of better sensitivity, except for brain MRIs. For brain MRIs, the obtained specificity and sensitivity results with the DWT-Gabor approach show improved true negative detection, but lower true positive performance. Finally, the obtained results reveal also that level-one DWT decomposition is preferable to level-two decomposition.

Finally, Table 5 compares the results obtained with the DWT-Gabor approach to other work that we surveyed. In many cases, the DWT-Gabor method yields higher classification rates, particularly for mammograms and retina. For the problem of brain MRI classification, our obtained performance is better than the results of [38], but less than what is reported in [6, 8]. However, these comparisons should be viewed with caution as not all the results stem from a common image database and the different authors use different sample and image sizes. Moreover, many authors use no cross-validation and simply perform a single arbitrary split of their data into training and test sets to obtain their accuracy results. Obviously, one cannot generalize or draw definite conclusion from such efforts, and comparisons between

works cannot be made other than in general terms. In this respect, it can only be concluded from our results that the DWT-Gabor for feature extraction is effective for obtaining high image classification accuracy by an SVM and that it may outperform other feature extraction and classification techniques reported in the literature, at least those based on DWT-only image decomposition. Unfortunately, a more definite conclusion is impossible without gaining access to the image databases used by the other authors.

5. Computational Complexity

Finally, the computational complexity of the DWT, Gabor, contourlet, and curvelet for an $N \times N$ image is, respectively, $O(N)$, $O(N^2 \times M^2)$ with M being the width of Gabor (Gaussian) mask filter, $O(N^2)$, and $O(N^2 \log(N))$. As a result, the computational complexity of the combination of the DWT and Gabor filter is $O(N) + O(N^2 \times M^2)$. In terms of features extraction processing time, the average time required to process a brain, a mammogram, and a retina image with the DWT approach (DWT-Gabor) was, respectively, 0.19 (0.31), 0.17 (0.32), and 0.15 seconds (0.35) using Matlab R2009a on a 1.5 GHz Core2 Duo processor.

6. Conclusion

We proposed a supervised system for biomedical images classification that uses statistical features obtained from the combination of the discrete wavelet transform and Gabor filter to classify normal images versus cancer images, using support vector machines as classifiers. Our experimental results show that such a hybrid processing model achieves higher accuracy in comparison to using DWT or Gabor filter banks alone. Therefore, the proposed image processing and features extraction approach seem to be very promising for the detection of certain pathologies in biomedical images.

For future works, it is recommended to consider a larger set of features and a selection process to identify the most discriminant ones. In addition, the Gabor parameters will be adjusted for each type of image separately to improve the accuracy. Furthermore, the DWT-Gabor will be directly compared to the dual-tree complex wavelet, curvelet, and contourlet using the same databases and images in order to draw general conclusions. Also, multilabels classifications will be considered in future works to investigate the discriminative power of our approach for each type of pathology.

TABLE 5: Comparison with the literature.

	Features	Classifier	Accuracy*
Mammograms			
[3]	Gabor	k-NN	80%
[4]	DT-CWT	SVM	88.64%
[5]	Contourlet	SVM	96.6%
Our approach	DWT-Gabor	SVM	96.67% ($\pm$0.05)
Retina			
[1]	DWT + GLCM	LDA	82.2%
[2]	Morphological + GLCM	Probabilistic boosting algorithm	81.3%–92.2% 71.7%–85.2%
[39]	Gabor	SVM	83%
Our approach	DWT-Gabor	SVM	100%
Brain			
[6]	DWT	SVM	98%
[8]	DWT + PCA	BPNN	100%
		SVM	90%
[38]	Voxels	Bayes	92%
		VFI	78%
Our approach	DWT-Gabor	SVM	97.36% ($\pm$0.02)

*Correct classification rate.

Finally, more experiments on the effect of kernel choice and its parameter on classification accuracy will be investigated.

References

[1] A. Khademi and S. Krishnan, "Shift-invariant discrete wavelet transform analysis for retinal image classification," *Medical and Biological Engineering and Computing*, vol. 45, no. 12, pp. 1211–1222, 2007.

[2] N. Lee, A. F. Laine, and T. R. Smith, "Learning non-homogenous textures and the unlearning problem with application to drusen detection in retinal images," in *Proceedings of the 5th IEEE International Symposium on Biomedical Imaging: From Nano to Macro (ISBI '08)*, pp. 1215–1218, Paris, France, May 2008.

[3] A. Dong and B. Wang, "Feature selection and analysis on mammogram classification," in *Proceedings of the IEEE Pacific Rim Conference on Communications, Computers and Signal Processing (PACRIM '09)*, pp. 731–735, Victoria, BC, Canada, August 2009.

[4] A. Tirtajaya and D. D. Santika, "Classification of microcalcification using dual-tree complex wavelet transform and support vector machine," in *Proceedings of the 2nd International Conference on Advances in Computing, Control and Telecommunication Technologies (ACT '10)*, pp. 164–166, Jakarta, Indonesia, December 2010.

[5] F. Moayedi, Z. Azimifar, R. Boostani, and S. Katebi, "Contourlet-based mammography mass classification using the SVM family," *Computers in Biology and Medicine*, vol. 40, no. 4, pp. 373–383, 2010.

[6] S. Chaplot, L. M. Patnaik, and N. R. Jagannathan, "Classification of magnetic resonance brain images using wavelets as input to support vector machine and neural network," *Biomedical Signal Processing and Control*, vol. 1, no. 1, pp. 86–92, 2006.

[7] Y. Zhang, S. Wang, and L. Wu, "A novel method for magnetic resonance brain image classification based on adaptive chaotic PSO," *Progress in Electromagnetics Research*, vol. 109, pp. 325–343, 2010.

[8] Y. Zhang, Z. Dong, L. Wu, and S. Wang, "A hybrid method for MRI brain image classification," *Expert Systems with Applications*, vol. 38, no. 8, pp. 10049–10053, 2011.

[9] M. E. Celebi, H. Iyatomi, G. Schaefer, and W. V. Stoecker, "Lesion border detection in dermoscopy images," *Computerized Medical Imaging and Graphics*, vol. 33, no. 2, pp. 148–153, 2009.

[10] Q. Abbas, M. E. Celebi, and I. F. García, "Skin tumor area extraction using an improved dynamic programming approach," *Skin Research and Technology*, vol. 18, pp. 133–142, 2012.

[11] Q. Li, F. Li, and K. Doi, "Computerized detection of lung nodules in thin-section CT images by use of selective enhancement filters and an automated rule-based classifier," *Academic Radiology*, vol. 15, no. 2, pp. 165–175, 2008.

[12] A. El-Bazl, M. Nitzken, E. Vanbogaertl, G. Gimel'jarb, R. Falfi, and M. Abo El-Ghar, "A novel shaped-based diagnostic approach for early diagnosis of lung nodules," in *Proceedings of the IEEE International Symposium in Biomedical Imaging (ISBI '11)*, pp. 137–140, Chicago, Ill, USA, 2011.

[13] M. T. Coimbra and J. P. S. Cunha, "MPEG-7 visual descriptors—contributions for automated feature extraction in capsule endoscopy," *IEEE Transactions on Circuits and Systems for Video Technology*, vol. 16, no. 5, pp. 628–636, 2006.

[14] B. Li and M. Q. H. Meng, "Texture analysis for ulcer detection in capsule endoscopy images," *Image and Vision Computing*, vol. 27, no. 9, pp. 1336–1342, 2009.

[15] C. K. Chui, *An Introduction to Wavelets*, Academic Press, San Diego, Calif, USA, 1992.

[16] M. Vetterli and C. Herley, "Wavelets and filter banks: theory and design," *IEEE Transactions on Signal Processing*, vol. 40, no. 9, pp. 2207–2232, 1992.

[17] J. G. Daugman, "Uncertainty relation for resolution in space, spatial frequency, and orientation optimized by two-dimensional visual cortical filters," *Journal of the Optical Society of America A*, vol. 2, no. 7, pp. 1160–1169, 1985.

[18] I. W. Selesnick, R. G. Baraniuk, and N. G. Kingsbury, "The dual-tree complex wavelet transform," *IEEE Signal Processing Magazine*, vol. 22, no. 6, pp. 123–151, 2005.

[19] E. Candès and D. Donoho, "Ridgelets: a key to higher-dimensional intermittency?" *Philosophical Transactions of the London Royal Society*, vol. 357, pp. 2495–2509, 1999.

[20] E. J. Candès and D. L. Donoho, "Continuous curvelet transform—I. Resolution of the wavefront set," *Applied and Computational Harmonic Analysis*, vol. 19, no. 2, pp. 162–197, 2005.

[21] M. N. Do and M. Vetterli, "The contourlet transform: an efficient directional multiresolution image representation," *IEEE Transactions on Image Processing*, vol. 14, no. 12, pp. 2091–2106, 2005.

[22] R. J. Ferrari, R. M. Rangayyan, J. E. L. Desautels, and A. F. Frère, "Analysis of asymmetry in mammograms via directional filtering with Gabor wavelets," *IEEE Transactions on Medical Imaging*, vol. 20, no. 9, pp. 953–964, 2001.

[23] Z. Cui and G. Zhang, "A novel medical image dynamic fuzzy classification model based on ridgelet transform," *Journal of Software*, vol. 5, no. 5, pp. 458–465, 2010.

[24] T. Gebäck and P. Koumoutsakos, "Edge detection in microscopy images using curvelets," *BMC Bioinformatics*, vol. 10, article 75, 2009.

[25] J. Ma and G. Plonka, "The curvelet transform: a review of recent applications," *IEEE Signal Processing Magazine*, vol. 27, no. 2, pp. 118–133, 2010.

[26] N. Kingsbury, "Complex wavelets and shift invariance," in *Proceedings of the IEEE Seminar on Time-Scale and Time-Frequency Analysis and Applications*, pp. 501–510, London, UK, 2000.

[27] Y. L. Qiao, C. Y. Song, and C. H. Zhao, "M-band ridgelet transform based texture classification," *Pattern Recognition Letters*, vol. 31, no. 3, pp. 244–249, 2010.

[28] F. Gómez and E. Romero, "Texture characterization using a curvelet based descriptor," *Lecture Notes in Computer Science*, vol. 5856, pp. 113–120, 2009.

[29] H. Shan and J. Ma, "Curvelet-based geodesic snakes for image segmentation with multiple objects," *Pattern Recognition Letters*, vol. 31, no. 5, pp. 355–360, 2010.

[30] R. Eslami and H. Radha, "New image transforms using hybrid wavelets and directional filter banks: analysis and design," in *Proceedings of the IEEE International Conference on Image Processing (ICIP '05)*, pp. 733–736, Genova, Italy, September 2005.

[31] O. O. V. Villegas and V. G. C. Sánchez, "The wavelet based contourlet transform and its application to feature preserving image coding," *Lecture Notes in Computer Science*, vol. 4827, pp. 590–600, 2007.

[32] S. Lahmir and M. Boukadoum, "Classification of brain MRI using the LH and HL wavelet transform sub-bands," in *Proceedings of the IEEE International Symposium on Circuits and Systems (ISCAS '11)*, pp. 1025–1028, Rio de Janeiro, Brazil, May 2009 2011.

[33] S. Lahmir and M. Boukadoum, "Brain MRI classification using an ensemble system and LH and HL wavelet Sub-bands Features," in *Proceedings of the IEEE Symposium Series on Computational Intelligence (SSCI '11)*, pp. 1–7, Paris, France, April 2011.

[34] S. Lahmir and M. Boukadoum, "Hybrid Cosine and Radon Transform-based processing for Digital Mammogram Feature Extraction and Classification with SVM," in *Proceedings of the 33rd IEEE Annual International Conference on Engineering in Medecine and Biology Society (EMBS '11)*, pp. 5104–5107, Boston, Mass, USA, 2011.

[35] S. Lahmir and M. Boukadoum, "DWT and RT-Based Approach for Feature Extraction and classification of Mammograms with SVM," in *Proceedings of the IEEE Biomedical Circuits and Systems Conference (BioCAS '11)*, pp. 412–415, San Diego, Calif, USA, November 2011.

[36] V. N. Vapnik, *The Nature of Statistical Learning Theory*, Springer, 1995.

[37] S. Lahmiri and M. Boukadoum, "Hybrid discret wavelet transform and Gabor filter banks processing for mammogram features extraction," in *Proceedings of the IEEE New Circuits and Systems (NEWCAS '11)*, pp. 53–56, Bordeaux, France, June 2011.

[38] L. M. Bruce and N. Shanmugam, "Using neural networks with wavelet transforms for an automated mammographic mass classifier," in *Proceedings of the 22nd Annual International Conference of the IEEE Engineering in Medicine and Biology Society*, pp. 985–987, Chicago, Ill, USA, July 2000.

[39] S. M. H. Jamarani, G. Rezai-rad, and H. Behnam, "A novel method for breast cancer prognosis using wavelet packet based neural network," in *Proceedings of the 27th Annual International Conference of the Engineering in Medicine and Biology Society (IEEE-EMBS '05)*, pp. 3414–3417, Shanghai, China, September 2005.

[40] http://medical-dictionary.thefreedictionary.com/circinate+retinopathy.

[41] C. I. O. Martins, F. N. S. Medeiros, R. M. S. Veras, F. N. Bezerra, and R. M. Cesar Jr., "Evaluation of retinal vessel segmentation methods for microaneurysms detection," in *Proceedings of the IEEE International Conference on Image Processing (ICIP '09)*, pp. 3365–3368, Cairo, Egypt, November 2009.

[42] C. Plant, S. J. Teipel, A. Oswald et al., "Automated detection of brain atrophy patterns based on MRI for the prediction of Alzheimer's disease," *NeuroImage*, vol. 50, no. 1, pp. 162–174, 2010.

[43] J. Meier, R. Bock, L. G. Nyúl, and G. Michelson, "Eye fundus image processing system for automated glaucoma classification," in *Proceedings of the 52nd Internationales Wissenschaftliches Kolloquium*, Technische Universität Ilmenau, 2007.

[44] E. Sakka, A. Prentza, I. E. Lamprinos, and D. Koutsouris, "Microcalcification detection using multiresolution analysis based on wavelet transform," in *Proceedings of the IEEE International Special Topic Conference on Information Technology in Biomedicine*, Ioannina, Greece, October 2006.

[45] S. G. Mallat, "Theory for multiresolution signal decomposition: the wavelet representation," *IEEE Transactions on Pattern Analysis and Machine Intelligence*, vol. 11, no. 7, pp. 674–693, 1989.

[46] J. Suckling, J. Parker, and D. R. Dance, "The mammographic image analysis society digital mammogram database," in *Proceedings of the the 2nd International Workshop on Digital Mammography*, A. G. Gale, S. M. Astley, D. D. Dance, and A. Y. Cairns, Eds., pp. 375–378, Elsevier, York, UK, 1994.

[47] H. J. Chiou, C. Y. Chen, T. C. Liu et al., "Computer-aided diagnosis of peripheral soft tissue masses based on ultrasound

imaging," *Computerized Medical Imaging and Graphics*, vol. 33, no. 5, pp. 408–413, 2009.

[48] J. K. Kim, J. M. Park, K. S. Song, and H. W. Park, "Adaptive mammographic image enhancement using first derivative and local statistics," *IEEE Transactions on Medical Imaging*, vol. 16, no. 5, pp. 495–502, 1997.

[49] H. S. Sheshadri and A. Kandaswamy, "Breast tissue classification using statistical feature extraction of mammograms," *Medical Imaging and Information Sciences*, vol. 23, no. 3, pp. 105–107, 2006.

[50] N. Cristianini and J. Shawe-Taylor, *Introduction to Support Vector Machines and Other Kernel-Based Learning Methods*, Cambridge University Press, Cambridge, UK, 2000.

[51] L. Chen, X. Mao, Y. Xue, and L. L. Cheng, "Speech emotion recognition: features and classification models," *Digital Signal Processing*, vol. 22, pp. 1154–1160, 2012.

[52] F. Palmieri, U. Fiore, A. Castiglione, and A. De Santis, "On the detection of card-sharing traffic through wavelet analysis and Support Vector Machines," *Applied Soft Computing*, vol. 13, no. 1, pp. 615–627, 2013.

[53] A. Azadeh, M. Saberi, A. Kazem, V. Ebrahimipour, A. Nourmohammadzadeh, and Z. Saberi, "A flexible algorithm for fault diagnosis in a centrifugal pump with corrupted data and noise based on ANN and support vector machine with hyperparameters optimization," *Applied Soft Computing*, vol. 13, no. 3, pp. 1478–1485, 2013.

[54] R. J. Martis, U. R. Acharya, K. M. Mandana, A. K. Ray, and C. Chakraborty, "Cardiac decision making using higher order spectra," *Biomedical Signal Processing and Control*, vol. 8, pp. 193–203, 2013.

[55] M. R. Mohammad, "Chi-square distance kernel of the gaits for the diagnosis of Parkinson's disease," *Biomedical Signal Processing and Control*, vol. 8, pp. 66–70, 2013.

[56] R. Vandenberghe, N. Nelissen, E. Salmon et al., "Binary classification of 18F-flutemetamol PET using machine learning: comparison with visual reads and structural MRI," *NeuroImage*, vol. 64, no. 1, pp. 517–525, 2013.

[57] http://marathon.csee.usf.edu/Mammography/Database.html.

[58] http://www.ces.clemson.edu/~ahoover/stare/.

[59] http://www.med.harvard.edu/aanlib/.

Spectroscopic Detection of Caries Lesions

Mika Ruohonen,[1] **Katri Palo,**[2] **and Jarmo Alander**[1]

[1] *Faculty of Technology, University of Vaasa, P.O. Box 700, 65101 Vaasa, Finland*
[2] *Dental Services of the City of Vaasa, Social and Health Administration, P.O. Box 241, 65101 Vaasa, Finland*

Correspondence should be addressed to Mika Ruohonen; mika.ruohonen@uwasa.fi

Academic Editor: Hengyong Yu

Background. A caries lesion causes changes in the optical properties of the affected tissue. Currently a caries lesion can be detected only at a relatively late stage of development. Caries diagnosis also suffers from high interobserver variance. *Methods.* This is a pilot study to test the suitability of an optical diffuse reflectance spectroscopy for caries diagnosis. Reflectance visible/near-infrared spectroscopy (VIS/NIRS) was used to measure caries lesions and healthy enamel on extracted human teeth. The results were analysed with a computational algorithm in order to find a rule-based classification method to detect caries lesions. *Results.* The classification indicated that the measured points of enamel could be assigned to one of three classes: healthy enamel, a caries lesion, and stained healthy enamel. The features that enabled this were consistent with theory. *Conclusions.* It seems that spectroscopic measurements can help to reduce false positives at *in vitro* setting. However, further research is required to evaluate the strength of the evidence for the method's performance.

1. Introduction

Minimally invasive dentistry is an approach that seeks to maintain the patient's oral health with preventive measures and to treat possible disturbances of health as early as possible and with as little intervention as possible [1]. This requires that caries is detected at an early stage of development and that its status can be monitored frequently [2]. However, the current methods for diagnosing caries are able to detect caries only at a relatively advanced stage. Accordingly, methods for early detection of caries have been researched for the past twenty years. Many of these methods still require extensive research before they can be used in clinical practice. Optical caries diagnosis methods are based on the fact that caries cause changes in the tooth's optical properties at an early stage of development [3].

This was a pilot study to investigate whether diffuse reflectance visible/near-infrared spectroscopy (VIS/NIR-S) can be used to detect dental caries lesions. Reflectance spectroscopy measures the intensity of light at several different wavelengths, that is, its spectra, after the light has reflected from the studied object. Diffuse reflectance refers to light that has been reflected from the inside of the object, rather than from its surface. In this study the intensity was measured at wavelengths in the visible range and at wavelengths in the near-infrared range, covering wavelengths in the range 420–1000 nanometers. Within this range, the intensity was measured at 2305 different wavelengths, so that the difference between consecutive wavelengths was approximately 0.25 nm. This study was limited to studying natural caries lesions that could be diagnosed with fiber-optic illumination, on smooth surfaces of extracted tooth.

A theory of caries diagnosis using near-infrared spectroscopy emerges from the previous studies of detecting caries lesions with near-infrared light [2–7]. According to this theory, the development of a caries lesion increases the porosity of the affected tissue, which in turn leads to an increased scattering of light in the lesion. Wavelengths in the near-infrared range are considered better than the wavelengths in the visible range, because the former can penetrate deeper into the tissue and are less affected by stains on the tooth. The purpose of this study was to provide additional evidence in support of this theory. More work on this topic can be found in [8–15].

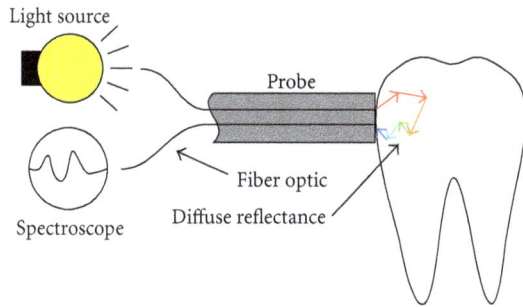

FIGURE 1: An illustration of the measurement setup.

2. Methods

2.1. Samples. The dental services of the City of Vaasa provided extracted human teeth for the study. The teeth were stored immersed in denatured alcohol in order to disinfect them and to keep them hydrated. Before inspection and measurements, the teeth were gently dried with a cue tip. The teeth were inspected by the first author with fiber-optic illumination, after the technique was introduced to him by the second author, in order to detect healthy areas of enamel and areas of enamel that contained caries lesions.

In total 21 teeth were used in the study. A total of 109 points of enamel were measured on the teeth, consisting of 69 points which were thought to represent healthy enamel and 40 points which were thought to represent caries lesions. Each measurement point produced a spectra, a sample for the rest of the analysis. In pattern recognition terminology the diagnosis of a given sample, as either healthy or carious, is called the label of the sample. The analysis of the samples tries to create a method which estimates the diagnosis, the label, of the sample based only on the measurements. The resulting estimates are called predictions.

2.2. Measurements. The measurement setup is presented in Figure 1. An optical fiber, placed in contact with the sample, conveys light from a light source to the sample. The light enters the sample and scatters to all directions inside of it. Another optical fiber is placed in contact with the sample at a small distance from the first fiber. Some fraction of the light which scatters inside the sample will eventually exit the sample so that it enters the second optical fiber. It then gets conveyed to a spectrometer, which measures the spectra of the reflected light. Properties of the sample material affect the measured spectra. Photonics describes the key properties with the absorption coefficient and the scattering coefficient of the material. The measured spectra is analyzed in order to deduce information about the sample material.

The measurements were made with a spectrometer HR4000 (Ocean Optics Inc., Dunedin, FL, USA) and with a general purpose transmission dip probe model T300-RT-VIS/NIR (Ocean Optics Inc., Dunedin, FL, USA). The probe contains two optical fibers, both with a diameter of $300\,\mu m$, housed in a stainless steel assembly with a diameter of 3.175 mm. The assembly is surrounded by a ferrule with a diameter of 6.35 mm. One of the fibers is connected to a light source and brings light to the sample. The light source used in this study was a tungsten halogen lamp HL-2000 (Ocean Optics Inc., Dunedin, FL, USA). The other fiber is connected to the spectrometer. It collects and transmits the diffusely reflected light. Construction of a custom probe for this study was deemed unfeasible. Thus, the study had to be carried out with a probe that was readily available in our laboratory. The selected probe is designed for measuring the transmission spectra of liquid samples. However, it was considered to be suitable for this study when the ferrule enclosing the inner assembly was removed, exposing the stainless steel assembly that houses the fiber optics.

The period of time for which the spectrometer collects light when it is making one measurement is called the integration time. In this study integration time was set to 20 milliseconds. A longer integration time produces better measurement results than a short one, because the intensity of the collected light increases at all wavelengths, yielding a better signal-to-noise ratio (SNR). Therefore, the integration time is typically set as long as possible. However, if the intensity of the collected light at a given wavelength exceeds the measurement range of the spectrometer, the spectrometer saturates. In that case the intensity cannot be measured, and we know only that it exceeded the maximum measurable value.

In order to make the measurement results comparable to results that would have been obtained with the same spectroscope using another light source or another integration time, the spectroscope has to be calibrated for these factors. This is done by measuring the smallest and the greatest intensity value that a measurement can produce with the given integration time for each wavelength and by scaling all other measurements to that range. This gives values between zero and one for all wavelengths. These scaled results are called normalized intensities. The lowest possible intensity values are obtained by measuring the so-called dark current, which is caused by thermal noise. Measuring a white reference sample produces the greatest possible intensity values. In this study, the integration time was set so that the white reference sample (a white reference tile WS-2, Avantes Inc., Eerbeek, The Netherlands) did not saturate at any wavelength. A spectrometer must also be calibrated for its detector, so that its measurement results are comparable to those obtained by other spectroscopes. This is done by measuring the spectra of a sample whose spectra is known. In this study the used spectroscope was calibrated for its detector by the manufacturer as part of its construction.

A spectroscopic measurement result contains many small random errors, which are collectively called (thermal) noise. These errors are caused by heat, or thermal energy, in the spectroscope. They follow a normal distribution with a given mean value. The dark current presents the mean value of the noise for each wavelength. When the dark current is subtracted from the spectra, the mean value of the effect caused by the noise is shifted to zero, and thus the effect of noise is observed as errors which have a normal distribution with a zero mean. In order to minimize the effect of noise in the samples, each point was measured one hundred times consecutively, and the resulting spectra were averaged. This meant that the probe needed to stay as motionless as possible for two seconds. However, a far shorter time period would have probably been sufficient.

2.3. Analysis. As a further measure against noise, the samples were smoothed by using the Savitzky-Golay method with a window length of 61 and sixth degree polynomials. This method selects the coefficients of a sixth degree polynomial so that the polynomial is the best possible approximation for the measurement result, that is, the spectra, for the 30 wavelengths before a given wavelength and for the 30 wavelengths after it. The value of the polynomial at the given wavelength replaces the measured intensity at that wavelength. This removes, or smoothens, fast and small changes in the spectra, which are mainly caused by noise.

A simple computational algorithm, based on exhaustive search, was then used to find a set of rules that could be used for detecting caries lesions. At this point, the goal was to classify the samples into two classes: points on healthy enamel (healthy samples) and points on caries lesions (carious samples). For this, a set of rules was searched for, so that every rule had the following format: if the sample's normalized intensity at a given wavelength λ is greater than (or smaller than) a given threshold I^*, the sample is classified as carious; otherwise, the sample is classified as healthy. Thus, each rule had three parameters: the wavelength λ, the intensity threshold I^*, and whether or not the threshold is an upper or lower limit for the intensity. If, and only if, one or more of the rules classified the sample as carious, the sample was classified as carious. If none of the rules considered the sample as carious, it was classified as healthy. A pseudocode for this step is given in Pseudocode 1. It was hoped that the algorithm would select a set of rules which resembles the results found in earlier studies on this subject.

A number of wavelengths were selected from the range of available wavelengths ($\approx$420–1000 nm) as options for parameter λ in the search, so that the intervals between the wavelengths were equal and the first and the last wavelength were always selected as options. A pseudo-code for this is given in Pseudocode 2. The search was done with different numbers of wavelengths. For each of the selected wavelengths, the algorithm sorted the samples' intensities at that wavelength and considered the midpoint between each two consecutive intensities as a possible threshold I^* in a rule. A pseudo-code for this is given in Pseudocode 3.

The algorithm calculated the classification accuracy on the training set for each of the pairs λ and I^* described above, using the threshold I^* first as an upper limit for classifying the sample as carious and then using it as a lower limit, and chose the values of λ and I^* and the type of threshold, which gave the best accuracy (see pseudo-code at Pseudocode 4). After a rule had been selected this way, the algorithm selected another rule with the same method, so that the new rule gave the best possible accuracy when used together with the previously selected rule(s). This was continued until the maximum allowed number of rules, here five rules, was reached, or until the classifier was unable to find a new rule

which would improve the classification accuracy. A pseudo-code for this logic is given in Pseudocode 5.

This algorithm, like every machine learning method, requires a set of samples which is used for searching for the rules and a separate set of samples which is used for evaluating the accuracy that is achieved with the resulting rules. The former set of samples is called the training set and the latter set is called the validation set. The number of samples available for this study was rather limited. This may cause problems for the machine learning method when the samples are divided into a training set and a validation set, because some types of samples may become overrepresented in the training set, misleading the learning method as it tries to recognize what discerns the two classes from each other.

In this study, this risk was alleviated by using a 4-fold cross-validation. In this method, the samples are divided into four groups, and one of them is used as the validation set while the other three groups form the training set. Each group in turn is used as the validation set, and the results from these four "folders" are averaged. This way each sample is a part of the training set in three folders and a part of the validation set in one folder. It is unlikely that the same types of samples would be overrepresented in all four training sets, unless the entire set of available samples has this problem. A single training set which has this problem would stand out from the others, and the skewed learning results from it would be corrected by the results from the other training sets. While the small set of samples may still give a skewed representation of the kinds of samples which are being studied, the cross-validation seeks to minimize this problem.

In this study the averaging was done so, that a median rule set was constructed from the rules which the algorithm selected for the folders, and all of the samples were then classified with the median rule set. Median of the numbers of rules in the folders determined the number of rules in the median rule set. Some manual deliberation was used when constructing the rules of the set. For each rule in the final set, a temporary rule set was composed by selecting one rule from each folder's rule set, so that the rules in the temporary set resembled each other, if that was possible given the available rules. The median of the wavelengths used in the rules in the temporary rule set determined the wavelength for the rule in the median rule set. The intensity threshold and the type of threshold were selected similarly for the rule in the median rule set. A pseudo-code for this is given in Pseudocode 6.

Each sample was diagnosed as either healthy or carious by the first author, and the selected rules estimated each sample to be either healthy or carious. Based on these two properties, the samples can be divided into four classes. Samples which were diagnosed as healthy and which were estimated to be healthy by the rules are called true negatives (TNs). Similarly, carious samples which were correctly estimated are called true positives (TPs). A healthy sample which was estimated to be carious is called false positive (FP) and a carious sample which was estimated to be healthy is called a false negative (FN). The sizes of these classes comprise a confusion matrix, or a contingency table. These four values can be used to calculate the following five values which describe the accuracy of the selected rules.

```
CLASSIFY(R, x̄)
(1) // Classify sample x̄ using the set of rules R
(2) for i = 1 to R.length
(3)     t = R[i].limitThreshold    // Threshold intensity I*
(4)     j = R[i].limitIndex    // Index of wavelength λᵢ
(5)     if R[i].limitType == UPPER and xⱼ > t
(6)         return (+1)
(7)     elseif R[i].limitType == LOWER and xⱼ < t
(8)         return (+1)
(9) return (−1)    // No rule indicated sample as positive
```

PSEUDOCODE 1: Pseudocode for classifying a sample. Samples in the positive class are carious, and samples in the negative class are healthy. A sample $\vec{x}$ is a vector, where each component x_i equals the normalized intensity at a given wavelength λ_i.

```
GET-WAVELENGTH-INDEX(X, i)
(1) // Get the ith wavelength option for a rule, given a set of samples X
(2) // For first wavelength, i = 1
(3) s = (X.maxWavelength − X.minWavelength)/WAVELENOPTIONCOUNT
(4) λ = X.minWavelength + s(i− 1)
(5) // Get the index of the measured wavelength λᵢ, which is closest to λ
(6) j = CLOSEST-INDEX(λ)
(7) return j
```

PSEUDOCODE 2: Pseudocode for computing the ith wavelength option for a rule.

(i) Positive predictive value (PPV) is the probability that the classifier, that is, the set of rules, is correct when it estimates a sample to be carious.

(ii) Negative predictive value (NPV) is the probability that a healthy estimate is correct.

(iii) Sensitivity is the fraction of all carious samples that were classified as carious.

(iv) Specificity is the fraction of healthy samples that were classified as healthy.

(v) Accuracy is the fraction of the samples which were correctly estimated, that is, where the rules gave the correct answer.

2.4. Two Hypotheses of Misdiagnosis. After the classification rules had been selected and the samples had been classified according to them, there were fifteen samples which the author had diagnosed as carious but which were classified as healthy (false negatives). The spectra of these samples were virtually indistinguishable from the spectra of the healthy samples (see Figure 3(a)), at least for the analysis methods used in this study. Thus, a hypothesis was made that these samples, the false negative cases, had been misdiagnosed by the author and subsequently mislabeled.

The rules that were selected by the algorithm suggested that a short wavelength, namely, 420 nm, was relatively useful in the diagnosis of caries. This was inconsistent with the theory on the optical diagnosis of caries. Therefore, another hypothesis was made, according to which a number of samples had been diagnosed by the author as carious while

in fact the measured points were only stained and were thus false positive cases of the diagnosis, even if they had been classified correctly by the classifier. A pair of rules was manually selected in order to detect such stained samples. These rules were $I(\lambda \approx 420) \leq 0.206 \wedge I(\lambda \approx 815) \leq 0.313$. Notation $I(\lambda)$ refers to the normalized intensity of the spectra at wavelength λ. In other words, the sample was thought to represent a stain if it had a small scattering coefficient at both a long wavelength (815 nm, which is in the near-infrared range) and a short wavelength (420 nm). Application of these rules identified eight samples as being misdiagnosed due to a stain.

3. Results

The samples, or the spectra of the measured points, are presented in Figure 2. The number of wavelengths which were selected as options for the rule's parameter λ, that is, parameter WAVELENGTHOPTIONCOUNT, had only a small effect on the accuracy of the resulting median rule set. When only the shortest wavelength ($\approx$420 nm) and the longest wavelength ($\approx$1000 nm) were available as options, the median rule set had an accuracy of 82%. With three wavelengths to choose from, the accuracy was 83%. When the number of options was between four and six, the accuracy was 85%. With greater numbers of wavelengths available, the accuracy was 84%.

The selected rules were very similar in all folders. This suggested that the rules depicted a phenomenon which was consistently present in all four folders. When the number of options for the rules' wavelengths was 15, the median rule

GET-THRESHOLDS($X, \lambda^{'}$)
(1) // Get the threshold options for a rule, given a set of samples X and
(2) // an index of wavelength.
(3) // Use local variables, arrays A and M
(4) **for** $i = 1$ **to** $X.sampleCount$
(5) $A[i] = X[i][\lambda^{'}]$ // Intensity at λ_i for sample $\bar{x}_i$
(6) $A = $ SORT(A)   // Ascending or descending
(7) **for** $i = 1$ **to** $A.length - 1$
(8) $M[i] = (A[i] + A[i + 1])/2$
(9) **return** M

PSEUDOCODE 3: Pseudocode for computing the threshold options for a rule at a given measured wavelength λ_i. The wavelength is defined by its index, $\lambda^{'} = i$.

FIND-NEW-RULE(R, X)
(1) // Select a new rule, given a set of rules R and a set of samples X
(2) // Use local variables, rules Q and B
(3) $b = 0.0$ // Best accuracy found so far
(4) **for** $i = 1$ **to** WAVELENOPTIONCOUNT $+ 1$
(5) $\lambda^{'} = $ GET-WAVELENGTH-INDEX(X, i)
(6) $Q.limitIndex = \lambda^{'}$ // Rule's wavelength λ, by index
(7) $T = $ GET-THRESHOLDS($X, \lambda^{'}$)
(8)   **for** $j = 1$ **to** $T.length$
(9) $Q.limitThreshold = T[j]$ // Rule's threshold intensity I^{*}
(10) $Q.limitType = $ UPPER
(11) $a = $ CLASSIFY-SAMPLES($R + Q, X$)   // Classification accuracy
(12)   **if** $a > b$
(13) $B = Q$
(14) $b = a$
(15) $Q.limitType = $ LOWER
(16) $a = $ CLASSIFY-SAMPLES($R + Q, X$)   // Classification accuracy
(17)   **if** $a > b$
(18) $B = Q$
(19) $b = a$
(20) **return** B // Best new rule found

PSEUDOCODE 4: Pseudocode for selecting a new rule.

SELECT-RULES(X)
(1) // Select the set of rules for given set of samples X
(2) // Use local variable, set of rules R
(3) $a = 0.0$ // Accuracy with current set of rules
(4) $R = \varnothing$ // Current set of rules
(5) **for** $i = 1$ **to** MAXRULECOUNT
(6) $B = $ FIND-NEW-RULE(R, X)
(7) $b = $ CLASSIFY-SAMPLES($R + B, X$)   // Classification accuracy
(8)   **if** $a \geq b$
(9)   **return** R // New rule did not help
(10) $R = R + B$ // Add new rule to set
(11) $a = b$
(12) **return** R

PSEUDOCODE 5: Pseudocode for selecting the set of rules. Here X is the set of training samples and MAXRULECOUNT $= 5$.

```
COMPOSE-MEDIAN-RULES(S)
(1) // Compose median rule set from given set of rule sets S
(2) // S = (R₁, R₂,..., Rₙ), n = FOLDERCOUNT
(3) // Use local variables, sets of rules M and T, and rule Q
(4) M = ∅
(5) N = MEDIAN(R₁.length, R₂.length,..., Rₙ.length)
(6) for i = 1 to N
(7)     // Compose temporary rule set, T = (T₁, T₂,···,Tₙ)
(8)     // If possible, have T₁ ≈ T₂ ≈ ... ≈ Tₙ
(9)     // Each rule in R ∈ S appears in at most one temporary rule set T
(10)    T = COMPOSE-TEMP-SET(S)
(11)    Q.limitIndex = MEDIAN(T₁.limitIndex,..., Tₙ.limitIndex)
(12)    Q.limitThreshold = MEDIAN(T₁.limitThreshold,..., Tₙ.limitThreshold)
(13)    Q.limitType = MEDIAN(T₁.limitType,..., Tₙ.limitType)
(14)    M = M + Q
(15) return M
```

PSEUDOCODE 6: Pseudocode for computing the median rule set. In this study FOLDERCOUNT = 4.

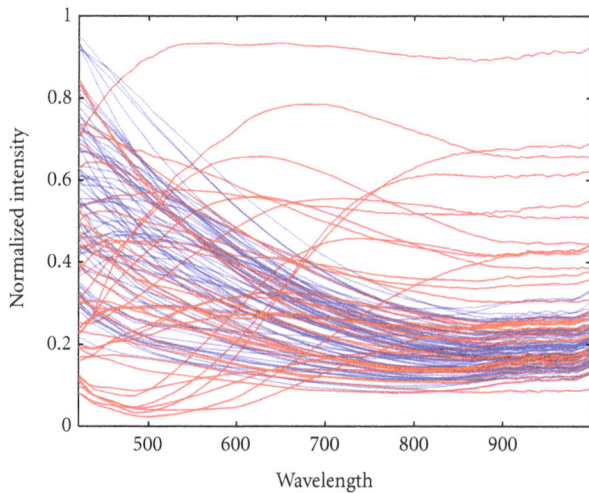

FIGURE 2: The samples, that is, the spectra of the measured points. The blue curves depict samples which were diagnosed as healthy and the red curves depict samples which were diagnosed as carious.

TABLE 1: The confusion matrix, or the contingency table, of the median rule set.

	Carious	Healthy	
Estimated carious	25 (TP)	2 (FP)	93% (PPV)
Estimated healthy	15 (FN)	67 (TN)	82% (NPV)
	63% (Sens.)	97% (Spec.)	84% (Acc.)

set indicated that a sample is carious if, and only if, $I(\lambda \approx 420) \leq 0.2642 \vee I(\lambda \approx 750) \geq 0.3502$. A confusion matrix of the classification accuracy that is achieved with these rules is presented in Table 1, showing that these rules reached an accuracy of 84%.

As can be seen in Figure 3(a) and in Table 1, there were fifteen carious samples which were classified as healthy (false negatives), and whose spectra was virtually indistinguishable from the spectra of the healthy samples. As explained in Section 2.4, this leads to a hypothesis that these samples had been misdiagnosed and subsequently mislabeled, and that they therefore represented healthy samples and were in fact classified correctly.

According to the theory on optical caries diagnosis, an elevated intensity in the near-infrared range is the best indication of a dental caries lesion. However, the rules selected

by the search algorithm indicated that a short wavelength, namely, 420 nm, was relatively useful in the diagnosis of caries. As explained in Section 2.4, another hypothesis was thus made, according to which a number of samples had been diagnosed as carious while in fact they were only stained. A pair of rules was manually selected in order to detect such stained samples.

Application of these rules identified eight samples as being misdiagnosed due to a stain. All samples that were identified as stained had been diagnosed and classified as carious, and thus appeared to be true positive cases. These suspected misdiagnoses had not lowered the apparent accuracy of the classification, but they may have caused the rule set to erroneously consider stains as caries lesions.

When the search algorithm was run again, giving 15 options for the parameter λ, after first relabeling the fifteen false negative cases as healthy samples (first hypothesis) and then relabeling the eight suspected stains as healthy samples (second hypothesis), the algorithm selected only one rule in every cross-validation folder. All rules set an upper limit for the normalized intensity at a wavelength in the near-infrared range. If the intensity was greater than this, the sample was classified as carious. The median of those rules was $I(\lambda \approx 791) \geq 0.3255$, which is consistent with the theory. The confusion matrix of this rule is presented in Table 2. This rule produced an accuracy of 97%.

(a)

(b)

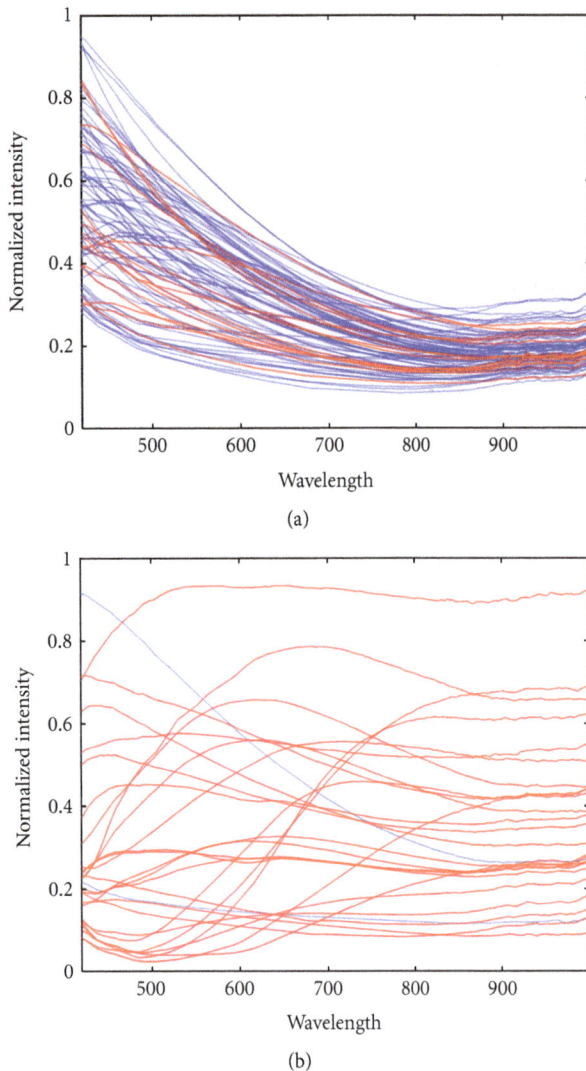

FIGURE 3: Samples which were classified (a) as healthy and (b) as carious by the median rule set. Blue curves represent healthy samples and red curves represent carious samples. The samples which were diagnosed as carious but classified as healthy (false negatives) are emphasized.

TABLE 2: The confusion matrix, or the contingency table, for the median rule (set) which was selected after relabeling the samples according to the two hypotheses of misdiagnosis.

	Carious	Healthy	
Estimated carious	14 (TP)	0 (FP)	100% (PPV)
Estimated healthy	3 (FN)	92 (TN)	97% (NPV)
	82% (Sens.)	100% (Spec.)	97% (Acc.)

4. Discussion

This study suffers from a small number of samples. Although the study used 109 measurements, they were taken from only 21 individual teeth. This fact is significant, because it is probable that samples taken from a single tooth resemble each other more than samples taken from different teeth or from different patients. Furthermore, the 109 measurements contained only 40 measurements from a caries lesion. Fifteen of those measurements were considered to be misdiagnosed by the first hypothesis, and further eight measurements were considered to be misdiagnosed by the second hypothesis. Therefore, further study is needed to increase the reliability of the accuracy estimate of this method.

The measurement results together with the theory on the topic suggest that many of the measurements which were supposedly made from a caries lesion are in fact made from healthy enamel, which was in some cases stained. When we make these suggested corrections to the labeling of the samples, the samples seem to fit well to the theory and the samples can easily be accurately classified. These kinds of diagnostic mistakes, or false positive diagnoses, are a credible explanation, because the diagnoses were made by a novice on the subject. However, such corrections also pose a risk that the measurement results are relabeled to make them fit the theory, which would inflate the accuracy of the method. Further study of the method might dispel such possibilities.

The composition of the dental tissues varies from tooth to tooth and between different sites of a given tooth [16]. As can be seen in Figure 2, the spectra of the different healthy samples vary quite a bit, especially at the visible wavelengths. This suggests that the threshold intensity or intensities for diagnosing a suspected lesion as carious might also vary similarly. In order to compensate for the inter-tooth and intra-tooth variance, we might consider measuring the average spectra for a given tooth by measuring several points on the tooth surface, that is, by scanning the surface and by evaluating how much the spectra of the suspected lesion differ from the tooth's average.

Unfortunately, this approach could potentially make this method less effective for its original purpose. The method is being developed for the detection of caries lesions at an early stage of development. Thus, the dentist does not necessarily notice all of the lesions which are detected by the device. If the inspected tooth surface contains several developing caries lesions, the average spectra of the surface could be something in between the healthy enamel and the carious enamel, making the lesions appear too similar to the average surface to be diagnosed as carious. A set of fixed thresholds would avoid this problem. The scanning method would also make it rather awkward to inspect several teeth per patient.

Quantitative Light-induced Fluorescence (QLF) and Laser-induced Fluorescence (LF) are two optical methods for the detection of caries lesions. They are based on fluorescence, or the phenomenon that when the tooth sample is illuminated with a light source, some of the light is absorbed in the sample, after which the sample emits light at a longer wavelength. For both methods the emitted wavelength falls within the range of measured wavelengths [3]. In this study the sample was considered carious if the measured intensity was greater than a fixed threshold, $I(\lambda \approx 791) \geq 0.3255$. The proposed explanation is that the increased scattering due to caries causes more light to be reflected to the measuring fiber optic. QLF expects to find a reduced intensity for carious samples at wavelengths $\lambda > 520$ nm because increased scattering due to caries interferes with the detection of the fluorescence, and LF expects to find increased intensity at the near-infrared range caused by fluorescence from organic molecules in the sample [3].

Since the samples in this study were stored in denatured alcohol, they were probably relatively free of organic molecules. Further study is required to determine whether the fluorescence from organic molecules, that is, the phenomenon measured by LF, interferes with the detection method outlined in this study, especially for *in vivo* measurements. If it does interfere, it probably makes the method more eager to label a sample as carious, thus increasing its sensitivity and reducing its specificity. This effect may be modified, at least in part, by selecting a new set of rules based on results from *in vivo* measurements. Incidentally, low specificity has been cited as a major weakness of the LF method [3]. In contrast, authors of this study felt that the method outlined in this paper helped them to increase specificity.

An ability to measure the amount of dental tissue lost to caries could be pursued by inducing caries *in vitro* to a tooth sample (see [6, 17, 18]) so that the amount of mineral dissolved from the tooth could be measured without destroying the sample, and by measuring the spectra of the sample at varying degrees of mineral loss. One possible method for this would be to cycle the tooth sample in de- and remineralization solutions and to measure the amount of mineral dissolved to the solutions with a mass spectrometer. This would have to be repeated with a sufficient number of samples. Finding a method to calculate the amount of the mineral loss from the spectra would be a regression problem.

5. Conclusions

It seems that spectroscopic measurements can help to reduce false positives at *in vitro* setting, including those caused by stains. This method may also give objective evidence of the presence of a caries lesion. However, the work reported in this paper was a pilot study, and further research is required to evaluate the strength of the evidence for the method's performance at *in vitro* setting and to extend the measurements to *in vivo* setting.

Acknowledgments

The authors would like to acknowledge the Field-NIRce project for funding this study and Professor Paul Geladi for his role in organizing the project. The project was funded by Bothnia-Atlantica, the European Union, Regional Council of Ostrobothnia, Region Västerbotten, and Provincial Government of Västerbotten. The authors would also like to acknowledge the support that this study has received from the Dental Services of the City of Vaasa, particularly the previous Chief Dental Officer Ph.D. Jukka Kentala. This study has greatly benefited from the help and guidance of Professor Jouni Lampinen, DSc Petri Välisuo, and Dr. Vladimir Bochko. Finally, the authors wish to thank the anonymous reviewers, whose comments helped to improve the quality of this paper.

References

[1] N. Wilson and A. Plasschaert, "Dental caries, minimally invasive dentistry and evidencebased clinical practice," in *Minimally Invasive Dentistry-the Management of Caries*, N. H. F. Wilson, Ed., pp. 1–6, Quintessence, 2007.

[2] R. S. Jones, G. D. Huynh, G. C. Jones, and D. Fried, "Near-infrared transillumination at 1310-nm for the imaging of early dental decay," *Optics Express*, vol. 11, no. 18, pp. 2259–2265, 2003.

[3] L. Karlsson, "Caries detection methods based on changes in optical properties between healthy and carious tissue," *International Journal of Dentistry*, vol. 2010, Article ID 270729, 9 pages, 2010.

[4] C. M. Bühler, P. Ngaotheppitak, and D. Fried, "Imaging of occlusal dental caries (decay) with near-IR light at 1310-nm," *Optics Express*, vol. 13, no. 2, pp. 573–582, 2005.

[5] R. S. Jones, *Near-Infrared Optical Imaging of Early Dental Caries*, University of California, San Francisco, Calif, USA, 2006.

[6] J. Wu and D. Fried, "High contrast near-infrared polarized reflectance images of demineralization on tooth buccal and occlusal surfaces at $\lambda = 1310$-nm," *Lasers in Surgery and Medicine*, vol. 41, no. 3, pp. 208–213, 2009.

[7] M. Staninec, C. Lee, C. L. Darling, and D. Fried, "In vivo near-IR imaging of approximal dental decay at 1,310 nm," *Lasers in Surgery and Medicine*, vol. 42, no. 4, pp. 292–298, 2010.

[8] D. Fried, J. D. B. Featherstone, C. L. Darling, R. S. Jones, P. Ngaotheppitak, and C. M. Bühler, "Early caries imaging and monitoring with near-infrared light," *Dental Clinics of North America*, vol. 49, no. 4, pp. 771–793, 2005.

[9] I. A. Pretty, "Caries detection and diagnosis: novel technologies," *Journal of Dentistry*, vol. 34, no. 10, pp. 727–739, 2006.

[10] C. Zakian, I. Pretty, and R. Ellwood, "Near-infrared hyperspectral imaging of teeth for dental caries detection," *Journal of Biomedical Optics*, vol. 14, no. 6, Article ID 064047, 2009.

[11] A. M. A. Maia, D. D. D. Fonseca, B. B. C. Kyotoku, and A. S. L. Gomes, "Evaluation of sensibility and specificity of NIR transillumination for early enamel caries detection—An *in vitro* study," in *Proceedings of the European Conference on Lasers and Electro-Optics and the European Quantum Electronics Conference (EQEC '09)*, p. 1, IEEE, June 2009.

[12] L. Karlsson, *Optical Based Technologies for Detection of Dental Caries*, Karolinska Institutet, 2009.

[13] W. A. Pena, *Optical Imaging of Early Dental Caries in Deciduous Teeth With Near-IR Light at 1310 nm*, University of California, San Francisco, Calif, USA, 2009.

[14] C. Lee, D. Lee, C. L. Darling, and D. Fried, "Nondestructive assessment of the severity of occlusal caries lesions with near-infrared imaging at 1310 nm," *Journal of Biomedical Optics*, vol. 15, no. 4, Article ID 047011, 2010.

[15] S. Chung, D. Fried, M. Staninec, and C. L. Darling, "Near infrared imaging of teeth at wavelengths between 1200 and 1600 nm," in *Lasers in Dentistry XVII*, vol. 7884 of *Proceedings of SPIE*, San Francisco, Calif, USA, January 2011.

[16] J. A. Weatherell, C. Robinson, and A. S. Hallsworth, "Variations in the chemical composition of human enamel," *Journal of Dental Research*, vol. 53, no. 2, pp. 180–192, 1974.

[17] M. Marquezan, F. N. P. Corrêa, M. E. Sanabe et al., "Artificial methods of dentine caries induction: a hardness and morphological comparative study," *Archives of Oral Biology*, vol. 54, no. 12, pp. 1111–1117, 2009.

[18] T. Aoba, "Solubility properties of human tooth mineral and pathogenesis of dental caries," *Oral Diseases*, vol. 10, no. 5, pp. 249–257, 2004.

Comparison of Respiratory Resistance Measurements Made with an Airflow Perturbation Device with Those from Impulse Oscillometry

J. Pan,[1,2] A. Saltos,[3] D. Smith,[3] A. Johnson,[2] and J. Vossoughi[1,2]

[1] *Fischell Department of Bioengineering, University of Maryland, College Park, MD 20742, USA*
[2] *Engineering and Scientific Research Associates, Olney, MD 20832, USA*
[3] *School of Medicine, University of Maryland, Baltimore, MD 21201, USA*

Correspondence should be addressed to A. Johnson; artjohns@umd.edu

Academic Editor: Chun-Yuh Charles Huang

The airflow perturbation device (APD) has been developed as a portable, easy to use, and a rapid response instrument for measuring respiratory resistance in humans. However, the APD has limited data validating it against the established techniques. This study used a mechanical system to simulate the normal range of human breathing to validate the APD with the clinically accepted impulse oscillometry (IOS) technique. The validation system consisted of a sinusoidal flow generator with ten standardized resistance configurations that were shown to represent a total range of resistances from 0.12 to 0.95 kPa·L^{-1}·s (1.2–9.7 cm H$_2$O·L^{-1}·s). Impulse oscillometry measurements and APD measurements of the mechanical system were recorded and compared at a constant airflow of 0.15 L·s^{-1}. Both the IOS and APD measurments were accurate in assessing nominal resistance. In addition, a strong linear relationship was observed between APD measurements and IOS measurements ($R^2 = 0.999$). A second series of measurements was made on ten human volunteers with external resistors added in their respiratory flow paths. Once calibrated with the mechanical system, the APD gave respiratory resistance measurements within 5% of IOS measurements. Because of their comparability to IOS measurements, APD measurements are shown to be valid representations of respiratory resistance.

1. Introduction

1.1. Respiratory Resistance. Respiratory resistance is proportional to the total opposition to breathing caused by frictional forces in the airway passages. At any given time, respiratory resistance is equal to the ratio of respiratory pressure to airflow, given by

$$R_{\text{res}} = \frac{\Delta P_{\text{res}}}{\dot{V}}, \tag{1}$$

where R_{res} is respiratory resistance, ΔP_{res} is the respiratory pressure gradient, and $\dot{V}$ is airflow [1, 2].

Increases in respiratory resistance are symptomatic of a number of restrictive and obstructive pulmonary conditions including COPD [3], asthma [4], and bronchiolitis [5]. Useful feedback during administration of endotracheal tubes [6], anesthesia [7], bronchodilator medications [8], and mechanical ventilation [1] can also be given by respiratory resistance measurements.

1.2. Measurement Methods. Respiratory resistance is composed of resistances of airways, lung tissue, and chest wall components. Spirometry, specifically peak expiratory flow (PEF) and forced expiratory volume in one second (FEV1), has traditionally been the bedside measurement of choice for diagnosing increased resistance. However, spirometric tests do not directly measure respiratory resistances, are dependent on effort and lung volume, and require complete subject cooperation [9, 10]. Furthermore, since spirometric measurements are principally made during forced exhalation, little insight is normally provided for diagnosing inhalation-related pathologies. Although it has been shown that children as young as 5 years of age are capable of

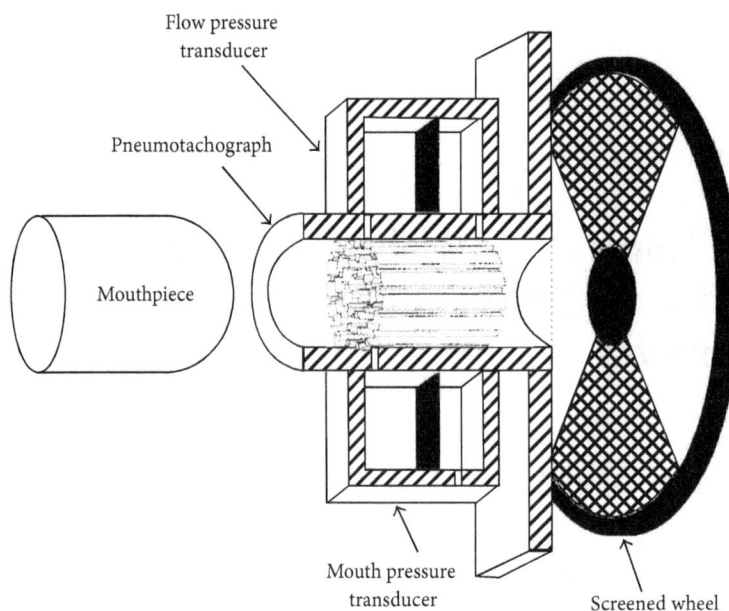

FIGURE 1: Cut-away diagram of airflow perturbation mechanism. The screened wheel rotates, allowing screened segments to briefly slow, or perturb, airflow passing through the pneumotachograph.

performing spirometry, appropriate coaching is required [11].

In contrast to spirometry, passive measurements of total respiratory resistance are a more attractive option for assessing unconscious or noncooperative patients because they require minimal cooperation on the part of the subject. One such method that relies on passive measurements is the forced oscillation technique (FOT). The use of FOT in young and uncooperative patients, for example, has been extensively studied since the 1980s [12], and its theoretical background is comprehensively described.

The FOT determines the mechanical properties of the entire respiratory system (all three components) by superimposing a loudspeaker-generated external pressure signal on the normal breathing pattern of a patient [13]. Impulse oscillometry (IOS) is a form of FOT that uses short duration pressure impulses as the external driver. These impulses are confined to the respiratory system by a parallel bias tube acting as an inertance element to allow patients to breathe regularly to the atmosphere. Dead volume added to the respiratory system by the bias tube can cause respiratory adjustments in airway caliber and depth of breathing. From the in-phase and out-of-phase components of the measured flows and mouth pressures, the IOS machine is able to compute respiratory resistance and reactance over a range of frequencies [14]. IOS has been shown to be effective in measuring lung function both of children [4, 15–18] and of adults [19, 20]. Because it measures total respiratory resistance, IOS is an obvious instrument with which to compare APD measurement

The APD is an instrument that is seeking clinical acceptance. Its advantages over other measurement techniques are

that it is small, lightweight, inexpensive, quick, has little external dead volume, and requires little patient cooperation [21–23]. Shown in Figure 1, the APD is composed of a pneumotachograph, two differential pressure transducers, a segmented wheel, motor, and a signal conditioning circuit. The device is held by the patient, while the patient breathes normally through a disposable mouthpiece. A periodic resistance is introduced into the flow path by the rotating wheel with open and screened segments. As the wheel rotates, perturbations are induced in both mouth pressure and flow. The depths of these perturbations depend on relative resistance inside the patient and resistance of the APD itself. Continuous measurement of mouth pressure and flow gives the resistance to the atmosphere of the APD device. From that, it is relatively simple to calculate the patient respiratory resistance as the depth of the pressure perturbation divided by the depth of the flow perturbation. It has been determined from actual measurements that the resistance measured by the APD includes airways, lung tissue, and chest wall components [21–23].

The APD is theoretically similar to both the FOT and the interrupter (INT) techniques in measuring respiratory resistance. Like FOT, the APD imposes a periodic perturbation on the breathing waveform. Unlike FOT, the APD determines resistance in the time domain rather than in the frequency domain, easily separates resistance during inhalation from resistance during exhalation, and requires patient breathing to obtain a signal. Some of these characteristics are also shared by INT, but the APD only partially obstructs airflow during tidal breathing. Both pressure and flow signal components are assessed at the same time, so the problem of lung accommodation time that has discredited INT measurements is avoided with the APD.

Comparison of Respiratory Resistance Measurements Made with an Airflow Perturbation Device with Those from
Impulse Oscillometry

25

APD measurements have been shown to be reproducible and sensitive to resistance changes in human subjects between the ages of 2 and 88. For example, respiratory resistances measured by the APD clearly show the expected reduction with age as children grow and also the differentiation expected between adult men and women [24]. Similarly, studies have shown the APD to be comparable to the esophageal balloon technique [23] and sensitive to resistance changes in a controlled excised sheep-lung respiratory model [22].

1.3. Validation of the APD by Comparing with IOS. The objective of this study was to compare respiratory resistance measurements made with the APD with those made using IOS. This comparison was made using two procedures: (1) in an artificial, nonbiological respiratory model in which resistances were controllable and known and (2) in human volunteers with external resistances added to their respiratory systems.

There are several important differences between the IOS and APD that influence validation procedures. First, frequency response of the respiratory system is measured by IOS, whereas time response is measured by the APD. Therefore, many measurements are taken by the APD in the time that it takes the IOS machine to take just one. Although the IOS impulse makes the IOS machine capable of quicker readings than the traditional forced oscillation technique, a series of repetitive readings with acceptable coherences is still required for the IOS machine to estimate resistance and reactance. By contrast, the APD is theoretically capable of making measurements as quickly as mouth pressure and airflow can be sampled. To reconcile this difference, IOS and APD measurements can be made more comparable by averaging APD measurements over a discrete duration, such as one minute. In addition, because the IOS impulse contains a range of frequencies, IOS can give resistance and reactance measurements at several different frequencies simultaneously. The commercially available IOS instrument used in the system described in this paper displayed resistance values at 5 Hz and 20 Hz as R5 and R20, respectively. Unless the APD wheel speed is changed, the APD is capable of expressing resistance at only the one wheel speed.

The second important difference is that the APD signal is developed across and through the varying resistance of the wheel. Consequently, there is no phase angle between mouth pressure and flow, so reactance cannot be measured as directly with the APD as with IOS. As a result, the system by which the devices are validated must be dominated by resistance and not reactance. Moreover, since the IOS machine produces its signal with a speaker-like transducer, its signal-to-noise (S/N) ratio is highest with no respiratory flow rate (no noise, only signal). On the other hand, the APD requires a respiratory flow rate in order to produce a signal, so the higher the respiratory flow rate, the larger is the perturbation. Thus, the APD S/N ratio is highest at peak flow. This difference must be compromised in the validation procedure.

Third, the APD can distinguish easily between resistance during inhalation and resistance during exhalation. The IOS

unit used with the system described in this paper did not have this capability, so the IOS resistance reading was compared to the average of inhalation and exhalation resistances from the APD.

Lastly, the IOS machine requires a large power input in order to produce its signal in addition to a separate computer to operate, while the APD does not require any more power than to rotate the wheel and is completely self-contained. Thus, the APD can be small, portable, and compact, while the IOS machine is a large instrument. Whereas this difference does not directly influence the validation procedure, it does influence the physical layout of the components and the interface with the human volunteers.

2. Methods

2.1. Artificial Respiratory Model System

2.1.1. System Configuration. The artificial respiratory model used for APD/IOS comparisons had three components analogous to a biological system: a sinusoidal flow source to model breathing, a compliance chamber to model lung compliance, and a controllable resistance to model respiratory resistance.

First, a motorized syringe pump (no. 17050-3) purchased from VacuMed (Ventura, CA, USA) was selected to be used as the flow generator. The syringe pump was powered by a motor that generated a smooth, continuous sinusoidal flow. The magnitude and frequency of the flow was controlled by adjusting the stroke volume of the pump and the rotational frequency of the motor. In order to mimic human resting breathing, the stroke volume was set to 0.5 L for all experiments, and the motor frequency was set to 18 RPM to give a volumetric flow, $\dot{V}$, of $0.15 \, \text{L} \cdot \text{s}^{-1}$.

Second, a glass cylindrical container (height = 44.5 cm, inner diameter = 14 cm) was installed at the outlet of the piston pump to act as a compliance chamber. Because a positive displacement pump delivers nearly the same output regardless of downstream resistance, its effective internal resistance is uncontrollable and very high. If the fluid being delivered by the pump is incompressible, then the pump delivery would be determined solely by the pump and not by downstream resistance. Air, however, is compressible, so there is some effect of downstream resistance as air pressure in the piston cylinder increases or decreases. As a result, adding sufficient compliance to the flow pathway so that the extra downstream resistance causes the air to compress in the compliance chamber rather than to be forced through the resistance is one method to make a piston pump sensitive to downstream resistance changes, as required by the APD.

Third, a series of resistances were installed following the compliance chamber. The resistance of the system was changed depending on the type and number of resistors installed. Two types of resistors were used to test two ranges of resistances. Fleisch no. 1 and no. 2 pneumotachographs (Phipps & Bird, Arlington, VA, USA) were used to provide a low range of resistance ($0.12-0.31 \, \text{kPa} \cdot \text{L}^{-1} \cdot \text{s}$ or $1.2-3.2 \, \text{cm}$ $H_2O \cdot L^{-1} \cdot s$), and Hans Rudolph standard flow resistors (Series 7100R2, Shawnee, KS, USA) were used to provide a

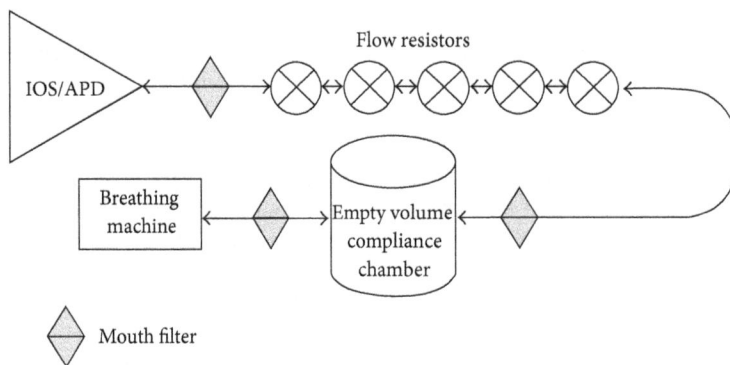

FIGURE 2: Schematic of the artificial respiratory model.

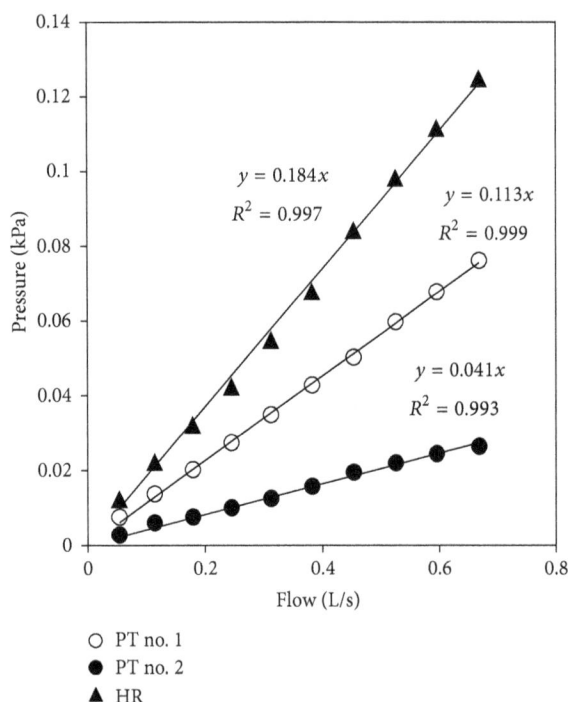

FIGURE 3: Pneumotachographs no. 1 and no. 2 (PT no. 1 and PT no. 2) and a Hans Rudolph standard flow resistor (HR) exhibit linear pressure and flow relationships at tested flows.

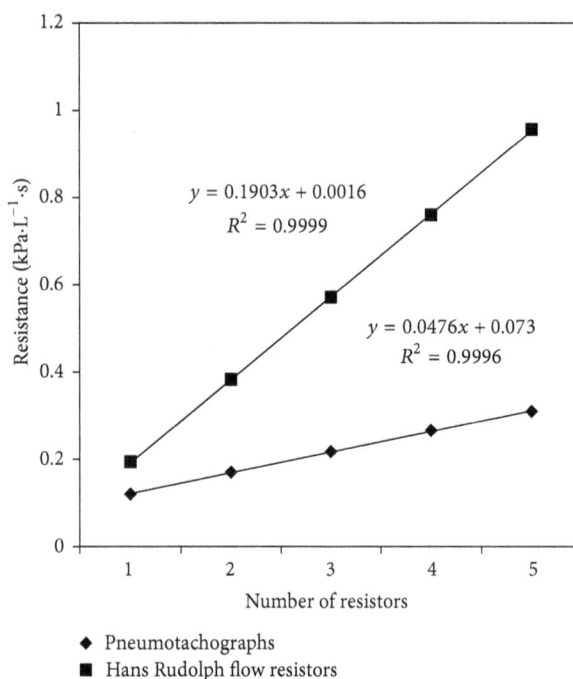

FIGURE 4: Increase in resistance in a series combination of resistors is linearly proportional to the number of resistors.

high range of resistance (0.19–$0.95\,\mathrm{kPa \cdot L^{-1} \cdot s}$ or 1.9–$9.7\,\mathrm{cm}$ $H_2O \cdot L^{-1} \cdot s$). Previous testing with the APD on people from ages 2 to 88 [24] has shown that this range incorporates the vast majority of expected resistance values.

Tubing with rubber end connectors was used to connect the components. Antibacterial filters were also used as connectors between the components to ensure tight fits. Pneumotachographs were placed end-to-end and sealed with rubber connectors cut from mouth pieces in the Pulmonary Function Filter Kit purchased from AllianceTech Medical, Inc. (Granbury, TX). Hans Rudolph standard flow resistors were also placed end-to-end and connected with Hans Rudolph small connectors (Series 7023, $22\,\mathrm{mm} \times 22\,\mathrm{mm}$). A schematic of the complete system is illustrated in Figure 2.

2.1.2. Calibration of the Resistors. The resistance of each individual resistor and each series combination of resistors was carefully measured from their steady-state flow-pressure relationships to provide a nominal value to compare with IOS and APD measurements. A steady flow of $0.16\,\mathrm{L \cdot s^{-1}}$ was applied for these measurements. For pressure measurements, a Dwyer Model 40–1 (Michigan City, IN, USA) manometer was used, and to monitor flow, a Gilmont 40453 (Pelham, NH, USA) Flowmeter was used. Both the manometer and flowmeter were carefully calibrated. The four no. 2 pneumotachographs were labeled $PT\#2_{1-4}$, and the five Hans Rudolph standard flow resistors were labeled HR_{1-5}. The no. 1 pneumotachograph was identified as PT#1.

The pressure/flow relationships of the pneumotachographs and Hans Rudolph Standard Flow Resistors were tested for linearity. The pressure drops across no. 1 and

Comparison of Respiratory Resistance Measurements Made with an Airflow Perturbation Device with Those from
Impulse Oscillometry

27

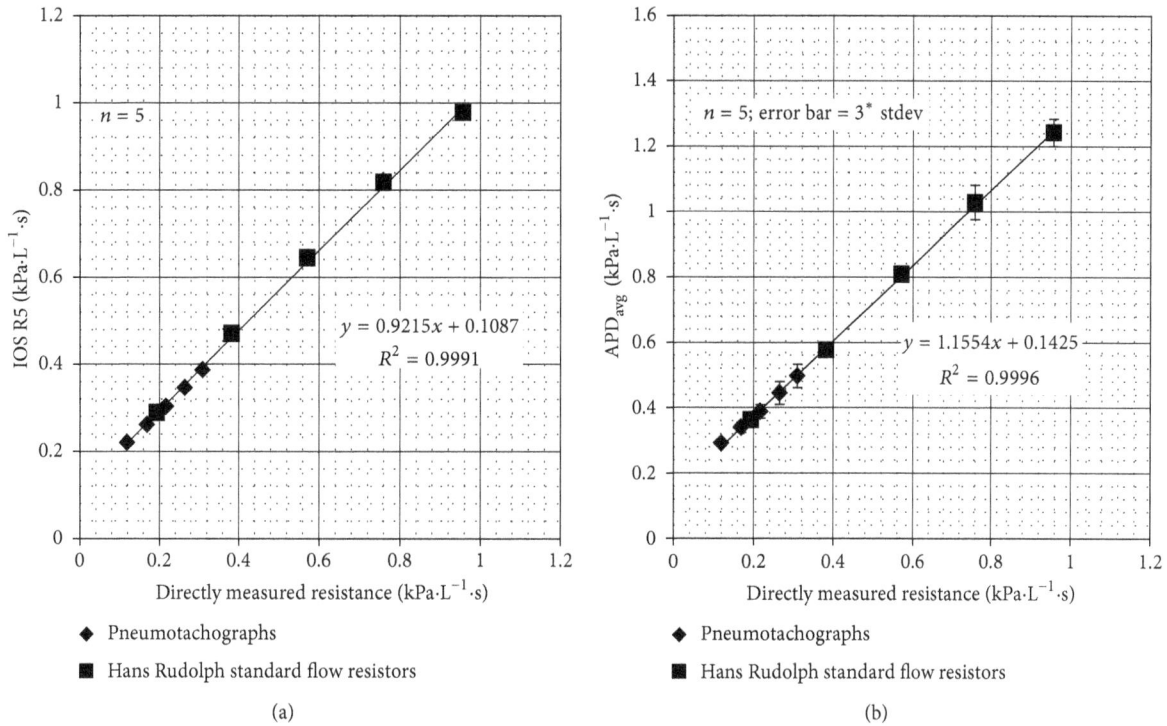

FIGURE 5: IOS R5 measurements of resistance of series combination of resistors slightly underestimated the directly measured resistance (a), whereas APD measurements of the same series combinations slightly overestimated the directly measured resistance (b). The offsets in both graphs are due to the unavoidable inherent resistance of the mechanical system without additional added resistances.

no. 2 pneumotachograph and Hans Rudolph standard flow resistor were recorded at various flows ranging from 0.05 to $0.70 \, \text{L} \cdot \text{s}^{-1}$.

Plastic tubing was fit by compression onto the ends of each pneumotachograph, and each plastic tube was tapped to measure static pressure in close proximity to the pneumotachograph. The pressure taps on the pneumotachographs themselves were plugged with masking tape. The pressure drop was measured as the difference in pressures at the tubing pressure taps. This was done to obtain the entire pressure drop across the device and to account for end effects in the pneumotachs.

For the Hans Rudolph Standard Flow Resistors, Hans Rudolph small connectors were fit by compression onto the ends of each flow resistor. The pressure drop was measured across the pressure taps on these connectors.

To account for possible resistance differences due to resistor placement, total pressure drops were also obtained for series combinations of resistors. For the pneumotachographs, the pressure drop was recorded across each series combination as the number of pneumotachographs was increased incrementally up to four no. 2 pneumotachographs and one no. 1 pneumotachograph. For the Hans Rudolph Standard Flow Resistors, the number of resistors was increased incrementally from one to five resistors.

2.1.3. IOS and APD Measurements. A CareFusion (San Diego, CA, USA) MasterScreen IOS machine was used following the manufacturer specifications to measure the resistance of

the validation system [25]. The IOS machine was attached to the output end of the system. Volume, temperature, and pressure calibrations were performed according to the manufacturer specifications. For each resistance configuration of the system, one IOS measurement of R5 was recorded and used for comparison. It was decided to use this R5 value as a good choice for comparison with APD measurements because R5 represents respiratory resistance over the entire respiratory system.

Unlike normal APD operating procedure, no previous APD flow or pressure calibrations were performed before usage. The perturbation frequency on the APD was set, as normally used, to 9.8 Hz, with two perturbations occurring with each rotation of the wheel. Consistent with the IOS procedure, five APD measurements were taken using the same APD unit. The reported inhalation and exhalation resistances were recorded as $R_{\text{APD,in}}$ and $R_{\text{APD,ex}}$, respectively. These values were averaged and reported as a single value, R_{APD}.

2.2. Measurements on Human Subjects. Ten healthy nonasthmatic male and female volunteers between the ages of 18 and 30 years completed tests of respiratory resistance by the APD and the IOS machine. The perturbation frequency of the APD was set to 9.8 Hz. IOS measurements were made using a CareFusion (San Diego, CA, USA) MasterScreen IOS machine following the manufacturer specifications to measure resistance [25]. For both devices, an antibacterial filter from the Pulmonary Function Filter Kit purchased from

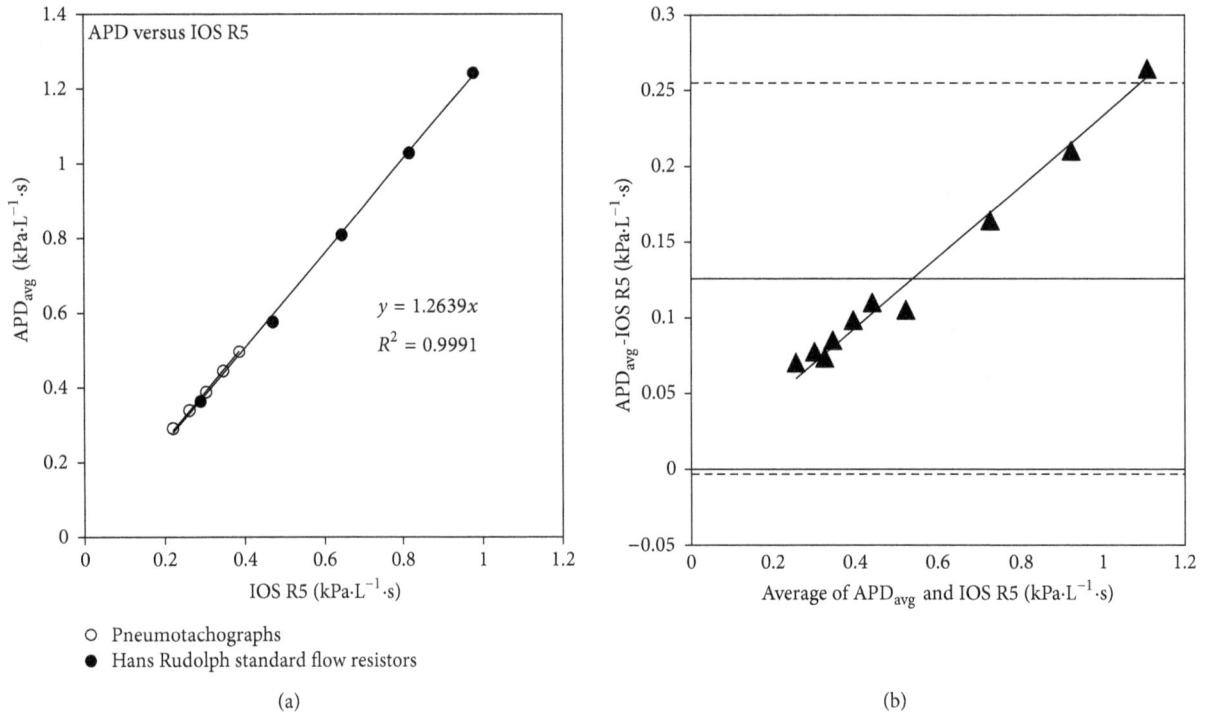

(a)   (b)

Figure 6: Strong linear correlation is observed between the IOS R5 measurements and the APD measurements ($R^2 = 0.999$) (a) as well as a strong proportional relation ($R^2 = 0.985$), shown in the Bland-Altman diagram (b). The dotted line represents ± 1.96 standard deviation; the solid line represents the mean of the differences.

AllianceTech Medical, Inc. (Granbury, TX, USA) was used as a mouthpiece for the test subjects. This protocol was approved by the University of Maryland Institutional Review Board.

Identical tests were performed on each subject using the APD and the IOS machines. Subjects performed four trials on the APD and one trial on the IOS machine. Inhalation and exhalation resistance values from the APD were averaged during each trial to give R_{APD}. IOS measurements of R5 were recorded following completion of APD measurements to compare against R_{APD}.

Respiratory resistance of the subjects was incrementally increased by adding various series combinations of Hans Rudolph Standard Flow Resistors (Series 7100R2) in between the test subject and the measuring device. The number of flow resistors was increased one at a time from one to four, giving a total range of 0.20–$0.78\,\text{kPa} \cdot \text{L}^{-1} \cdot \text{s}$ (2.00–$8.00\,\text{cmH}_2\text{O} \cdot \text{L}^{-1} \cdot \text{s}$) as indicated by the manufacturer, and measurements of respiratory resistance were taken at every increment. Resistors were connected by compression fitting using Hans Rudolph small connectors, and rubber adapters were used to connect resistor combinations to mouthpieces and devices.

3. Results

3.1. Calibration of the Resistors

Individual Resistors. The resistance of each resistor was calculated using the relationship between pressure and flow,

Table 1: Measured resistance values of each resistor.

Pneumotachographs	Resistance ($\text{kPa} \cdot \text{L}^{-1} \cdot \text{s}$)	Hans Rudolph standard flow resistors	Resistance ($\text{kPa} \cdot \text{L}^{-1} \cdot \text{s}$)
PT no. 1	0.109	HR_1	0.182
PT no. 2_1	0.044	HR_2	0.179
PT no. 2_2	0.051	HR_3	0.182
PT no. 2_3	0.051	HR_4	0.179
PT no. 2_4	0.049	HR_5	0.182

$R = \Delta P / \dot{V}$. The resistance values were found to be consistent across all no. 2 pneumotachographs and across all Hans Rudolph Standard Flow Resistors. The resistance of the no. 1 pneumotachograph was found to be approximately as twice as that of a single no. 2 pneumotachograph, which is consistent with the manufacturer specifications. Table 1 summarizes these findings.

Good linearity between pressure and flow was observed for the no. 1 and no. 2 pneumotachographs and the Hans Rudolph Standard Flow Resistors, for flow rates between 0.05 and $0.70\,\text{L} \cdot \text{s}^{-1}$. Dividing the pressure over the flow corresponded to the slope of the regression lines, giving a measure of the resistance value. Figure 4 shows that the slopes of the regression lines (forced through zero) were within 5% agreement with the measured resistance values of each resistor listed in Table 1, with the exception of the no. 2 pneumotachograph, which was within 15% of the average

Comparison of Respiratory Resistance Measurements Made with an Airflow Perturbation Device with Those from
Impulse Oscillometry

29

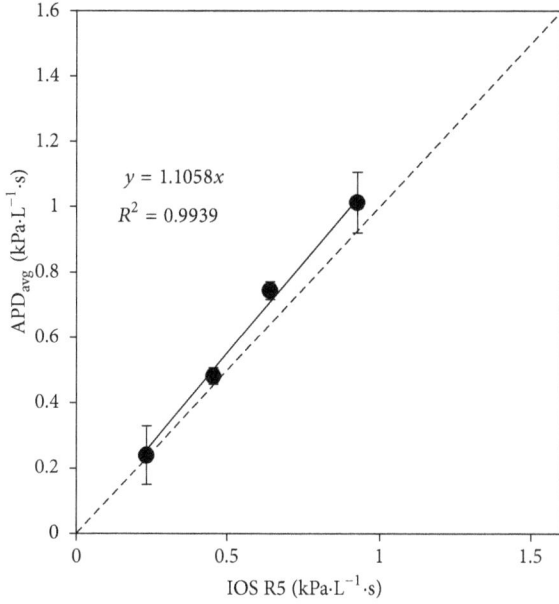

FIGURE 7: Uncalibrated APD-measured respiratory resistance at 9.8 Hz compared to IOS-measured respiratory resistance at 5 Hz for one typical subject, illustrating the high degree of linearity between the two devices. The linear regression was forced through zero. The dotted line indicates the line of identity. Points are averages of four trials. Error bars are three times the standard deviation.

TABLE 2: Measured resistance values of series combinations of resistors.

Pneumotachographs	Resistance $(kPa \cdot L^{-1} \cdot s)$	Hans Rudolph standard flow resistors	Resistance $(kPa \cdot L^{-1} \cdot s)$
PT no. 1	0.119	HR_1	0.194
PT no. 1 + PT no. 2_1	0.169	HR_{1-2}	0.382
PT no. 1 + PT no. 2_{1-2}	0.217	HR_{1-3}	0.571
PT no. 1 + PT no. 2_{1-3}	0.265	HR_{1-4}	0.759
PT no. 1 + PT no. 2_{1-4}	0.309	HR_{1-5}	0.956

$(0.041 \, kPa \cdot L^{2-1} \cdot s$ compared to $0.048 \, kPa \cdot L^{-1} \cdot s$, or 0.49 cm $H_2O \cdot L^{-1} \cdot s$).

Resistors in Series. Resistors were coupled together in series in single unit increments for the pneumotachographs and the Hans Rudolph Standard Flow Resistors, and their resistances calculated from their pressure/flow relationships. Resistances of series combinations were approximately (but not exactly) equivalent to the algebraic sum of the individual resistors. The differences between the sum of individual resistor values and their combinations are likely due to small interfacing resistances incurred when connecting them together. Table 2 summarizes these findings.

For both the series combination of pneumotachographs and the series combination of Hans Rudolph Standard Flow Resistors, the resistance was linearly proportional to the number of resistors in the system. Moreover, the proportionality for the series combination of both types of resistors was close to the averaged resistance of the individual resistors. This is represented by the slope of the linear regression shown in Figure 3. Specifically, for the series combination of pneumotachographs, the slope of the linear regression was equal to $0.047 \, kPa \cdot L^{-1} \cdot s$ per pneumotachograph, compared to $0.048 \pm 0.003 \, kPa \cdot L^{-1} \cdot s$ (0.49 cm $H_2O \cdot L^{-1} \cdot s$), the calibration average of the individual resistances of the no. 2 pneumotachographs. Similarly, for the series combination of Hans Rudolph Flow Resistors, the slope of the linear regression was equal to $0.190 \, kPa \cdot L^{-1} \cdot s$ per Hans Rudolph Flow Resistor, compared to $0.181 \pm 0.002 \, kPa \cdot L^{-1} \cdot s$ (1.85 cm $H_2O \cdot L^{-1} \cdot s$), the average calibration of the individual resistances of the Hans Rudolph Flow Resistors.

3.2. IOS and APD Measurements of the Physical System. IOS Measurements. IOS measurements of R5 of the system were recorded for various series combinations of the pneumotachographs and the Hans Rudolph Standard Flow Resistors. In Figure 5(a), the resistance measured at 5 Hz, R5, was plotted against the directly measured resistance values (see Table 2), and a linear regression was performed. The data fit the regression curve exceptionally well ($R^2 = 0.999$). Moreover, the slope of the curve (0.921), an indication of how well IOS resistance measurements agree with directly measured resistances, suggested only slight underestimation by the IOS.

APD Measurements. Because the APD is capable of both inhalation and exhalation measurements of resistance, these measurements were recorded and averaged to produce the APD-measured resistance. In Figure 5(b), the APD-measured resistance values were plotted against the directly measured resistance values (see Table 2), and a linear regression was performed. Consistent with the IOS machine, the APD data also fit the regression curve exceptionally well ($R^2 = 0.999$). In contrast to the IOS machine, the slope of the curve (1.155) indicated slight overestimation by the uncalibrated APD.

Despite the slight under and overestimations by the two devices, Figure 6 illustrates that a strong correlation exists between the two methods when one is compared against the other. A Bland-Altman diagram is typically used to compare two clinical measurements when the properties to be measured exhibit significant variation over their range such that correlation may not necessarily equate to agreement. In other words, two measures may be highly correlated yet exhibit substantial differences across their range of values. Here, the Bland-Altman diagram in Figure 6(b) illustrates that the difference between APD R_{APD} and IOS R5 depends highly on the magnitude of the resistance. This is indicative of a proportional relationship and conforms to the conclusions indicated by the data in Figure 5.

3.3. APD and IOS Measurements of Human Subjects. Correlations between APD R_{APD} and IOS R5 were linear for all subjects breathing through several external resistors. Figure 7 shows representative data from a typical subject, highlighting the high degree of linearity. Table 3 lists the linear correlation factors as well as the goodness of fit, R^2. Linear regressions

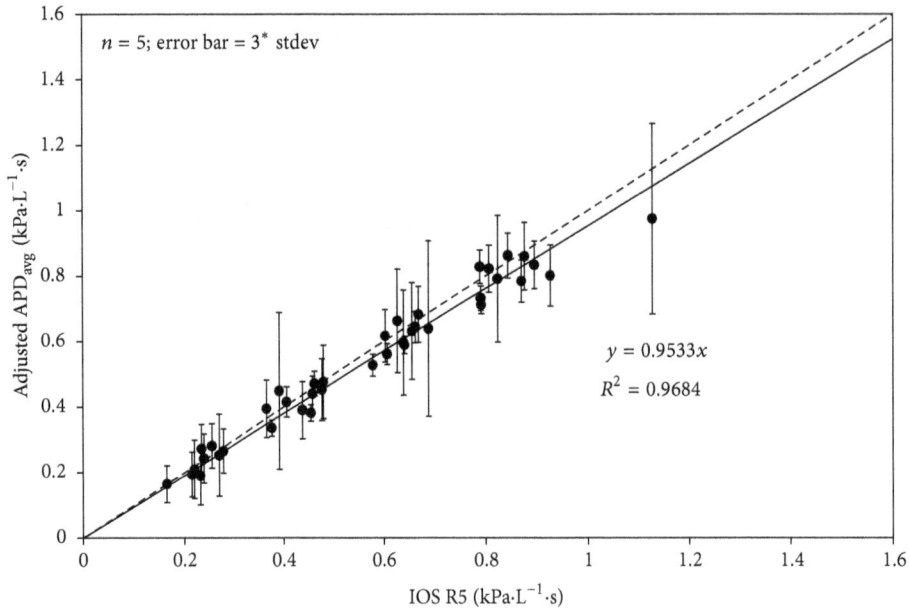

FIGURE 8: Average data of calibrated APD-measured respiratory resistances from the cohort of all ten test subjects are nearly identical to IOS-measured respiratory resistances. The linear regression was forced through zero. The dotted line indicates the line of identity. Points are averages of four trials. Error bars are three times the standard deviation.

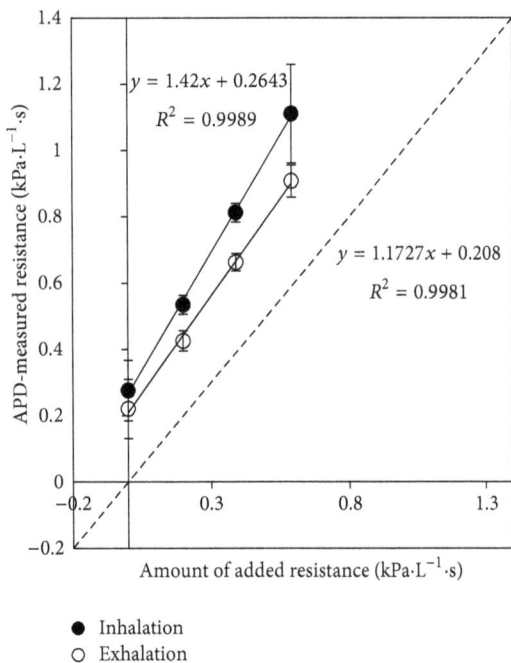

● Inhalation
○ Exhalation

FIGURE 9: APD-measured respiratory resistance indicates higher respiratory resistance in the inhalation direction compared to the exhalation direction for one particular subject. The APD used for this measurement was uncalibrated. The dotted line indicates the line of identity. Points are averages of four trials. Error bars are three times the standard deviation. The offsets for the subject data are respiratory resistances that are located inside the respiratory system of the subject.

TABLE 3: All subjects showed high linearity between APD-measured respiratory resistance and IOS-measured respiratory resistance, as indicated by strong goodness of fit coefficients. In addition, all subjects also showed slight overestimation by the APD compared with IOS R5.

	Correlation factor	Goodness of fit (R^2)
Subject no. 1	1.306	0.997
Subject no. 2	1.198	0.997
Subject no. 3	1.105	0.993
Subject no. 4	1.265	0.993
Subject no. 5	1.160	0.912
Subject no. 6	1.142	0.998
Subject no. 7	1.149	0.998
Subject no. 8	1.312	0.993
Subject no. 9	1.247	0.992
Subject no. 10	1.197	0.977

were forced through zero. The average correlation factor (1.208 ± 0.072) indicates slight overestimation by the APD compared with IOS. This is consistent with the previous results reported above that found a similar overestimation by the APD when measuring the resistance of the mechanical respiratory model.

The APD was calibrated to the IOS using the calibration factor (1/1.263) obtained from the comparison made on the mechanical system, and the data from the ten subjects were plotted in Figure 8. Both the calibrated APD and IOS gave nearly identical resistance measurements, with less than 5%

Comparison of Respiratory Resistance Measurements Made with an Airflow Perturbation Device with Those from
Impulse Oscillometry

31

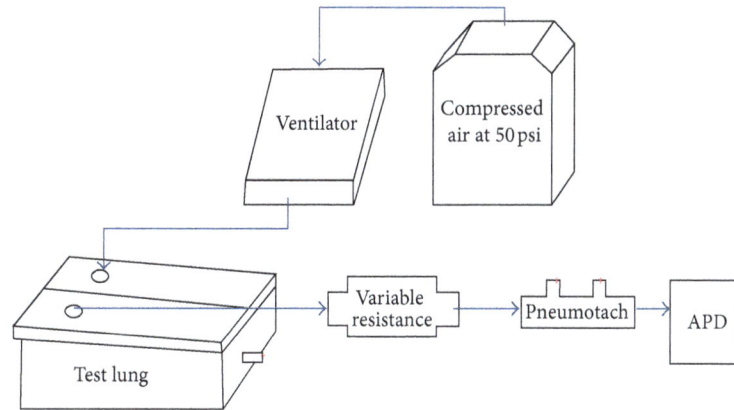

FIGURE 10: Airflow through the ventilator test lung system. Arrows indicate flexible plastic tubing connections.

difference between the two. This difference is less than the natural variation inherent in the respiratory system [26].

4. Discussion

Here, we presented data that illustrate that the APD is capable of making airflow resistance measurements comparable to those made by the commercially-available IOS instrument. This was shown on a nonbiological, mechanical respiratory system as well as on human subjects. The APD produced measurements as consistently as those made by IOS.

The APD used in these experiments was not calibrated for prior pressure or flow accuracy. Normally, it would be expected that each individual APD would require calibration either against the IOS system or against the calibrated resistor combination to assure consistency among devices. This experiment has provided the means to do so. In the past, different APD devices gave slightly different respiratory resistance measurements due to variations in individual component parts. This no longer needs to be the case. The ratio of APD measurement values to IOS measurement values provides a calibration factor that can be used to calibrate APD average resistances. The fact that this factor is nearly the same for human subjects as it is for a mechanical system means that the highly reproducible and reliable mechanical system can be used for this purpose. The coefficient of variation of the correlation factors for all human subjects, an indicator of how random the amount of overestimation is, was 5.92%, suggesting that measurements on human beings are probably not the best to use for calibration purposes. A study by Johnson et al. [26] showed that there is a substantial variation in the average respiratory measurements that comes from the human respiratory system and not from the APD. Using the calibration factor obtained from the measurements on the mechanical system gave agreement within 5% of human subject respiratory resistances obtained from both devices. Considering that resistance measurements on humans can vary by much more than this, about 10%, the agreement by the calibrated APD with the IOS is certainly acceptable.

Both APD and IOS measurements correlated well with directly measured resistances of the series combinations, but a slight underestimation by the IOS machine and a slight overestimation by the APD were observed. For the IOS machine, this error may be due to the compressive nature of air in the compliance chamber. Although the inclusion of a compliance component also makes the physical system similar to a human respiratory system, the intent of the compliance chamber was to provide a volume in which the downstream resistance would force air to compress during every stroke of the piston pump. In this way, airflow becomes sensitive to downstream resistance and not just the cycling of the pump, allowing for the measurements of the resistance with the APD. However, since the impulse that the IOS machine emits is itself a pressure wave, it may become attenuated by the air within the compliance chamber, resulting in a lower measurement of R5. In fact, when no resistors were attached, the IOS machine gave no reading at all, presumably because the entire impulse was being attenuated (data not shown).

Additionally, differences between inhalation respiratory resistance and exhalation respiratory resistance can give insight into diagnosing certain respiratory pathologies. One powerful feature of the APD is its ability to resolve these directions. For example, Figure 9 illustrates a sample test subject who appears to exhibit higher respiratory resistance in the inhalation direction than the exhalation direction. It would be expected that exhalation resistance should be higher than inhalation resistance because of the distensible airways and high external pressures during exhalation [27]. This subject may have a respiratory problem, or the difference may be due to the fact that the APD used in these tests was uncalibrated.

Some might suggest alternate means to fabricate a mechanical system to be used for calibration purposes. A previous attempt at APD validation was based upon a ventilator test lung system (Dual Adult Training and Test Lung, Michigan Instruments, Grand Rapids, MI, USA) that relied on the compliance of the test lung to soften the flow source. One of the two parallel test lungs was supplied through a ventilator with compressed air at 50 psi and set to 10 BPM with a 3 second inspiration time to mimic human resting breathing. This raised the second lung mechanically coupled to the first. The flow from the second test lung was subsequently fed into the APD through interchangeable orifices (Pneuflo Rp5 and Rp20, Michigan Instruments) used as resistances. Figure 10 shows a diagram of the airflow path through the system.

The problems that were encountered with this system were the inconsistency between trials (probably due to flow rate dependence of the orifice resistances) and APD resistance values that differed considerably from the resistance values supplied by the manufacturer. By contrast, the current system diagrammed in Figure 2 gave very consistent measurements.

Based on the results given in this paper, we can conclude that the APD compares well with the IOS method, and it can be an accurate method to assess respiratory resistance in patients.

Acknowledgment

This work was supported by the National Institutes of Health under the Grant no. R44HL078055.

References

[1] D. C. Grinnan and J. D. Truwit, "Clinical review: respiratory mechanics in spontaneous and assisted ventilation," *Critical Care*, vol. 9, no. 5, pp. 472–484, 2005.

[2] A. T. Johnson, *Biomechanics and Exercise Physiology: Quantitative Modeling*, Taylor & Francis, Boca Raton, Fla, USA, 2007.

[3] A. M. G. T. Di Mango, A. J. Lopes, J. M. Jansen, and P. L. Melo, "Changes in respiratory mechanics with increasing degrees of airway obstruction in COPD: detection by forced oscillation technique," *Respiratory Medicine*, vol. 100, no. 3, pp. 399–410, 2006.

[4] F. M. Ducharme, G. M. Davis, and G. R. Ducharme, "Pediatric reference values for respiratory resistance measured by forced oscillation," *Chest*, vol. 113, no. 5, pp. 1322–1328, 1998.

[5] M. E. Wohl, L. C. Stigol, and J. Mead, "Resistance of the total respiratory system in healthy infants and infants with bronchiolitis," *Pediatrics*, vol. 43, no. 4, pp. 495–509, 1969.

[6] T. J. Gal and P. M. Suratt, "Resistance to breathing in healthy subjects following endotracheal intubation under topical anesthesia," *Anesthesia and Analgesia*, vol. 59, no. 4, pp. 270–274, 1980.

[7] P. Pelosi, M. Croci, E. Calappi et al., "The prone positioning during general anesthesia minimally affects respiratory mechanics while improving functional residual capacity and increasing oxygen tension," *Anesthesia and Analgesia*, vol. 80, no. 5, pp. 955–960, 1995.

[8] H. K. Kil, G. A. Rooke, M. A. Ryan-Dykes, and M. J. Bishop, "Effect of prophylactic bronchodilator treatment on lung resistance after tracheal intubation," *Anesthesiology*, vol. 81, no. 1, pp. 43–48, 1994.

[9] M. R. Miller, J. Hankinson, V. Brusasco et al., "Standardisation of spirometry," *European Respiratory Journal*, vol. 26, no. 2, pp. 319–338, 2005.

[10] S. Kano, D. L. Burton, C. J. Lanteri, and P. D. Sly, "Determination of peak expiratory flow," *European Respiratory Journal*, vol. 6, no. 9, pp. 1347–1352, 1993.

[11] H. Eigen, H. Bieler, D. Grant et al., "Spirometric pulmonary function in healthy preschool children," *American Journal of Respiratory and Critical Care Medicine*, vol. 163, no. 3, pp. 619–623, 2001.

[12] P. Lebecque, K. Desmond, Y. Swartebroeckx, P. Dubois, J. Lulling, and A. Coates, "Measurement of respiratory system resistance by forced oscillation in normal children: a comparison with spirometric values," *Pediatric Pulmonology*, vol. 10, no. 2, pp. 117–122, 1991.

[13] A. B. DuBois, S. Y. Botelho, and J. H. Comroe Jr., "A new method for measuring airway resistance in man using a body plethysmograph: values in normal subjects and in patients with respiratory disease," *The Journal of Clinical Investigation*, vol. 35, pp. 327–335, 1956.

[14] H. J. Smith, P. Reinhold, and M. D. Goldman, "Forced oscillation technique and impulse oscillometry," *European Respiratory Monograph*, vol. 31, pp. 72–105, 2005.

[15] W. Buhr, R. Jörres, D. Berdel, and F. J. Làndsér, "Correspondence between forced oscillation and body plethysmography during bronchoprovocation with carbachol in children," *Pediatric pulmonology*, vol. 8, no. 4, pp. 280–288, 1990.

[16] A. B. Bohadana, R. Peslin, S. E. Megherbi et al., "Dose-response slope of forced oscillation and forced expiratory parameters in bronchial challenge testing," *European Respiratory Journal*, vol. 13, no. 2, pp. 295–300, 1999.

[17] C. Delacourt, H. Lorino, M. Herve-Guillot, P. Reinert, A. Harf, and B. Housset, "Use of the forced oscillation technique to assess airway obstruction and reversibility in children," *American Journal of Respiratory and Critical Care Medicine*, vol. 161, no. 3, pp. 730–736, 2000.

[18] P. Lebecque and D. Stǎnescu, "Respiratory resistance by the forced oscillation technique in asthmatic children and cystic fibrosis patients," *European Respiratory Journal*, vol. 10, no. 4, pp. 891–895, 1997.

[19] T. Chinet, G. Pelle, I. Macquin-Mavier, H. Lorino, and A. Harf, "Comparison of the dose-response curves obtained by forced oscillation and plethysmography during carbachol inhalation," *European Respiratory Journal*, vol. 1, no. 7, pp. 600–605, 1988.

[20] J. Hellinckx, M. Cauberghs, K. de Boeck, and M. Demedts, "Evaluation of impulse oscillation system: comparison with forced oscillation technique and body plethysmography," *European Respiratory Journal*, vol. 18, no. 3, pp. 564–570, 2001.

[21] C. G. Lausted and A. T. Johnson, "Respiratory resistance measured by an airflow perturbation device," *Physiological Measurement*, vol. 20, no. 1, pp. 21–35, 1999.

[22] A. T. Johnson and M. S. Sahota, "Validation of airflow perturbation device resistance measurements in excised sheep lungs," *Physiological Measurement*, vol. 25, no. 3, pp. 679–690, 2004.

[23] D. C. Coursey, S. M. Scharf, and A. T. Johnson, "Comparing pulmonary resistance measured with an esophageal balloon to resistance measurements with an airflow perturbation device," *Physiological Measurement*, vol. 31, no. 7, pp. 921–934, 2010.

[24] A. T. Johnson, W. H. Scott, E. Russek-Cohen et al., "Resistance values obtained with the airflow perturbation device," *International Journal of Medical Implants and Devices*, vol. 1, pp. 137–151, 2005.

Comparison of Respiratory Resistance Measurements Made with an Airflow Perturbation Device with Those from
Impulse Oscillometry

33

[25] E. Oostveen, D. MacLeod, H. Lorino et al., "The forced oscillation technique in clinical practice: methodology, recommenda-
tions and future developments," *European Respiratory Journal*,
vol. 22, no. 6, pp. 1026–1041, 2003.

[26] A. T. Johnson, S. C. Jones, J. J. Pan, and J. Vossoughi, "Variation of respiratory resistance suggests optimization of airway
caliber," *IEEE Transactions on Biomedical Engineering*, vol. 59,
pp. 2355–2361, 2012.

[27] E. R. Lopresti, A. T. Johnson, F. C. Koh, W. H. Scott, S. Jamshidi,
and N. K. Silverman, "Testing limits to airflow perturbation
device (APD) measurements," *BioMedical Engineering Online*,
vol. 7, article 28, 2008.

4

A MATLAB-Based Boundary Data Simulator for Studying the Resistivity Reconstruction Using Neighbouring Current Pattern

Tushar Kanti Bera and J. Nagaraju

Department of Instrumentation and Applied Physics, Indian Institute of Science Bangalore, Bangalore, Karnataka 560012, India

Correspondence should be addressed to Tushar Kanti Bera; tkbera77@gmail.com

Academic Editor: Nicusor Iftimia

Phantoms are essentially required to generate boundary data for studying the inverse solver performance in electrical impedance tomography (EIT). A MATLAB-based boundary data simulator (BDS) is developed to generate accurate boundary data using neighbouring current pattern for assessing the EIT inverse solvers. Domain diameter, inhomogeneity number, inhomogeneity geometry (shape, size, and position), background conductivity, and inhomogeneity conductivity are all set as BDS input variables. Different sets of boundary data are generated by changing the input variables of the BDS, and resistivity images are reconstructed using electrical impedance tomography and diffuse optical tomography reconstruction software (EIDORS). Results show that the BDS generates accurate boundary data for different types of single or multiple objects which are efficient enough to reconstruct the resistivity images for assessing the inverse solver. It is noticed that for the BDS with 2048 elements, the boundary data for all inhomogeneities with a diameter larger than 13.3% of that of the phantom are accurate enough to reconstruct the resistivity images in EIDORS-2D. By comparing the reconstructed image with an original geometry made in BDS, it would be easier to study the inverse solver performance and the origin of the boundary data error can be identified.

1. Introduction

Electrical impedance tomography (EIT) [1, 2] reconstructs the spatial distribution of electrical conductivity or resistivity of a closed conducting domain (Ω) from the surface potentials developed by a constant current injection through the surface electrodes surrounding the domain to be imaged. Before carrying out the practical measurements on patients, it is advised to test an EIT system with a tissue mimicking model of known properties [3] called practical phantoms [4–10]. Hence, phantoms are often required to assess the performance of EIT systems for their validation, calibration, and comparison purposes. Two-dimensional (2D) EIT (2D-EIT) assumes that the electrical current flows in a 2D space which is actually three-dimensional inside real volume conductors. Hence, the development of a perfect 2D practical phantom is a great challenge as the real electrodes always have a definite surface area, and hence the injected current signal cannot be confined in a 2D plane in bathing solution [5]. Researchers have developed a number of practical phantoms which are three-dimensional objects, and those phantoms are designed and developed, generally, for their own EIT systems. Practical phantoms containing electrolyte (or other conducting medium) [4–10] are three-dimensional in shape and hence they will have some data error due to the three dimensional current conduction. Also, the phantoms containing electrolytes (e.g., NaCl solution or saline) [5, 7, 8] are difficult to transport and are prone to errors since the evaporation of the water gives rise to changes in conductivity [9]. In addition, temperature variations have a marked effect on the conductivity because the temperature coefficient is large [11]. Therefore, the practical phantoms will have a poor stability and a gradually increasing data error over time. Network or mesh phantoms [12, 13] are compact, more stable, rugged, portable, easy to move, consistent over time, and less temperature dependent. But these phantoms need a huge number of identical electronic components properly designed in a mesh mimicking the conductivity distribution of a practical biological tissue. Furthermore, for a large tissue structure, a mesh phantom requires a huge number of very precision components. The reproduction of these kinds of phantoms having different properties is often time-consuming [14]. The option for changing the position and property of an inhomogeneity is limited by the phantom structure and the number of elements in mesh phantom but the practical phantoms allow us to put several types of object in different positions in the bathing solution, but they produce several errors contributing to the poor signal to noise ratio (SNR) in boundary data.

Reconstructed image quality in impedance tomography depends on the errors associated with practical phantom, electronic hardware, and inverse solver performance. Image quality is largely affected by the practical phantom design parameters such as phantom geometry, electrode geometry, electrode materials, and the nature and behavior of the inhomogeneity and bathing solution. SNR is also reduced by the error contributed by current injector, data acquisition system, and signal conditioner circuits. In practical phantoms, the voltage data developed by a three-dimensional current conduction are collected form surface electrodes connected to an analog instrumentation. Therefore, it is quite confusing to identify the source of the errors responsible for poor image quality in a 2D-EIT system. In order to overcome the difficulties and limitations of practical and mesh phantoms, a MATLAB-based boundary data simulator (BDS) is developed to generate accurate 2D boundary data for assessing the EIT inverse solvers. BDS is an absolute 2D data simulator which is required to generate the errorless 2D boundary data to study and modify the inverse solver of a 2D EIT system. As the BDS is a computer program, it is free from the instrumentation errors and allows us to generate voltage profile with different types of phantom geometry, inhomogeneity and background conductivity profile, and inhomogeneity geometry (shape, size, and position). Moreover, it is absolutely stable, compact, easy to use, and easy to handle and modify for further development. Boundary data for different phantom geometries are generated in BDS, and resistivity images are reconstructed in standard reconstruction algorithm. BDS is studied to conform its suitability to use for boundary data generation with different phantom configurations which are required to assess the EIT inverse solvers.

2. Methods

2.1. Mathematical Modelling of EIT. EIT image reconstruction is a nonlinear inverse problem [15] in which the electrical conductivity distribution of a closed domain (Ω) in a volume conductor is reconstructed from the surface potential data developed at the boundary ($\partial\Omega$) by injecting a constant current signal. A low frequency and low magnitude constant sinusoidal current is injected through an array of electrodes attached to the boundary, and the boundary potentials are measured using a data acquisition system. The voltage data collected from surface electrodes are then used by an image reconstruction algorithm [15] which reconstructs the conductivity distribution of the domain under test (DUT). The reconstruction algorithm computes the boundary potential

for a known current injection and known conductivity values and tries to compute the conductivity distribution for which the difference between the measured boundary potential (V_m) and the calculated (V_c) is minimum. The reconstruction algorithm is developed with two parts: forward solver (FS) [5, 15–17] and inverse solver (IS) [15–17]. Forward solver calculates the boundary potential data for a known current injection and known conductivity values. Inverse solver computes the conductivity distribution for which the boundary voltage difference ($\Delta V = V_m - V_c$) becomes minimum.

The DUT will have the distinct conductivity values at each points defined by their corresponding coordinates (x, y). Due to a constant current injection, a potential profile is developed within DUT, and its potential profile without any internal energy sources depends on the conductivity profile. Hence, a relationship, called EIT governing equation, between the electrical conductivity (σ) of the points within the DUT and their corresponding potential values (Φ) can be established. The governing equation in EIT [1, 2] can be derived from the Maxwell's equation and can be represented as

$$\nabla \cdot \sigma \nabla \Phi = 0. \qquad (1)$$

To calculate the domain potential developed for a constant current injected to the DUT with a known conductivity distribution, the above equation is essentially to be solved. As the EIT governing equation is a nonlinear partial differential equation, the direct or analytical technique fails to solve it. Therefore, to calculate the domain potential, the equation is solved by developing a mathematical model called "forward model" which is derived from (1) using a numerical technique like finite element method (FEM) [18].

The EIT governing equation has an infinite number of solutions, and hence the FEM formulation of the EIT technique is essentially required to be provided by some boundary conditions [18–20] to restrict its solutions space. The boundary conditions are imposed into the FEM formulation of EIT by specifying the value of certain parameters (voltage or current). The parameters defining the boundary conditions may be either the potentials at the surface or the current density crossing the boundary or mixed conditions.

The boundary conditions, in which the parameters are the potential at the surface, are called the *Dirichlet* boundary conditions and are represented as [1, 5, 19, 20]

$$\Phi = \Phi_i, \qquad (2a)$$

where $i = 1,\ldots,N$ are the measured potentials on the electrodes.

The boundary conditions, in which the parameters are current density crossing the boundary, are known as the *Neumann* boundary conditions [1, 5, 19, 20] which are given by

$$\int_{\partial\Omega} \sigma \frac{\partial\Phi}{\partial n} = \begin{cases} +I & \text{on the source electrode} \\ -I & \text{on the sink electrode} \\ 0 & \text{otherwise,} \end{cases} \qquad (2b)$$

where $\partial\Omega$ is the boundary, and n is the outward unit normal vector on an electrode surface.

In EIT, the FEM technique is used to derive the forward model from the governing equation in the form of a matrix equation establishing the relationship between the injected current and the developed potential within a DUT. The relationship can be assumed as the transfer function of the system which is mathematically represented as a matrix called global stiffness matrix (GSM) [18] or transformation matrix constructed with the elemental conductivities (σ) and nodal coordinates (x, y). In EIT, FEM discretizes the DUT by a finite element mesh containing finite number of elements of defined geometry and finite number of node. FEM applied on the governing equation to derive the forward model of a DUT in the form of a matrix equation using the σ and nodal coordinates. In the EIT forward model, the relationship established between the current injection matrix $[C]$ (matrix of the applied signal) and the nodal potential matrix $[\Phi]$ (matrix of the developed signal) through the transformation matrix $[K(\sigma)]$ is mathematically represented as

$$[\Phi] = [K(\sigma)]^{-1}[C]. \qquad (3)$$

Now, in FEM formulation in EIT, when the current matrices $[C]$ and $[K(\sigma)]$ are known, and the nodal potential matrix $[\Phi]$ is unknown, the forward model or the mathematical problem is termed as the "forward problem". The procedure of calculating the $[\Phi]$ by solving the forward problem (3) with known $[K(\sigma)]$ and known $[C]$ is termed as "forward solution". In EIT, the forward solver first computes the potential distribution with the assumed initial conductivity distribution (σ_0) with a known constant current simulation, and then the inverse solver reconstructs the conductivity distribution from the measured boundary potential data for a same constant current injection through surface electrodes. The EIT reconstruction algorithm tries to mathematically find the elemental conductivity values (conductivity distribution) for which the difference between the estimated nodal potentials (V_c) computed in the FS and the potentials measured (V_m) on the surface electrodes (for a same current injection values) becomes minimum.

The inverse solver of the EIT reconstruction algorithm is developed with a mathematical minimization algorithm (MMA) [19–22] such as Gauss-Newton-based mathematical minimization algorithm (GN-MMA). In GN-MMA, the conductivity update vector ($[\Delta\sigma]$) is calculated and the boundary data mismatch vector ($\Delta V = V_m - V_c$) is minimized by an iteration technique like the modified Newton-Raphson iteration technique (NRIT) [19–22]. The $[\Delta\sigma]$ matrix is the desired variation in the elemental conductivity values in $[\sigma]$ matrix for which the forward solver calculates the boundary potentials more similar to the measured value in next iteration using NRIT. Therefore, the algorithm starts with an initial elemental conductivity vector ($[\sigma_0]$), and it is then updated to ($[\sigma_1] = [\sigma_0] + [\Delta\sigma]$) in the next iteration. Using this $[\sigma_1]$, FS calculates a new potential distribution in DUT and a new voltage mismatch vector $[\Delta V_1]$ is thus obtained and compared with the previous voltage mismatch vector $[\Delta V_0]$. If the ΔV_1 is not found as the minimum, the iteration process is continued till the kth iteration using the conductivity update vector ($[\Delta\sigma_k]$) developed by GN-MMA.

Using, NRIT the $[\sigma]$ matrix is iteratively updated to $[\sigma_{k+1}] = [\sigma_k] + [\Delta\sigma_k]$ and repetitively tries to find out the minimum value of $[\Delta V]$.

Hence, in the EIT inverse solver, it is understood that the desired elemental conductivity matrix is obtained by a minimization algorithm (MMA) which is composed of Gauss-Newton method and Newton-Raphson iteration in which the technique iteratively tries to find out an optimum conductivity distribution $[\sigma_k]$ for which the voltage mismatch vector is minimized $[\Delta V]$. At a particular iteration in this MMA, the elemental conductivity matrix is calculated when the current matrices $[C]$ and $[\Phi]$ or $[\Delta V = V_m - V_c]$ are known. This process is logically an opposite process to the forward problem. Thus, when the current matrices $[C]$ and $[\Phi]$ are known, and the elemental conductivity matrix $[\sigma]$ is unknown, the model or the problem is called the "inverse problem." The procedure of calculating the $[\sigma]$ or $[\Delta\sigma]$ using with known $[\Delta V]$ and the known $[C]$ is termed as "inverse solution."

2.2. Image Reconstruction with GN-MMA and NRIT. Electrical conductivity imaging is a highly nonlinear and ill-posed inverse problem [19–22]. In EIT, a minimization algorithm is used to obtain an optimized elemental conductivity value $[\sigma]$ for which the voltage mismatch vector $[\Delta V]$ becomes minimum. In the image reconstruction process, the minimization algorithm [17, 18] first defines an objective function (s) from the computational predicted data $[V_c]$ and the experimental measured data $[V_m]$ and runs iteratively to minimize it. Generally, in the EIT image reconstruction algorithm, the inverse solver searches for a least square solution of the minimized object the function (s) using by a Gauss-Newton method and the NRIT-based iterative approximation techniques.

If f is a function mapping a t-dimensional (t is the number of element in the FEM mesh) impedance distribution into a set of M (number of the experimental measurement data ($[V_m]$) available) approximate measured voltages, then the Gauss-Newton-method-based minimization algorithm [19–26] tries to find a least square solution of the minimized object function (s) [19–26] which is defined as:

$$s = \frac{1}{2}\|V_m - f\|^2 = \frac{1}{2}(V_m - f)^T(V_m - f). \qquad (4)$$

Now, differentiating (4) with respect to the conductivity σ, it reduces to

$$s' = -[f']^T[V_m - f] = -J^T\Delta V, \qquad (5)$$

where the matrix $J = f'$ is known as Jacobin matrix [19–22], which may be calculated by a method as described in [19, 22] or by the adjoint method [23] represented by (6)

$$J = \oint_\Omega \nabla\Phi_s \cdot \nabla\Phi_d d\Omega, \qquad (6)$$

where Φ_s is the forward solution for a particular source location, and Φ_d is the forward solution for the adjoint source location (source at the detector location and detector at the source location).

Differentiating (5) with respect to σ again, the equation reduces to

$$s'' = \left[f'\right]^T \left[f'\right] - \left[f''\right]^T \left[V_m - f\right]. \tag{7}$$

By Gauss-Newton method, the conductivity update vector $[\Delta\sigma]$ is given by

$$\Delta\sigma = -\frac{s'}{s''} = \frac{J^T \Delta V}{\left[f'\right]^T \left[f'\right] - \left[f''\right]^T \Delta V}. \tag{8}$$

Thus, the conductivity update vector is given by

$$\Delta\sigma = \left[\left[f'\right]^T \left[f'\right] - [H]^T \Delta V\right]^{-1} J^T \Delta V, \tag{9}$$

where the higher-order term $H = [f'']$ is known as the *Hessian matrix* [24]. In (9) by neglecting H, the update conductivity vector reduces to

$$\Delta\sigma = \left[\left[f'\right]^T \left[f'\right]\right]^{-1} J^T [\nabla V]. \tag{10}$$

In general, using NRIT method, the conductivity update vector expressed as in (10) can be represented for kth iteration (where k is a positive integer) as

$$\Delta\sigma_k = \left[\left[J_k\right]^T \left[J_k\right]\right]^{-1} \left[J_k\right]^T \left[\Delta V_k\right], \tag{11}$$

where $[\Delta V_k]$ and $[J_k]$ are the voltage mismatch matrix and Jacobian matrix, respectively.

The $\left[f'\right]^T$ matrix in (11) is always ill conditioned [19–24], and hence small measurement errors will make the solution of (11) changes greatly. In order to make the system well posed, the regularization method [19–26] is incorporated into the reconstruction algorithm by redefining the object function [19–26] with regularization parameters as

$$s_r = \frac{1}{2}\|V_m - f\|^2 + \frac{1}{2}\lambda\|G\sigma\|^2, \tag{12}$$

where s_r is the constrained least-square error of the regularized reconstructions, G is the regularization operator, and λ (the positive scalar) is called the regularization coefficient [19–26]

$$s_r = \frac{1}{2}(V_m - f)^T (V_m - f) + \frac{1}{2}\lambda(G\sigma)^T (G\sigma). \tag{13}$$

Differentiating the inject function in (12) with respect to the elemental conductivity: the following relations are obtained

$$s_r' = -\left(f'\right)^T (V_m - f) + \lambda(G)^T (G\sigma), \tag{14}$$

$$s_r'' = \left(f'\right)^T \left(f'\right) - \left(f''\right)^T (V_m - f) + \lambda G^T G. \tag{15}$$

Now, using Gauss-Newton- (GN-) method-based minimization process, the conductivity update vector $[\Delta\sigma]$ is obtained as

$$\Delta\sigma = \frac{s_r'}{s_r''} = \frac{\left(f'\right)^T (V_m - f) - \lambda(G)^T (G\sigma)}{\left(f'\right)^T \left(f'\right) - \left(f''\right)^T (V_m - f) + \lambda G^T G}. \tag{16}$$

Neglecting the *Hessian matrix* [24] in (15)

$$\Delta\sigma = \frac{s_r'}{s_r''} = \frac{\left(f'\right)^T (V_m - f) - \lambda(G)^T (G\sigma)}{\left(f'\right)^T \left(f'\right) + \lambda G^T G}. \tag{17}$$

Replacing f' by J and $G^T G$ by I (identity matrix) (21) reduces to

$$\Delta\sigma = \frac{J^T (V_m - f) - \lambda I\sigma}{J^T J + \lambda I}, \tag{18}$$

where the matrix $J = f'$ is the Jacobin as stated earlier.

Thus, the conductivity update vector $([\Delta\sigma])$ is found as

$$\Delta\sigma = \left(J^T J + \lambda I\right)^{-1} \left(J^T (V_m - f) - \lambda I\sigma\right). \tag{19}$$

Sometimes, the last term $(\lambda I\sigma)$ is neglected [22], and the conductivity update vector $[\Delta\sigma]$ is calculated as

$$\Delta\sigma = \left(J^T J + \lambda I\right)^{-1} J^T (V_m - f). \tag{20}$$

In general, the EIT image reconstruction algorithm provides a solution of the conductivity distribution of the DUT for the kth iteration as

$$\sigma_{k+1} = \sigma_k + \left(\left(J^T J + \lambda I\right)^{-1} \left(J^T (V_m - f) - \lambda I\sigma\right)\right)_k. \tag{21}$$

The EIT algorithm starts with the solution of FP obtained from the EIT governing equation, and the $[V_c]$ is calculated for a known current injection matrix $[C]$ and an initial guess (known or assumed) conductivity matrix $[\sigma_0]$. The voltage mismatch matrix $[\Delta V]$ is estimated, and then it is used to calculate the conductivity update matrix $[\Delta\sigma]$ using GN-MMA and is added to the initial conductivity matrix $([\sigma_o])$ to update it to a new conductivity matrix $[\sigma_1 = \sigma_o + \Delta\sigma]$ using NRIT. New update matrix $[\sigma_1]$ is used in forward solver to obtain a new calculated boundary data matrix $[V_{c1}]$ which provides a new voltage mismatch matrix $[\Delta V_1]$. Therefore, the NRIT algorithms iteratively calculate the $[\Delta\sigma]$ using GN-MMA to find out an optimized $[\sigma]$ matrix for which the $[\Delta V]$ reaches its minimum value. Thus, the EIT reconstruction algorithm is found to work in the following sequences:

(1) forward solver calculates the boundary potential matrix $[V_c]$ for a known current injection matrix $[C]$ and an initial guess (known) conductivity matrix $[\sigma_0]$,

(2) measured voltage data matrix $[V_m]$ is compared with $[V_c]$ to estimate the $[\Delta V]$ as $[\Delta V = V_c - V_m]$,

(3) Jacobian (J) is computed,

(4) conductivity update vector $[\Delta\sigma]$ is calculated by Gauss-Newton-based minimization algorithm,

(5) $[\sigma_o]$ matrix is updated to a new conductivity matrix $[\sigma_1 = \sigma_o + \Delta\sigma]$ by adding $[\Delta\sigma]$ to $[\sigma]$ using Newton-Raphson iteration technique (NRIT),

(6) new update matrix $[\sigma_1]$ is used in forward solver to calculate the new voltage mismatch matrix $[\Delta V_1]$,

(7) check whether the $[\Delta V_1]$ is minimum or not or compare the $[\Delta V]$ with a specified error limit (ε) if provided,

(8) stop the algorithm if $\Delta V \leq \varepsilon$ condition is achieved, otherwise repeat the steps 1 to 7 until the specified stopping criteria ($\Delta V \leq \varepsilon$) is achieved.

2.3. Boundary Data Simulator (BDS). A two-dimensional boundary data simulator (BDS) is developed in MATLAB R2010a [27] using finite element method (FEM) [15] to generate accurate boundary data for studying the EIT reconstruction algorithms. The MATLAB-based BDS is developed as an absolute 2D data simulator for EIT image reconstruction studies, and it is used suitably to generate the errorless 2D boundary data to study and modify the inverse solver of a 2D EIT system. As BDS is developed in a computer software, it is found free from errors produced by the EIT instrumentation and phantom. BDS also allows us to generate boundary potential data for different type of phantom geometry, inhomogeneity geometry (shape, size, and position), inhomogeneity conductivity profiles, and background conductivity profiles. Moreover, it is developed as a compact, absolutely stable, and easy to use and handle for EIT studies. It is developed in such a way that it can be modified for further modifications.

BDS is developed with MATLAB-based computer program consisting of four-part imaging domain simulator (IDS), EIT model developer (EMD), current injection simulator (CIS), and boundary data calculator (BDC). Imaging domain simulator (IDS) in BDS simulates a domain with inhomogeneity with their corresponding conductivity distributions. EIT model developer (EMD) derives a mathematical model of the forward solver by applying FEM on the governing equation of the DUT in the form of a matrix equation. Current injection simulator (CIS) simulates a constant current injection through the definite points at the domain boundary with neighbouring current injection protocol [1, 2, 28–30]. The boundary data calculator (BDC) solves the governing equation by solving the forward model and calculates the potentials at all electrodes at the domain boundary.

Imaging domain simulator (IDS) first defines a DUT with a desired area (A_D) defined by a required diameter and defined with a particular coordinate system. Imaging domain simulator applies the FEM to discretize the domain with a 2D finite element mesh containing finite element of triangular elements (t) and finite number of nodes (n). In IDS, a circular domain (Ω) to be imaged is defined with a required radius (R_p) using the Cartesian coordinate system (Figure 1(a)), and the domain is discretized with a finite element (FE) mesh (Figure 1(b)). The mesh is symmetrically composed of the first-order triangular elements with linear shape functions [18, 31]. The FE mesh is generated with the *pdetool* of MATLAB R2010a in such a way that it can be refined further to increase the number of elements as per the requirement. All the coordinates and parameters assigned to the finite elements and the nodes are stored in corresponding matrices. Boundary nodes are identified, and the sixteen nodes among

the boundary nodes are assigned as the electrodes called the electrode nodes. Inside the domain one (or more) smaller region (regions) is (are) defined as the inhomogeneity (inhomogeneities) positioned at a particular place. The center point (P) of the inhomogeneity with the required shape and size is positioned inside the phantom domain by defining its center with a polar coordinate (r, θ) as shown in Figure 1(a). Single or multiple inhomogeneities are defined with their desired areas (A_I) inside the DUT, and elements within the inhomogeneity and the background are identified. The background area is defined as the area of the domain surrounding the inhomogeneity ($A_B = A_D - A_I$), and the elements within the background area (A_B) are identified. The elements within the inhomogeneity are assigned with a particular conductivity called inhomogeneity conductivity, (σ_i) while the rest of the elements are assigned with a different conductivity called background conductivity (σ_b) as shown in Figure 1(b). The assigned conductivity values of all the elements are assumed to be featured at their corresponding centroids.

EIT model developer (EMD) develops the mathematical model of the forward solver by applying FEM on the governing equation and derive the forward model of a DUT in the form of a matrix equation (3) using the elemental conductivities and nodal coordinates. The EMD establishes a relationship between the current injection matrix, $[C]$ (matrix of the applied signal), and the nodal potential matrix, $[\Phi]$ (matrix of the developed signal), through the transformation matrix $[K(\sigma)]$ which is mathematically represented by (3). The global stiffness matrix $[K(\sigma)]$ in EIT is actually an admittance matrix [23] that is formed [16] using the nodal coordinates of all the elements with their corresponding conductivities. Thus, the $[K(\sigma)]$ inforward model represents the transfer function of the EIT system obtained from governing equation by FEM formulation [19].

The current injection simulator (CIS) is used to simulate a constant current injection through the sixteen nodes called simulated electrodes (SE) on the domain boundary with neighbouring current injection protocol. The CIS works in a *"for"* loop to execute all the projections [1, 28, 30, 32] of current injection process. In BDS, a constant current injection is simulated into the DUT surrounded by the sixteen simulated current electrodes (SE_I) with all the possible combination of SE_I pairs, and the potential data are calculated on all the electrodes called voltage electrodes (SE_V) in BDC. The current injection through a particular current electrode pair (say SE_{I1} and SE_{I2}) and corresponding voltage data collection from all the possible voltage electrodes (SE_{V1}, SE_{V2}, SE_{V3}, SE_{V4}, SE_{V15}, SE_{V16}, SE_{V7}, SE_{V8}, SE_{V9}, SE_{V10}, SE_{V11}, SE_{V12}, SE_{V13}, SE_{V14}, SE_{V15} and SE_{V16}) is known as a simulated current projection (SCP). Hence, in an N-electrode EIT system, there will be N-different current projections each of which will inject current through a particular current electrode pair and collect m voltage (differential/grounded) data where m may be either equal to N or less than N depending on the EIT data collection strategy called the current pattern [1, 28, 30, 32]. Therefore, a complete scan (containing all the current projections) conducted on the DUT yields $N \times m$ voltage data. As the BDS is studied for sixteen electrode system, the CIS runs for sixteen times and provides sixteen

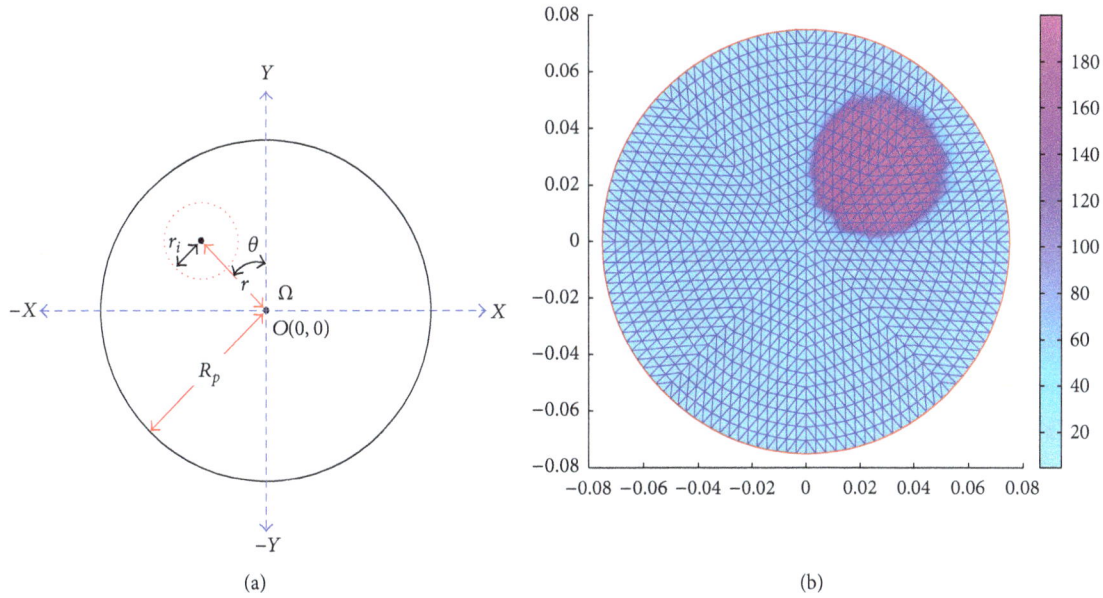

FIGURE 1: (a) A circular phantom domain (Ω) with an inhomogeneity defined by polar coordinate (r and θ); (b) a phantom domain (discretized by an FE mesh with 2048 elements and 1089 nodes) with a circular inhomogeneity ($R_p = 75$ mm, $r_i = 25$ mm, $r = 37.5$ mm, $\theta = 45°$, $\sigma_i = 0.005$ S/m, and $\sigma_b = 0.21$ S/m).

current projections (SCP_{V1}, SCP_{V2}, SCP_{V3}, SCP_{V4}, SCP_{V15}, SCP_{V16}, SCP_{V7}, SCP_{V8}, SCP_{V9}, SCP_{V10}, SCP_{V11}, SCP_{V12}, SCP_{V13}, SCP_{V14}, SCP_{V15}, and SCP_{V16}). Therefore, a complete data collection procedure (called a complete scan) in the BDS collects m voltage data from the voltage electrodes or voltage electrode pairs in all the sixteen current projections and computes $16 \times m$ voltage data.

Boundary data calculator (BDC) calculates the potentials (developed for a constant current injection by CIS) at all electrode points (electrode nodes) at the domain boundary in each current projection for a particular current pattern. The current injection matrix [32] is formed in CIS using the *Neumann type boundary conditions*, and the potential matrix is calculated from (3) using the matrix inversion technique working on *L-U* factorization [33] process. The BDS is developed to run in an another "*for*" loop for m times to calculate the m electrode potentials from voltage electrodes or voltage electrode pairs at each of the steps of the loop. This second "*for*" loop runs within the first "*for*" loop for m times and collects m voltage data for each step of first "*for*" loop and hence collects $16 \times m$ voltage data as first "*for*" loop runs for sixteen times. Moreover, as the EIT reconstruction process needs a complete scan, the BDS runs in each current projection and computes sixteen electrode potentials at each projection. The domain potential is calculated from the forward model (3), and the potential values of all the nodes are stored in a nodal potential matrix [33, 34] denoted by $[M_{\text{NP}}]$. Boundary potential data are separated from $[M_{\text{NP}}]$ and stored in a different matrix called boundary potential matrix $[M_{\text{BP}}]$. The electrode potential data are extracted from the nodal potential matrix $[M_{\text{NP}}]$ and are stored in a separate matrix called electrode potential matrix $[M_{\text{EP}}]$. In sixteen electrode EIT system, the $[M_{\text{EP}}]$ is formed as a column matrix

and contains the $16 \times m$ electrode potentials (differential or grounded) obtained for all the projections.

2.4. Neighbouring or Adjacent Current Injection Method. In neighbouring or adjacent current injection method, first reported by Brown and Segar [35], the current is applied through two neighbouring or adjacent electrodes, and the differential voltages is measured successively from all other adjacent electrode pairs excluding the pairs containing one or both of the current electrodes. For a sixteen electrode EIT system with domain under test surrounded by equally spaced sixteen electrodes (E_1, E_2, E_3, E_4, E_5, E_6, E_7, E_8, E_9, E_{10}, E_{11}, E_{12}, E_{13}, E_{14}, E_{15}, and E_{16}), the neighbouring method injects current through the current electrode pairs for sixteen current projections (Figure 2), and the differential voltages are measured across the voltage electrode pairs using four electrode method in each projection.

As shown in Figure 2(a) in the first current projection (P_1) of adjacent method, the current is injected through electrode 1 (E_1) and electrode 2 (E_2), and the thirteen differential voltage data ($V_1, V_2, V_3, \ldots, V_{13}$) are measured successively between the thirteen electrode pairs E_3-E_4, E_4-E_5, $\ldots$, and E_{15}-E_{16}, respectively (Figure 2(a)). As reported by Brown and Segar, in neighbouring current injection method, the current density within the DUT is found highest between the current electrodes (E_1 and E_2 for P_1); the current density then decreases rapidly as a function of distance [35]. Similarly, in current projection 2 (P_2), the current signal is injected through electrodes 2 (E_2) and 3 (E_3), and an another set of thirteen differential voltage data ($V_1, V_2, V_3, \ldots, V_{13}$) are collected between the thirteen electrode pairs E_4-E_5, E_5-E_6, $\ldots$, E_{16}-E_1, and so on. Lastly, in the current projection 16 (P_{16}), the last set of thirteen differential voltage

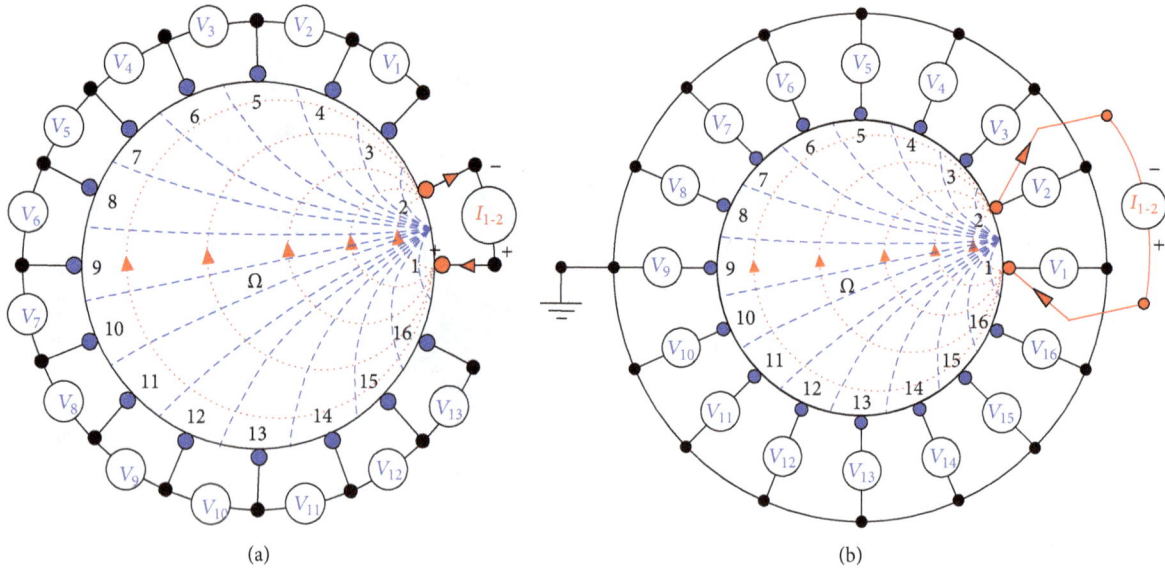

FIGURE 2: Current injection and boundary data collection in neighbouring current injection method; (a) data collection method as suggested by Brown and Segar and (b) data collection strategy as suggested by Cheng et al.

data $(V_1, V_2, V_3, \ldots, V_{13})$ are collected between the thirteen-electrode pairs E_2-E_3, E_3-E_4, ..., and E_{14}-E_{15} by injecting the current through the electrodes E_{16} and E_1. Thus, the neighbouring current injection method in a sixteen electrode EIT system data collection procedure consists of sixteen current projections ($P_1, P_2, P_3, \ldots, P_{15}$, and P_{16}), and each of the current projection yields thirteen differential voltage data $(V_1, V_2, V_3, \ldots, V_{13})$. Therefore, a complete data collection scan with the neighbouring current injection method in a sixteen electrode EIT system yields $16 \times 13 = 208$ voltage measurements.

Though in neighbouring method, EIT boundary data are not collected across the electrode pairs containing one or two current electrode for contact impedance problem [35], but sometimes it is advantageous to collect the boundary data from all the electrodes including the current electrodes to obtain the greatest sensitivity to the resistivity changes in the domain as reported by Cheng et al. [36]. In the present study, the boundary potentials are calculated at all the electrodes (Figure 2(b)) with respect to a virtual ground point selected within the DUT. Hence, in a complete data collection scan, the potentials on all the electrodes are collected in all the sixteen current projection and are stored in $[M_{EP}]$. Therefore, the $[M_{EP}]$ is found as a column matrix containing 16×16 voltage data all collected with respect to the virtual ground point of the DUT. Hence, in the present study, with neighbouring current injection method, the $[M_{EP}]$ is found as a 256×1 matrix containing 256 electrode potentials. In the present study, 1 mA current injection is simulated through the electrodes of the simulated domain containing sixteen nodal electrodes using adjacent or neighboring current injection protocol (Figure 2(b)). The potentials on all the sixteen electrodes are calculated using boundary data calculator (BDC) for all the current projections, and the electrode potential matrix $[M_{EP}]$ is used as the calculate boundary potential

matrix $[V_c]$ to reconstruct the conductivity distribution of DUT.

The BDS is designed in such a way that a huge number of voltage data sets can be generated using different types of phantoms with their different design parameters. Boundary potential data $[V_c]$ are generated for different type of phantom configurations, and the boundary data have been tested with electrical impedance tomography and diffuse optical tomography reconstruction software (EIDORS) [37, 38] for 2D-EIT. A large number of data sets are generated by changing the values of one or more phantom parameters like: phantom diameter ($D = 2R_p$), inhomogeneity radius (r_i), inhomogeneity geometry (shape, size, and position), inhomogeneity number (N_i), bathing solution conductivity (σ_b), and inhomogeneity conductivity (σ_i). 1 mA current injection is simulated to the domain boundary, and corresponding boundary data sets are used for image reconstruction in EIDORS. Data generation in BDS and image reconstruction in EIDORS are studied for different inhomogeneity geometries in DUT. Reconstruction is also studied for different iterations and for multiple inhomogeneity reconstruction to evaluate the BDS.

3. Results and Discussion

Image reconstruction quality in EIT depends on the boundary data accuracy which is dependent on the geometric accuracy of the inhomogeneity developed in BDS. Dimensional accuracy of the inhomogeneity depends on the number of finite elements in the FE mesh or mesh refinement number (N_{mr}) as shown in Figure 3. As the N_{mr} increases, the number of elements in the FE mesh is increased, and hence the geometric accuracy of the inhomogeneity increases which gives more accurate boundary data and better image reconstruction (Figure 3). But the BDS with a highly refined

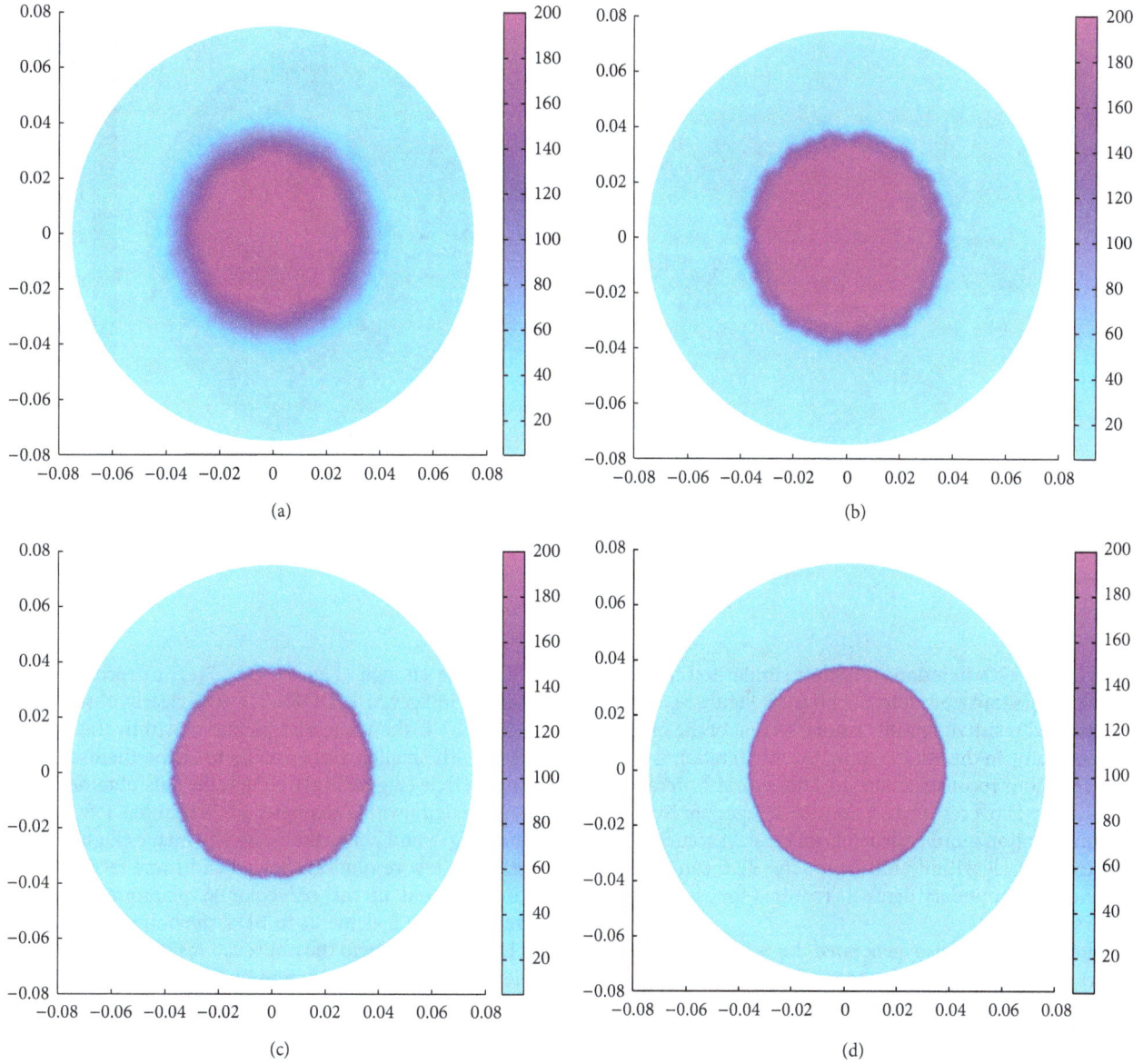

FIGURE 3: Circular inhomogeneity (R_p = 75 mm, r = 0, r_i = 37.5 mm, σ_i = 0.005 S/m, and σ_b = 0.21 S/m) with FEM mesh with different number of finite elements: (a) 512 elements and 289 nodes, (b) 2048 elements and 1089 nodes, (c) 8192 elements and 4225 nodes, and (d) 32768 elements and 16641 nodes.

mesh needs a high PC memory and large computation time. In this paper, the mesh refinement is found suitable as N_{mr} = 4 as per the configuration of the PC (2.4 GHz/1.5 GBRAM/ P-IV) used. It is observed that the FE mesh with N_{mr} = 4 (containing 2048 elements and 1089 nodes) gives almost an accurate geometry (Figure 3) to the desired inhomogeneity and generates a reconstructible data set in less than 10 seconds. EIDORS reconstructs the resistivity images from the BDS data sets using regularized image reconstruction technique.

Results show that the resistivity or conductivity can be successfully reconstructed from the boundary data generated by our BDS using a circular domain (R_p = 75 mm) with a circular inhomogeneity (r = 37.5 mm, r_i = 25 mm, θ = 45°,

σ_i = 0.005 S/m, and σ_b = 0.21 S/m) in the 9th iteration (Figure 4). It is also observed that the reconstructed shape of the inhomogeneity is similar to that of the original one (Figure 4(a)), and the reconstructed conductivity profile in Figure 4(b) is almost similar to that of the original object in Figure 4(a).

Iteration studies shows that in different reconstruction steps called iterations (Figure 5), the reconstructed images become more localized from iteration to iteration and the reconstruction errors (appeared by the red color at phantom periphery) are gradually reduced (Figure 5).

It is observed that the resistivity is successfully reconstructed from the boundary data in the 9th iteration (Figures 5(i) and 5(j)), though the shape of all the reconstructed

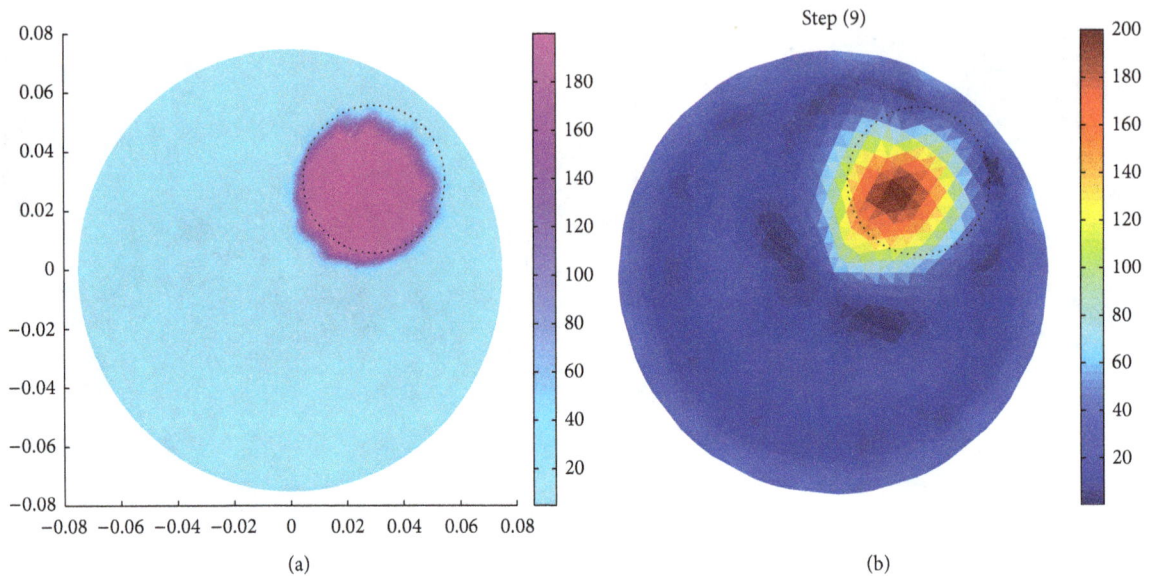

FIGURE 4: (a) Simulated domain with a circular object ($R_p = 75$ mm, $r = 37.5$, $\theta = 45°$, $r_i = 25$ mm, $\sigma_i = 0.005$ S/m, and $\sigma_b = 0.21$ S/m); (b) reconstructed image of (a).

images in 9th–12th iterations is almost similar to that of the original one (shown by dotted circles in Figure 5). As the reconstructed resistivity profile similar to that of the original is obtained only in the 9th iteration, the 9th iteration is taken as the optimum reconstruction. In 13th and 14th iterations, the resistivity is overestimated, and the images are lost. The optimum iteration number depends on the data accuracy and reconstruction algorithm, and hence the BDS can be used to generate the boundary data sets required for assessing the inverse solver in EIT.

Voltage data are also generated for a domain ($R_p = 75$ mm) with the circular inhomogeneities ($r_i = 25$ mm, $\sigma_i = 0.005$, S/m, and $\sigma_b = 0.21$ S/m) positioned at different places using the BDS (Figure 6). It is observed that the reconstructed image is more circular for an inhomogeneity positioned at the phantom centre where $r = 0$ and $\theta = 0°$ (Figure 6(a)). On the other hand, for $r \neq 0$, that is, for the inhomogeneities near domain boundary (Figure 6(b)), reconstructed images are not perfectly circular because of the comparatively less accurate shape of the original object obtained for $r \neq 0$. For a less number of mesh refinements, the geometry of the original side objects is not exactly circular itself (Figure 4), and hence the corresponding boundary data have lower accuracy. An FE mesh with large N_{mr} can easily produce an accurate geometry for the boundary objects (objects near domain boundary) with proper shape, which gives a boundary data without geometric error and automatically improves the image shape.

Boundary data sets are also generated with a circular domain ($R_p = 75$ mm and $\sigma_b = 0.21$ S/m) with a circular inhomogeneity ($\sigma_i = 0.005$ S/m) with different diameters ($2r_i$) and all positioned at the phantom center ($r = 0$). The boundary data are calculated and used for reconstructing the resistivity images. Results show that for the domain discretized with $N_{mr} = 4$, the data sets, generated with a diameter larger than 13.3% of the phantom diameter,

are accurate enough (Figures 7(a)–7(f)) to reconstruct the resistivity images in EIDORS-2D. It is clearly observed that for $N_{mr} = 4$, the triangular elements within the inhomogeneity with smaller r_i are unable to shape themselves into a proper circle (Figure 7(g)). Hence, the data obtained for the inhomogeneity with a diameter of 20 mm has low accuracy (Figure 7(g)), and hence the resistivity image (Figure 7(h)) is found with low resolution showed and some reconstruction error (appeared in the red color at phantom periphery). Increasing the FE elements in BDS, the boundary data error can be minimized, and the improved resistivity image can be achieved even for smaller inhomogeneities with a diameter less than 13.3% of R_p.

Boundary potential data are also generated for domains ($R_p = 75$ mm) containing multiple circular inhomogeneities ($r_i = 25$ mm, $r = 37.5$ mm, $\sigma_i = 0.005$ S/m, and $\sigma_b = 0.21$ S/m) placed at different positions inside the domain (Figure 8). Figure 8(a) shows a domain with two circular inhomogeneities (180° apart from each other) which are placed at a central distance (r) of 37.5 mm. Similarly, another domain with three circular inhomogeneities (120° apart from each other) placed inside the phantom domain is shown in Figure 8(c). All the inhomogeneities in both the domains are positioned at a central distance (r) of 37.5 mm. 1 mA current is simulated with the neighbouring current pattern, and the boundary data are collected for resistivity reconstruction. It is noticed that the resistivity images (Figures 8(b) and 8(d)) of inhomogeneities in both the domains are reconstructed successfully.

Results show that the boundary data simulator can be efficiently used to generate boundary potential data for a huge number of phantom configurations in less than 10 seconds. BDS is software-based virtual EIT phantom, and hence it has a number of advantages over the practical and mesh phantoms. The literatures [39–41] presenting the

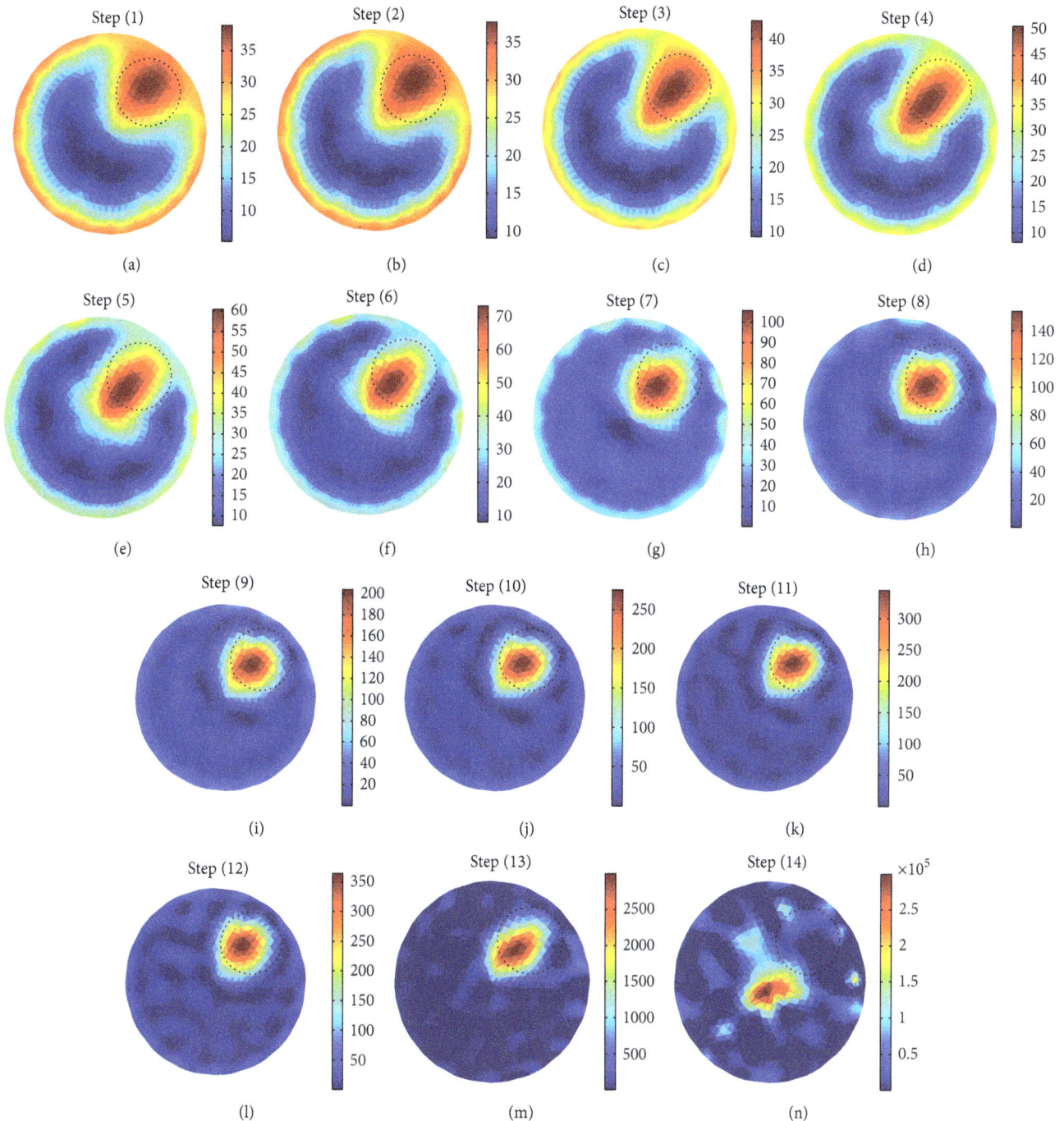

FIGURE 5: Reconstructed images of a simulated domain with a circular inhomogeneity (R_p = 75 mm, r = 37.5, θ = 45°, r_i = 25 mm, σ_i = 0.005 S/m, and σ_b = 0.21 S/m) for different number of iterations in inverse solver in EIDORS-2D: (a) 1st iteration, (b) 2nd iteration, (c) 3rd iteration, (d) 4th iteration, (e) 5th iteration, (f) 6th iteration, (g) 7th iteration, (h) 8th iteration, (i) 9th iteration, (j) 10th iteration, (k) 11th iteration, (l) 12th iteration, (m) 13th iteration, and (n) 14th iteration.

phantom simulations are limited, and they only discuss the software phantoms developed for their own systems. BDS is a software-based versatile boundary data simulator which generates boundary data suitable for studying the reconstruction algorithm required for several EIT systems, and hence it is better suited for assessing the performance of the inverse solver of 2D electrical impedance tomography.

4. Conclusions

A MATLAB boundary data simulator (BDS) is developed for studying the resistivity reconstruction in inverse solvers of 2D-EIT. BDS is developed with four parts: imaging domain simulator (IDS), EIT model developer (EMD), current injection simulator (CIS), and boundary data calculator (BDC).

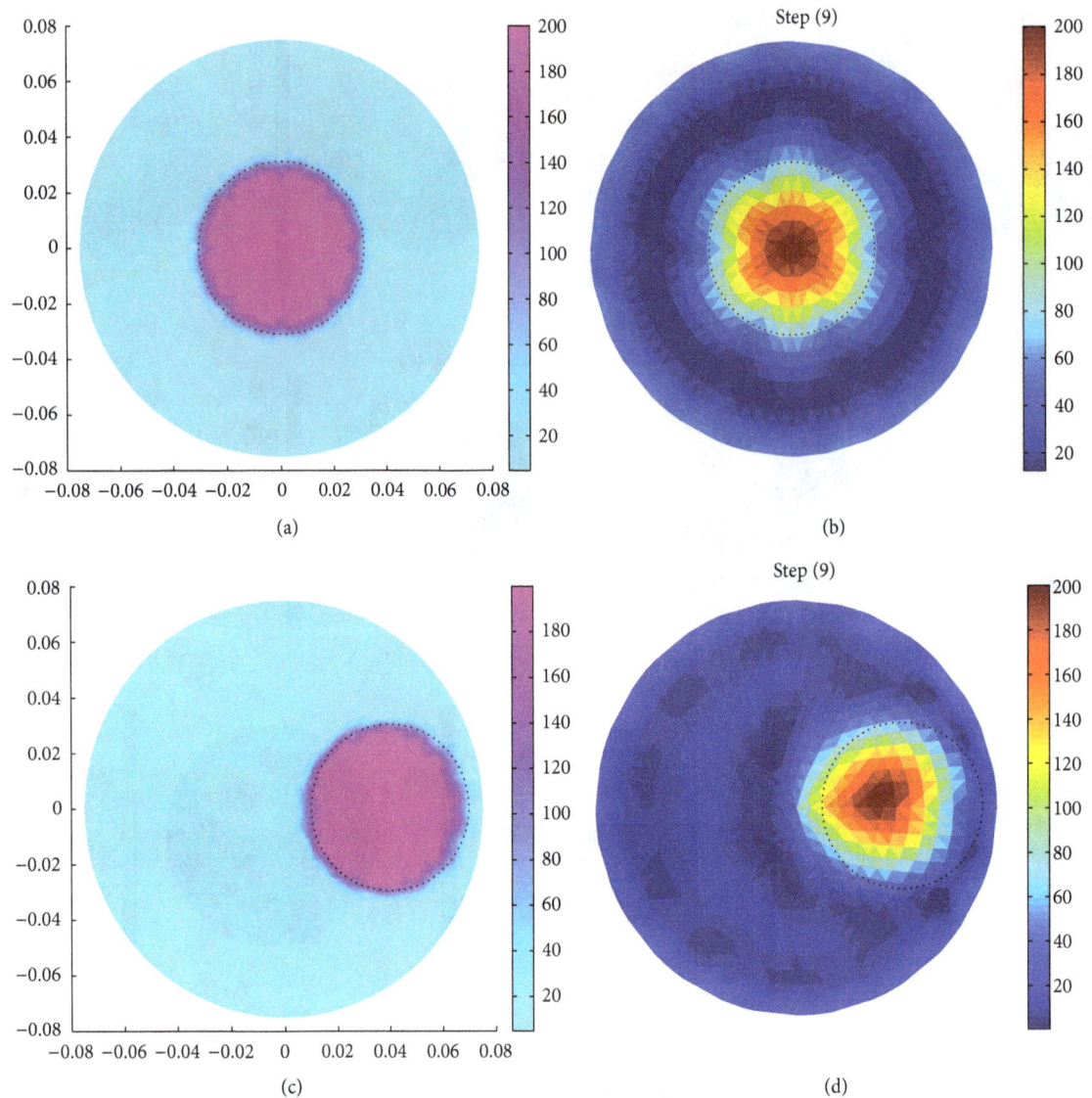

FIGURE 6: Image reconstruction of a simulated domain with a circular inhomogeneity (R_p = 75 mm, r_i = 30 mm, σ_i = 0.005 S/m, and σ_b = 0.21 S/m) at different positions (r, θ): (a) simulated domain with inhomogeneity at r = 0 mm and θ = 0°, (b) reconstructed image of the domain shown in Figures (a) and (c) simulated domain with inhomogeneity at r = 37.5 mm and θ = 0°, and (d) reconstructed image of the domain shown in Figure (c).

Imaging domain simulator (IDS) simulates a domain with single or multiple inhomogeneities of different geometries defined with their corresponding conductivity distributions, whereas the EIT model developer (EMD) derives a forward model using FEM to solve the governing equation of the DUT. Current injection simulator (CIS) simulates a constant current injection through the simulated electrodes positioned at the domain boundary with the neighbouring current injection protocol. The boundary data calculator (BDC) solves the forward model to solve the governing equation and calculates the potentials at all the simulated electrodes. Boundary data are generated with different type of domains simulated in BDS by changing its input parameters. Resistivity images are reconstructed from the boundary data

using standard EIT reconstruction software called EIDORS, and the BDS is evaluated. It is observed that the BDS with FE mesh with 2048 elements can simulate an inhomogeneity of desired geometry with suitable accuracy. The BDS with 2048 elements suitably generates the boundary data for simulated domains containing the objects with different geometries which are found efficient for image reconstruction in EIDORS. Results also show that the conductivity or resistivity profiles of the domains simulated in BDS are successfully reconstructed from their corresponding boundary data generated for different type of single and multiple inhomogeneities. By changing the inhomogeneity position, diameter, and number in BDS, boundary data are successfully generated as well as the resistivity images are reconstructed

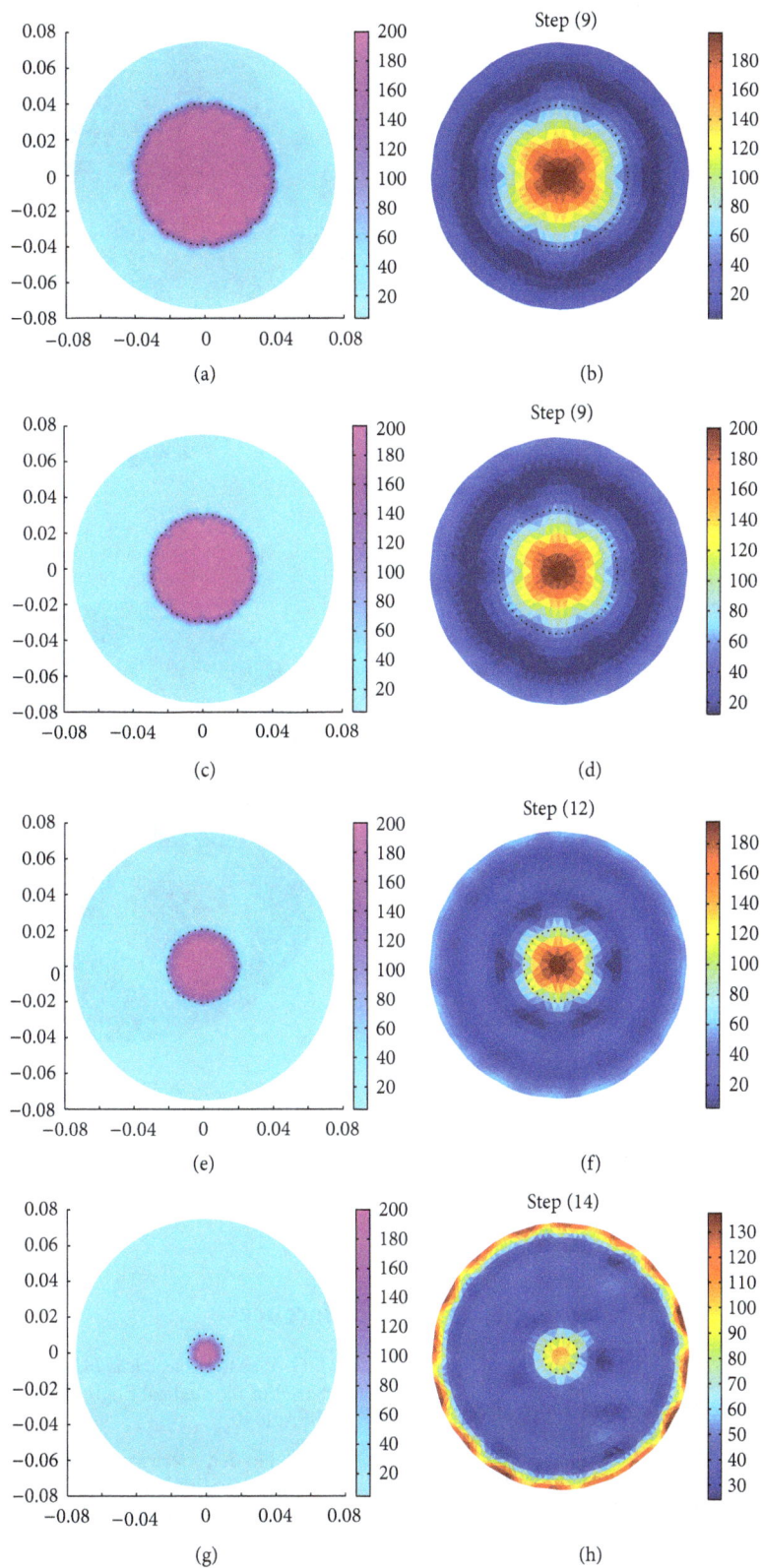

FIGURE 7: Image reconstruction of circular inhomogeneities (R_p = 75 mm, r = 0, σ_i = 0.005 S/m, and σ_b = 0.21 S/m) with different diameters: (a) original object with r_i = 40 mm, (b) reconstructed image of the object shown in (a), (c) original object with r_i = 30 mm, (d) reconstructed image of the object shown in (c), (e) original object with r_i = 20 mm, (f) reconstructed image of the object shown in (e), (g) original object with r_i = 10 mm, and (h) reconstructed image of the object shown in (g).

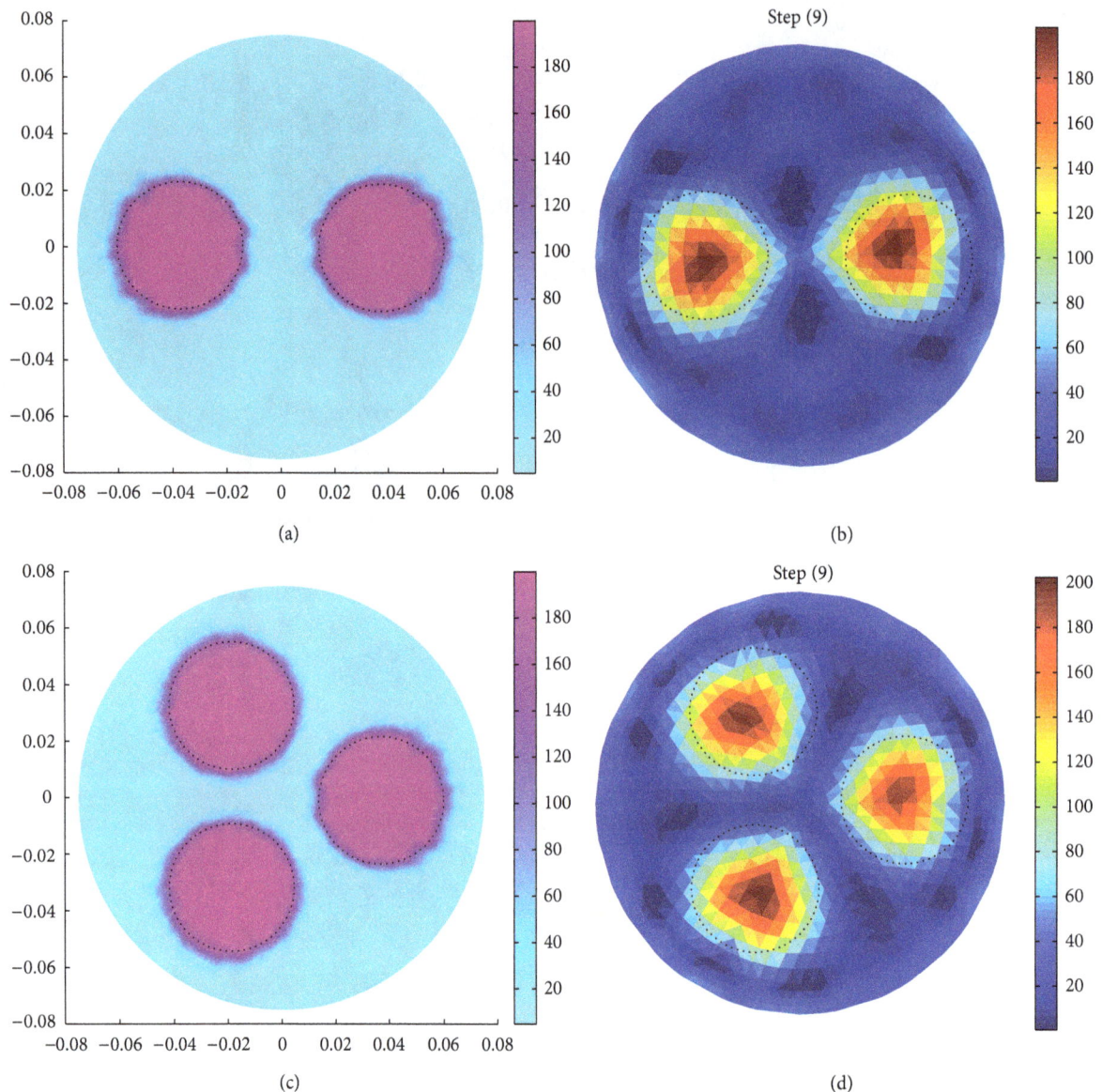

FIGURE 8: Image reconstruction of multiple inhomogeneities ($\sigma_b = 0.21$ S/m and $\sigma_i = 0.005$ S/m): (a) simulated domain with two circular objects ($r_i = 25$ mm, 180° apart), (b) reconstructed image of the domain shown in (a), (c) simulated domain with three circular objects ($r_i = 25$ mm, 120° apart), and (d) reconstructed image of the domain shown in (c).

successfully. Multiple inhomogeneity imaging shows that the BDS suitably generates boundary data with the desired accuracy, and the boundary data are found efficient for resistivity reconstruction in EIDORS. Results also show that for the simulated domains discretized with $N_{mr} = 4$, the boundary data sets generated for circular inhomogeneity with a diameter larger than 13.3% of the phantom diameter are accurate enough to reconstruct the resistivity images in EIDORS. Increasing the FE elements in BDS, the boundary data error can further be minimized, and the improved resistivity image reconstruction can be obtained even for smaller inhomogeneities. Hence, it is concluded that the BDS generated a number of boundary data sets which can suitably be used for inverse solver assessment in EIT.

References

[1] J. G. Webster, *Electrical Impedance Tomography*, Adam Hilger Series of Biomedical Engineering, Adam Hilger, New York, NY, USA, 1990.

[2] D. S. Holder, *Electrical Impedance Tomography: Methods, History and Applications*, Series in Medical Physics and Biomedical Engineering, Institute of Physics Publishing Ltd., Bristol, UK, 1st edition, 2005.

[3] H. Griffiths and Z. Zhang, "A dual-frequency electrical impedance tomography system," *Physics in Medicine and Biology*, vol. 34, no. 10, pp. 1465–1476, 1989.

[4] T. K. Bera and J. Nagaraju, "A multifrequency constant current source suitable for Electrical Impedance Tomography (EIT)," in *Proceedings of the IEEE International Conference on Systems*

in Medicine and Biology (ICSMB '10), pp. 278–283, Kharagpur, India, December 2010.

[5] T. K. Bera and J. Nagaraju, "A reconfigurable practical phantom for studying the 2D Electrical Impedance Tomography (EIT) using a FEM based forward solver," in *Proceedings of the 10th International Conference on Biomedical Applications of Electrical Impedance Tomography (EIT '09)*, School of Mathematics, The University of Manchester, Manchester, UK, June 2009.

[6] D. S. Holder, Y. Hanquan, and A. Rao, "Some practical biological phantoms for calibrating multifrequency electrical impedance tomography," *Physiological Measurement*, vol. 17, no. 4, pp. A167–A177, 1996.

[7] T. K. Bera and J. Nagaraju, "Studying the 2D-Image Reconstruction of Non Biological and Biological Inhomogeneities in Electrical Impedance Tomography (EIT) with EIDORS," in *Proceedings of the International Conference on Advanced Computing, Networking and Security (ADCONS' 11)*, pp. 132–136, NITK-Surathkal, Mangalore, India.

[8] T. K. Bera and J. Nagaraju, "A stainless steel electrode phantom to study the forward problem of Electrical Impedance Tomography (EIT)," *Sensors & Transducers Journal*, vol. 104, no. 5, pp. 33–40, 2009.

[9] H. Griffiths, "A phantom for electrical impedance tomography," *Clinical Physics and Physiological Measurement*, vol. 9, supplement A, pp. 15–20, 1988.

[10] T. K. Bera and J. Nagaraju, "Resistivity imaging of a reconfigurable phantom with circular inhomogeneities in 2D-electrical impedance tomography," *Measurement*, vol. 44, no. 3, pp. 518–526, 2011.

[11] H. Griffiths, Z. Zhang, and M. Watts, "A constant-perturbation saline phantom for electrical impedance tomography," *Physics in Medicine and Biology*, vol. 34, no. 8, pp. 1063–1071, 1989.

[12] H. Gagnon, A. E. Hartinger, A. Adler, and R. Guardo, "A resistive mesh phantom for assessing the performance of EIT systems," in *Proceedings of the International Conference on Biomedical Applications of Electrical Impedance Tomography (EIT '09)*, Manchester, UK, 2009.

[13] H. Gagnon, Y. Sigmen, A. E. Hartinger, and R. Guardo, "An active phantom to assess the robustness of EIT systems to electrode contact impedance variations," in *Proceedings of the International Conference on Biomedical Applications of Electrical Impedance Tomography (EIT '09)*, Manchester, UK, 2009.

[14] G. Hahn, A. Just, and G. Hellige, "Determination of the dynamic measurement error of EIT systems," in *Proceedings of the 13th International Conference on Electrical Bioimpedance and the 8th Conference on Electrical Impedance Tomography (ICEBI '07)*, IFMBE Proceedings 17, pp. 320–323, September 2007.

[15] W. R. B. Lionheart, "EIT reconstruction algorithms: pitfalls, challenges and recent developments," *Physiological Measurement*, vol. 25, pp. 125–142, 2004.

[16] T. K. Bera, S. K. Biswas, K. Rajan, and J. Nagaraju, "Improving image quality in Electrical Impedance Tomography (EIT) using projection error propagation-based regularization (PEPR) technique: a simulation study," *Journal of Electrical Bioimpedance*, vol. 2, pp. 2–12, 2011.

[17] T. K. Bera, S. K. Biswas, K. Rajan, and J. Nagaraju, "Improving conductivity image quality using block matrix-based multiple regularization (BMMR) technique in EIT: a simulation study," *Journal of Electrical Bioimpedance*, vol. 2, pp. 33–47, 2011.

[18] J. N. Reddy, *An Introduction to the Finite Element Method*, TATA McGraw-Hill Publishing Company Ltd., New Delhi, India, 3rd edition, 2006.

[19] T. J. Yorkey, *Comparing reconstruction methods for electrical impedance tomography [Ph.D. thesis]*, University of Wisconsin at Madison, Madison, Wis, USA, 1986.

[20] *Vauhkonen Marko Electrical Impedance Tomography and Prior Information*, Kuopio University Publications, Natural and Environmental Sciences, 1997.

[21] T. J. Yorkey, J. G. Webster, and W. J. Tompkins, "Comparing reconstruction algorithms for electrical impedance tomography," *IEEE Transactions on Biomedical Engineering*, vol. 34, no. 11, pp. 843–852, 1987.

[22] C. J. Grootveld, *Measuring and modeling of concentrated settling suspensions using electrical impedance tomography [Ph.D. thesis]*, Delft University of Technology, Delft, The Netherlands, 1996.

[23] B. M. Graham, *Enhancements in Electrical Impedance Tomography (EIT) image reconstruction for 3D lung imaging [Ph.D. thesis]*, University of Ottawa, 2007.

[24] M. C. Kim, K. Y. Kim, S. Kim, H. J. Lee, and Y. J. Lee, "Electrical impedance tomography technique for the visualization of the phase distribution in an annular tube," *Journal of Industrial and Engineering Chemistry*, vol. 8, no. 2, pp. 168–172, 2002.

[25] T. K. Bera, S. K. Biswas, K. Rajan, and J. Nagaraju, "Improving the image reconstruction in Electrical Impedance Tomography (EIT) with block matrix-based Multiple Regularization (BMMR): a practical phantom study," in *Proceedings of the IEEE World Congress on Information and Communication Technologies*, pp. 1346–1351, Mumbai, India, 2011.

[26] T. K. Bera, S. K. Biswas, K. Rajan, and J. Nagaraju, "Image reconstruction in Electrical Impedance Tomography (EIT) with projection error propagation-based regularization (PEPR): a practical phantom study," in *Advanced Computing, Networking and Security*, vol. 7135 of *Lecture Notes in Computer Science*, pp. 95–105, Springer, 2012.

[27] *MATLAB: The Language of Technical Computing, Version R2010a*, The MathWorks, Inc., Natick, Mass, USA, 2010.

[28] J. Malmivuo and R. Plonsey, *Bioelectromagnetism: Principles and Applications of Bioelectric and Biomagnetic Fields*, chapter 26, section 26.2.1, Oxford University Press, New York, NY, USA, 1995.

[29] T. K. Bera and J. Nagaraju, "Studying the resistivity imaging of chicken tissue phantoms with different current patterns in Electrical Impedance Tomography (EIT)," *Measurement*, vol. 45, pp. 663–682, 2012.

[30] T. K. Bera and J. Nagaraju, "Studying the 2D resistivity reconstruction of stainless steel electrode phantoms using different current patterns of Electrical Impedance Tomography (EIT), biomedical engineering," in *Proceeding of the International Conference on Biomedical Engineering (ICBME '11)*, pp. 163–169, Narosa Publishing House, Manipal, India, 2011.

[31] T. K. Bera and J. Nagaraju, "A FEM-based forward solver for studying the forward problem of Electrical Impedance Tomography (EIT) with a practical biological phantom," in *Proceedings of the IEEE International Advance Computing Conference (IACC '09)*, pp. 1375–1381, Patiala, India, March 2009.

[32] T. K. Bera and J. Nagaraju, "A study of practical biological phantoms with simple instrumentation for Electrical Impedance Tomography (EIT)," in *Proceedings of the IEEE Instrumentation and Measurement Technology Conference (I2MTC '09)*, pp. 511–516, Singapore, May 2009.

[33] E. Kreyszig, *Advanced Engineering Mathematics*, chapter 18, section 18.2, John Wiley & Sons, 8th edition, 1999.

[34] T. K. Bera and J. Nagaraju, "Studying the boundary data profile of a practical phantom for medical electrical impedance tomography with different electrode geometries," in *Proceedings of the World Congress on Medical Physics and Biomedical Engineering*, IFMBE Proceedings 25/II, pp. 925–929, Munich, Germany, September 2009.

[35] B. H. Brown and A. D. Segar, "The Sheffield data collection system," *Clinical Physics and Physiological Measurement*, vol. 8, supplement A, pp. 91–97, 1987.

[36] K. S. Cheng, S. J. Simske, D. Isaacson, J. C. Newell, and D. G. Gisser, "Errors due to measuring voltage on current-carrying electrodes in electric current computed tomography," *IEEE Transactions on Biomedical Engineering*, vol. 37, no. 1, pp. 60–65, 1990.

[37] N. Polydorides and W. R. B. Lionheart, "A Matlab toolkit for three-dimensional electrical impedance tomography: a contribution to the Electrical Impedance and Diffuse Optical Reconstruction Software project," *Measurement Science and Technology*, vol. 13, no. 12, pp. 1871–1883, 2002.

[38] M. Vauhkonen, W. R. B. Lionheart, L. M. Heikkinen, P. J. Vauhkonen, and J. P. Kaipio, "A MATLAB package for the EIDORS project to reconstruct two-dimensional EIT images," *Physiological Measurement*, vol. 22, no. 1, pp. 107–111, 2001.

[39] M. Kuzuoplu, M. Moh'dSaid, and Y. Z. Ider, "Analysis of three-dimensional software EIT (electrical impedance tomography) phantoms by the finite element method," *Clinical Physics and Physiological Measurement*, vol. 13, supplement A, pp. 135–138, 1992.

[40] R. P. Patterson and J. Zhang, "Evaluation of an EIT reconstruction algorithm using finite difference human thorax models as phantoms," *Physiological Measurement*, vol. 24, no. 2, pp. 467–475, 2003.

[41] R. Davalos and B. Rubinsky, "Electrical impedance tomography of cell viability in tissue with application to cryosurgery," *Journal of Biomechanical Engineering*, vol. 126, no. 2, pp. 305–309, 2004.

5

Preliminary Deformational Studies on a Finite Element Model of the Nasal Septum Reveals Key Areas for Septal Realignment and Reconstruction

Kyrin Liong,[1] Shu Jin Lee,[2] and Heow Pueh Lee[1]

[1] *Department of Mechanical Engineering, National University of Singapore, Singapore 117576*
[2] *Division of Plastic, Reconstructive and Aesthetic Surgery, National University Hospital, Singapore 119074*

Correspondence should be addressed to Kyrin Liong; kyrin.jo.liong@nus.edu.sg

Academic Editor: Rad Zdero

Background. With the current lack of clinically relevant classification methods of septal deviation, computer-generated models are important, as septal cartilage is indistinguishable on current imaging methods, making preoperative planning difficult. *Methods.* Three-dimensional models of the septum were created from a CT scan, and incremental forces were applied. *Results.* Regardless of the force direction, with increasing force, the septum first tilts (type I) and then crumples into a C shape (type II) and finally into an S shape (type III). In type I, it is important to address the dislocation in the vomer-ethmoid cartilage junction and vomerine groove, where stress is concentrated. In types II and III, there is intrinsic fracture and shortening of the nasal septum, which may be dislocated off the anterior nasal spine. Surgery aims to relieve the posterior buckling and dislocation, with realignment of the septum to the ANS and possible spreader grafts to buttress the fracture sites. *Conclusion.* By identifying clinically observable septal deviations and the areas of stress concentration and dislocation, a straighter, more stable septum may be achieved.

1. Introduction

Nasal septal deviation is a common nasal deformity. It can be a congenital disorder or a consequence of nasal trauma. Deviation of the bony or cartilaginous component of the nasal septum from the midline leads to its deviation. This results in external nasal deformity, internal nasal obstruction due to nasal airway constriction, or a combination [1–3].

Presently, septal deviation classification has largely been descriptive, based on nasal septal geometry and relationships between the bony and cartilaginous septa [4–7]. Jang et al. [6] presented a simplified classification of nasal deviation and the associated treatment outcome into five types based on the orientation of the bony pyramid and the cartilaginous vault. Jin et al. [7] presented a four-category classification of septal deviation based on the morphology, site, severity, and its influence on the external nose. Buyukertan et al. [4] reported a morphometric study of nasal septal deviation by separating the nasal septum into 10 segments. They

concluded that the system would constitute a new, objective, simple, and practical classification system. I. Baumann and H. Baumann [8] argued that the existing nomenclatures of septal deviation only dealt with nasal septum deformation exclusively and were rarely used in routine clinical work. They instead presented a method for the classification of septal deviations based upon the anatomical structures of the nasal septum and common clinical concepts. However, the most observable nasal septal deviation classification system was proposed by Rohrich et al. [9]. Therefore, for simplicity, nasal septal deviations will be classified according to that proposed by Rohrich et al. [9].

In order to improve the clinical outcome of septoplasty, a greater understanding of the etiopathogenesis of nasal septal deviation is necessary. This requires a basic understanding of the biomechanics of its formation. We aim to apply incremental force to a computer-generated septal model using structural modal analysis, which has also been utilized by Laura et al. [10], who previously described a simple method

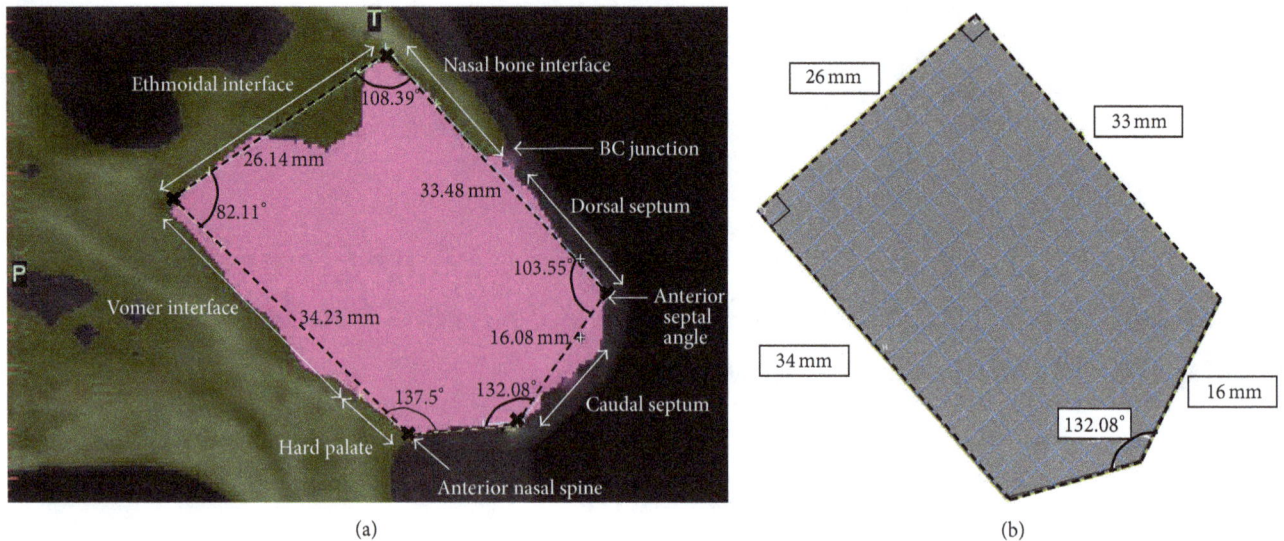

FIGURE 1: (a) Sagittal CT scan of the nasal septum with the indicated anatomical features and their measurements. (b) Idealized cartilaginous septum model created with the indicated measurements.

for determining the fundamental mode of a vibrating ulna to approximate its dynamic response. The objective of this study is to identify areas of high-stress and septal deformation patterns. Clinically, this may assist surgeons in the delineation of key areas for septal realignment and reconstruction.

2. Materials and Methods

2.1. Generation of Cranial Computed Tomography (CT) Scan and Finite Element Models. Cranial CT scans were obtained from a patient who possessed normal features—a straight nasal septum, normal occlusion and a perceivably symmetric face (Figure 1(a)). This study was performed in accordance with the guidelines of the institutional review board (IRB) and conforms to the Helsinki's Declaration. The patient had not previously undergone septoplasty or rhinoplasty, nor subject to nasal injury. Superposition of the CT images to create a three-dimensional (3D) model was conducted with Mimics software (Materialise Technologies, Leuven, Belgium). An idealized model (Figure 1(b)) and a patient-specific finite element model were generated for the study.

From the CT scans, we observed that the cartilaginous nasal septum was present in five slices. We chose to base the idealized model on the middle slice (Figure 1(a)) and measured the significant features of the nasal septum. We then utilized these measurements to create an idealized model (Figure 1(b)) in the Finite Element Analysis software, ABAQUS (Dassault Systèmes Technologies, Providence, RI, United States of America (USA)), where the idealized model was, subsequently, meshed. We recognize the thickness variation present in the septum. However, to simplify analyses and gain an estimate of nasal deformation, models were prescribed with a uniform thickness of 2 mm [11], which is an approximate average septum thickness, as reported previously [11–13]. To ensure mesh accuracy, convergence studies were carried out on the model.

To create a more realistic representation of the septum, which incorporated thickness variation, a 3D patient-specific model was created from the same CT scan utilizing Mimics software (Materialise Technologies, Leuven, Belgium) and meshed with Hypermesh (Altair HyperWorks, Troy, MI).

2.2. Material Properties. Cartilage exhibits a "nonhomogenous, anisotropic, nonlinear, viscoelastic behaviour" [14]. For deformations below 20% [14], however, no significant changes occur within the cartilage, and it is therefore sufficiently accurate to model cartilage as a homogenous, linearly elastic material in our analyses [15]. Mau et al. [11] utilized a similar homogenous, linear elastic material property to simulate septal L-strut deformation.

To define the linear elastic model of the cartilage, the Young's modulus, E, and Poisson's ratio, v, are required. However, the tensile and compressive Young's moduli are vastly different due to the structure of cartilage. According to Lee et al. [16], the tensile modulus ranges from 2.62 MPa to 10.6 MPa, the compressive modulus ranges from 0.40 MPa to 0.83 MPa, and the Poisson's ratio ranges from 0.26 to 0.38.

Specimen density is also required in the analysis. Cartilage is approximately 75% water, while the other 25% consists mainly of type-two collagen fibrils and proteoglycan molecules [17]. The density of water is 1000 kg/m^3, while the other components are highly dense structures. Therefore, the density of cartilage was estimated to be 2000 kg/m^3.

As the relative displacement within the septum is the main area of concern in this analysis, and since material properties affect the absolute and not the relative displacement of the septum, the average values of the elastic modulus and Poisson's ratio and an estimated value of the density were used. The elastic modulus was assigned a value of 5 MPa, Poisson's ratio was 0.32, and the density was 2000 kg/m^3.

Preliminary Deformational Studies on a Finite Element Model of the Nasal Septum Reveals Key Areas for Septal
Realignment and Reconstruction

51

2.3. Boundary Conditions

2.3.1. Bony Interfaces. As the bony interfaces with the nasal septum—ethmoidal, vomer, hard palate, and nasal bone interfaces (Figure 1(a))—are much stiffer than the septal cartilage, most of any applied force will be absorbed by the cartilaginous septum, leaving the bony septum uninjured [18]. It is therefore reasonable to assume these interfaces as rigidly fixed [15].

The nasal bone length overlapping the cartilaginous septum may affect the degree of nasal deformation and normally ranges from 3 to 15 mm [19]. However, to simplify analyses, a candidate that displayed a length that fell within this range—in this case, 14 mm—was considered, so that a typical deformation pattern could be observed.

2.3.2. Nasal Tip. In vivo, the nasal tip lies anterior to the anterior septal angle (ASA) where the lower lateral cartilages (LLCs) meet, although this may vary. However, due to the small distance between the ASA and the nasal tip, and for simplification in this analysis, the ASA was assumed to be the nasal tip.

According to Lee et al. [16], the nasal tip cartilages may be thought of as a spring and a cantilever, as they exhibit deformation recoil and elasticity. A cantilever is a result of the unequal stability in the tripod formed by the medial crura and paired LLCs, and a spring results from the LLCs, which produce an upward force that is in the form of stored elastic potential energy [16]. Therefore, the nasal tip may be modeled as spring supported.

The spring-stiffness constant, k, may be defined by (1) [20], and a spring-stiffness constant of 20 kN/m is applied in the three orthogonal axes.

$$k = \frac{E(\text{width} \times \text{height})^3}{4(\text{length})^3}. \tag{1}$$

A free nasal tip was prescribed as a preliminary step. A spring-supported nasal tip boundary condition, where the spring was connected between the two orange points on the nasal tip (Figure 2), was then applied to compare the effect of different boundary conditions. The dorsal and caudal septa were prescribed a free boundary condition.

2.4. Loading Conditions.

As "frontal force to the septum causes damage ranging from simple fracture of the nasal bones to severe flattening of the nasal bones and the septum" [21], two forms of frontal loading were applied—anteroposterior and dorsal and caudal septa in-plane loading. The force and pressure applied are estimates and are inconsequential to the relative displacements of the septum. As the present intention is to determine the Eigen modes, or the most likely deformed patterns of the septum, only the possible in-plane loading which will affect the resulting Eigen modes will be considered.

In the case of anteroposterior loading (Figure 2), a couple of forces of 1N each in both the vertical and horizontal axes were applied to the nasal tip to simulate a direct frontal punch

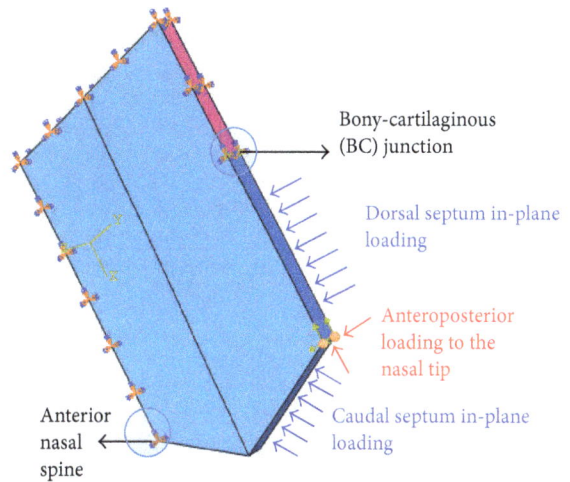

FIGURE 2: Three-dimensional idealized finite element model of the nasal septum with anteroposterior loading to the nasal tip, dorsal and caudal septa in-plane loading, and indicated anatomical features.

at an angle such that the forces on the nasal tip are the most significant.

In the case of dorsal and caudal septa in-plane loading (Figure 2), a uniform pressure of 2000 Pa was applied to both the dorsal and caudal septa. This was to simulate a frontal punch at an angle such that both the dorsal and caudal septa components are equally significant.

2.5. Eigen Modes and Finite Element Simulation.

Every object, including the septum, has a set of Eigen modes, depending on its structure and composition [22]. In each mode, all parts of the system vibrate with the same distinct frequency, which is referred to as the system's Eigen value at that mode. The lowest frequency is referred to as the fundamental frequency [23]. Since lower modes have lower frequencies and energies, they are more likely to occur. Hence, only the first 10 modes of the nasal septum were analyzed.

ABAQUS (Dassault Systèmes Technologies, Providence, RI) was used to obtain the Eigen mode shapes of the septum under the various loading conditions. A general, static step is created, in which one of the two loading conditions is applied. Thereafter, a linear perturbation, frequency step is created, in which the natural frequency and the corresponding mode shape will be extracted.

3. Results

The patterns of nasal septal deviation were similar to those described by Rohrich et al. [9]. In our study, the deviation patterns were therefore classified into three groups, each with its specific sites of high stress and dislocation and possible surgical corrective procedures (Table 1). Through observation of all deformation patterns, we were also able to identify the intrinsic points of fatigue within the cartilaginous septum—the BC junction, anterior nasal spine (ANS), vomer-ethmoidal cartilage junction (VEJ), and a single or

TABLE 1: Classification of septum deviation pattern based on sites of dislocation and possible surgical corrective procedures.

Type	Septum deviation pattern	Sites of dislocation	Surgical corrective procedures
I	Tilted in one piece	(i) Vomer-ethmoidal-cartilaginous (VEC) buckling (ii) Lower edge of septum dislocated off vomerine groove (iii) ANS attachment may be intact	(i) Submucous resection (ii) ±Septal reset to ANS (iii) ±Septal extension grafts for tip support
II	C-shaped	(i) Vomer-ethmoidal-cartilaginous (VEC) buckling (ii) Lower edge of septum dislocated off vomerine groove (iii) ANS attachment may be intact (iv) Septal fracture in a single site	(i) Septal reset to the ANS (ii) Submucous resection (iii) Inclusion of spreader grafts to buttress the septal fracture (iv) Possible septal extension grafts for vertical septal fracture
III	S-shaped	(i) Vomer-ethmoidal-cartilaginous (VEC) buckling (ii) Lower edge of septum dislocated off vomerine groove (iii) ANS attachment may be intact (iv) Septal fracture possibly in two sites, forming a septal concertina	(i) Septal reset to the ANS (ii) Submucous resection (iii) Inclusion of spreader grafts to buttress the septal fracture (iv) Inclusion of septal extension grafts to restore the support that is lost with septal shortening

couple of cracks in the quadrangular cartilage that lead to C-shaped and S-shaped nasal deformations, respectively. These points could lead to the septum levering off the vomerine groove and, in the latter two cases, a shortening of the septum (Figure 3).

For an idealized model, the slanted (Figure 4(a)), C-shaped (Figure 4(b)), and S-shaped (Figure 5(a)) deviation patterns were all observed (Table 2). In some modes, the system vibrates in-plane and therefore lacks a resultant deformation shape. In such cases, a dash is indicated. However, due to the lack of restriction on the nasal tip, it moves relatively freely, which may not represent in the vivo conditions.

In the following idealized model, the nasal tip is now constrained by a spring. While displaying similar patterns of deviation with a free nasal tip model, the spring-supported nasal tip model exhibits decreased displacement due to its prescribed restriction.

A patient-specific model was then analysed. By observation of the previous idealized models, it became apparent that the two forms of loading produced almost identical results. Therefore, only anteroposterior loading was prescribed to this model. The patient-specific model exhibited similar deformation patterns as the idealized nasal septal models (Table 2 and Figure 5).

4. Discussion

4.1. General Findings. The nasal septum is of utmost importance in the support of the "distal nose and for the maintenance of the bilateral nasal airway" [24]. A straight septum exists where there is force equilibrium [25], which may be disturbed in fracture, resulting in warping of the septal cartilage [18, 26]. Depending on the sustained trauma, the septum may deform in a myriad of patterns. Presently, however, studies have reported that septal deformation patterns may be categorized in a number of broad categories, regardless

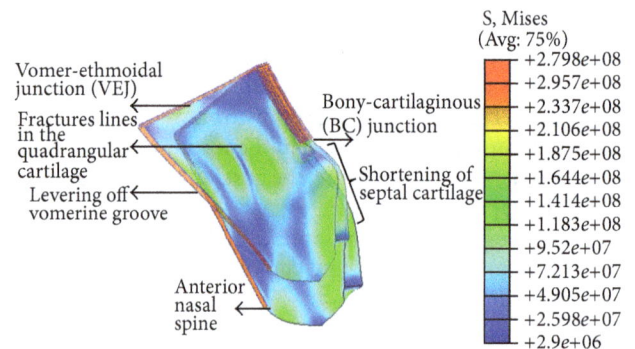

FIGURE 3: Intrinsic points of fatigue in the idealized cartilaginous septum model and the resultant shortening of the cartilage and levering off the vomerine groove. Note the stresses shown are Von Mises stresses. The ghost image that illustrates the shortening and levering of the septum is for illustrative purposes only.

of the trauma and/or injuries sustained. Guyuron et al. [5], Rohrich et al. [9], and Rhee et al. [24] categorized nasal deformities broadly into a septal tilt, anteroposterior or cephalocaudal C-shaped and an S- or reverse-S-shaped deformities. Unfortunately, these studies have not correlated these deviation patterns with degrees of force. Through the correlation of septum deformation patterns with increasing degrees of force, as well as with areas of dislocation and fracture, preoperative planning and septoplasty may be improved. The prompt identification and management of septal fractures are necessary to avoid nasal obstruction and posttraumatic septal deformity [24].

In our analyses, we were able to identify clinically observable nasal septal deviations and the aforementioned high-stress areas that would require stress-relief and the possible dislocation sites (Figure 3).

Preliminary Deformational Studies on a Finite Element Model of the Nasal Septum Reveals Key Areas for Septal Realignment and Reconstruction

53

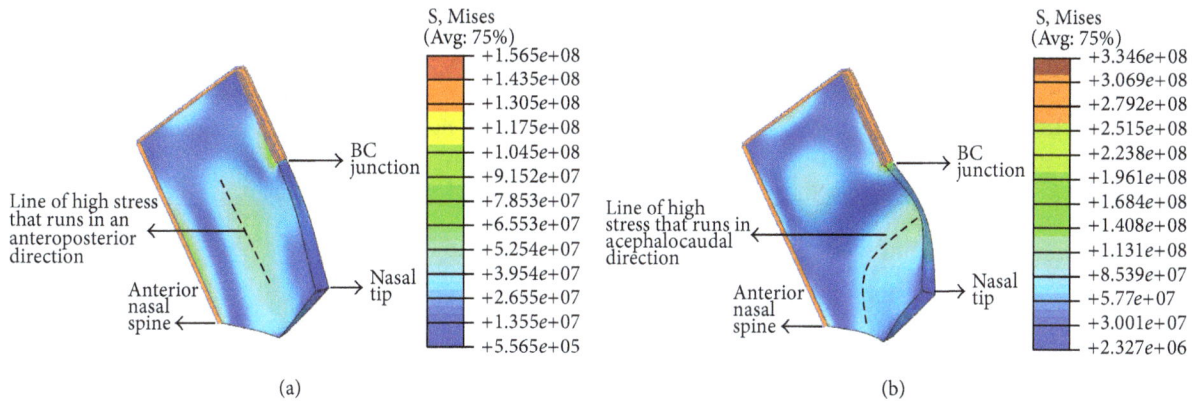

FIGURE 4: (a) Idealized nasal septum model with an unrestricted nasal tip in mode 3, displaying a C-shaped septum, with a line of high stress running through the anteroposterior direction. (b) Idealized nasal septum model with an unrestricted nasal tip in mode 4, displaying a C-shaped septum, with a line of high stress running through the cephalocaudal direction. Note that the stresses shown are Von Mises stresses.

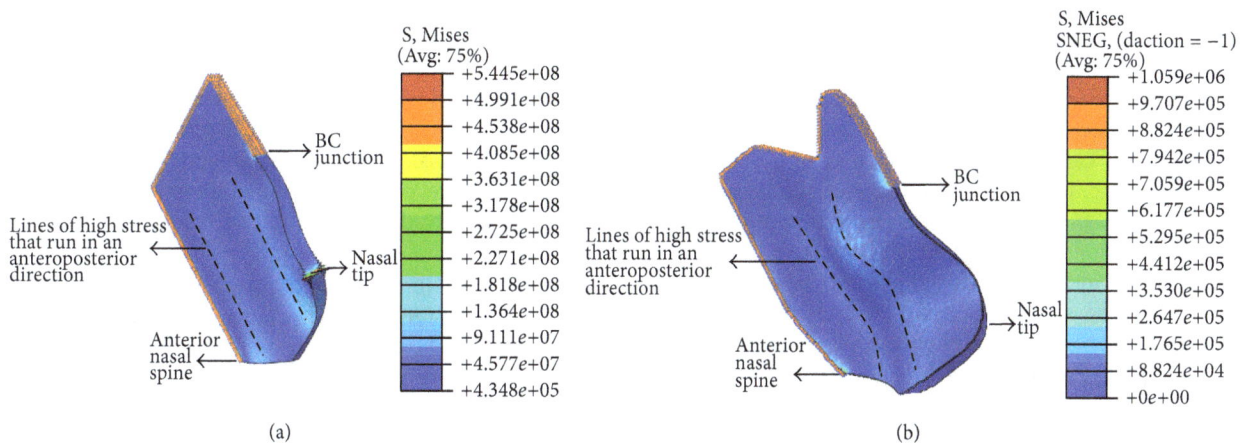

FIGURE 5: (a) Idealized nasal septum model with a spring-supported nasal tip in mode 6, displaying an S-shaped septum, with two lines of high stress running in the anteroposterior direction. (b) Patient-specific nasal septum model with a spring-supported nasal tip in mode 6, displaying an S-shaped septum, with relatively greater displacements than the corresponding idealized model. Note that the stresses shown are Von Mises stresses.

4.2. Prevalence of Modal Shapes and Concentrated Stress Zones in Idealized Models. We observed that regardless of force direction, with increasing force, the septum first tilts (type I) and then crumples into a C-shape (type II) and finally into an S-shape (type III). This was observed through the prevalence of the tilted septum in lower modes, while the "C-shaped" followed by the "S-shaped" deformation shapes occurring in relatively higher modes, respectively. As lower modes require less energy to manifest, they occur more frequently. Therefore, the lower the mode in which the deviation pattern is observed, the smaller the force required to cause this deformation, and consequently, the greater the probability of observing this pattern. Our findings are in agreement with clinical experience. According to Guyuron et al. [5], in a sample size of 93 patients who had undergone primary septoplasty, 40% had a septal tilt, 32% had a C-shape anteroposterior septum, 4% had a C-shape cephalocaudal septum, 9% had an S-shape anteroposterior septum, and 1% had an S-shape cephalocaudal septum.

In type I, when a tilted septum is observed, the highest stress concentration occurs at the BC junction and ANS. These high-stress areas were also reported by Lee et al. [21]. This suggests that with a low to moderate force, the septum dislocates en-mass from the midline vomerine groove (Figure 3) and levers off the BC junction to a tilted position. This may be observed on CT and MRI scans, and nasoendoscopy, where posterior buckling is frequently observed at the VEJ. Through submucous resection, the septum may be repositioned onto the groove [5], with prior resection of the cartilage tongue in the nasal floor. The septum may then be reset to the ANS (Table 1).

With a higher moderate force, a "C-shaped" deformation is likely due to a central line of stress in the septum, bending it into two pieces. The line of high stress may run through the anteroposterior (Figure 4(a)) or cephalocaudal (Figure 4(b)) directions. We propose that with significant loading, intrinsic septal fractures occur by breaking the cartilaginous septum into two, leading to the clinical morphology of a C-shaped

TABLE 2: Resultant modal shapes for idealized septum models with an unrestricted and spring-supported nasal tip and in the patient-specific nasal septum model with a spring-supported nasal tip.

| | Idealized nasal septum models | | | | | Patient-specific nasal septum model | |
| | Unrestricted nasal tip | | | Spring-supported nasal tip | | Spring-supported nasal tip | |
Mode	Anteroposterior loading	Dorsal and caudal in-plane loading	Mode	Anteroposterior loading	Dorsal and caudal in-plane loading	Mode	Frontal point loading
1	I	I	1	I	I	1	I
2	II	II	2	II	II	2	II
3	II	II	3	I	I	3	II
4	II	II	4	II	II	4	II
5	II	II	5	II	II	5	II
6	III	III	6	III	III	6	III
7	—	—	7	—	—	7	II
8	III	III	8	III	III	8	—
9	II	II	9	II	II	9	III
10	III	III	10	III	III	10	III

nose and shortening of the septum. In addition, the septum will be displaced off its vomerine groove and/or the ANS and will likely buckle at the VEJ (Figure 3). This is clinically significant as it cannot be observed on CT or MRI scans due to the invisibility of the septum in such modalities [27]. Hence, a clinical observation of a C-shaped septum may be the only indication. In addition to the corrective procedures mentioned previously, spreader grafts may be required on both sides of the septum, to assist in its straightening by providing the necessary nasal support that was relinquished when the septum fractured, thereby allowing the septum time to heal (Table 1).

With a force of higher magnitude, an "S-shaped" deformation may result due to multiple lines of stress leading to a septal concertina and shortening of the septum into a minimum of three overlapping pieces [28]. This is due to two lines of stress running in the anteroposterior direction (Figure 5). In addition to being shortened, the septum might be displaced from the vomerine groove and ANS (Figure 3). As with a C-shaped deformed septum, a clinical observation is its sole indication [27]. In addition to the aforementioned corrective procedures, longer spreader grafts will be required to brace both deformed sites to support the septum [5] and allow it to heal (Table 1).

Therefore, regardless of the deviation pattern, by relieving stress in these specific strips of concavity, in combination with the aforementioned surgical procedures, we propose that a more stable, straight septum may be achieved.

4.3. Comparison of Various Loading and Boundary Conditions.
Despite different loading conditions, the nasal septum deviates in a relatively constant pattern of a septal tilt, C- and S-shaped deviations with insignificant differences between the resultant modal shapes.

The free nasal tip and spring-supported nasal tip models responded differently to the loading conditions, specifically in mode three. A septal tilt is observed in the free nasal tip model, while a C-shaped deformation is observed in the latter

model. As the C-shape deviation is noted to occur with higher energy and septal tilt deviation with lower energy, this finding suggests that the spring of the LLCs acts to insulate and constrain the nasal tip and septum against deformation. The protective interrelationship of the LLCs to the nasal septum should be preserved during surgery.

4.4. Comparison of the Patient-Specific and Idealized Models.
The prevalence of modal shapes in patient-specific and idealized septal models, subject to frontal point-loading, is almost identical (Table 1). Slight deviations, such as those in mode three, are expected, due to the difference in shape between the models. Despite the patient-specific model exhibiting greater relative movement than the idealized model, this difference is insignificant as the basic modal shape remains (Figure 5). The similarities observed between these two models are a testament to the accuracy of the idealized model.

4.5. Limitations of the Study.
It is imperative to note that nasal septal deviations are secondary to the bony vault and cartilaginous changes. For the purpose of this study, the focus is on septal cartilage deformation patterns. Future research aims to combine the study of the deformations of the bony and cartilaginous septa. Due to the inherent collagen fibrils and the consequent anisotropy within the cartilaginous septum, we recognize that the prescription of a linearly elastic material model to the nasal septum material properties may not be fully representative of in vivo cartilage. In spite of this, an understanding of the relative displacement that occurs within the different models in different Eigen modes remains beneficial in aiding surgeons to correct a deviated nasal septum. No physical model was mechanically tested to validate the computational model in this preliminary study, which means that absolute stresses and relative stress patterns should be considered cautiously. Such an experimental validation study would typically make use of strain gages, but also infrared thermography, and global stiffness measurements [29, 30].

Preliminary Deformational Studies on a Finite Element Model of the Nasal Septum Reveals Key Areas for Septal
Realignment and Reconstruction

55

5. Conclusion

The purpose of this study was to gain a greater understanding of the septal deformation biomechanics. We found that despite different loading directions, the septum deformed consistently into only three shapes—a tilted position, a C-shaped septum, and an S-shaped septum. These patterns are in agreement with clinical observations of septal deformation patterns. The tilted septum is seen with the least force, C shape with moderate force, and S shape with high force. This suggests an intrinsic fracture of the septum into increased number of overlapping fragments with escalating force. Clinically, this is important information that provides insight into predictable patterns of internal septal fractures that need to be realigned and reconstructed to create a straight septum.

Disclosure

This study was performed in accordance with the guidelines of the institutional review board (IRB) and conforms to Helsinki's Declaration.

Conflict of Interests

The authors would like to declare that there is no issue related to conflict of interests for this study.

Acknowledgment

The authors would like to acknowledge the support by a grant from the Swiss-based CMF Clinical Priority Program of the AO Foundation under the Project no. C-09-2L.

References

[1] L. Bernstein, "Submucous operations on the nasal septum," *Otolaryngologic Clinics of North America*, vol. 6, no. 3, pp. 675–692, 1973.

[2] N. Edwards, "Septoplasty: rational surgery of the nasal septum," *Journal of Laryngology and Otology*, vol. 89, no. 9, pp. 875–897, 1975.

[3] P. McKinney and R. Shively, "Straightening the twisted nose," *Plastic and Reconstructive Surgery*, vol. 64, no. 2, pp. 176–179, 1979.

[4] M. Buyukertan, N. Keklikoglu, and G. Kokten, "A morphometric consideration of nasal septal deviations by people with paranasal complaints; a computed tomography study," *Rhinology*, vol. 41, no. 1, pp. 21–24, 2003.

[5] B. Guyuron, C. D. Uzzo, and H. Scull, "A practical classification of septonasal deviation and an effective guide to septal surgery," *Plastic and Reconstructive Surgery*, vol. 104, no. 7, pp. 2202–2209, 1999.

[6] Y. J. Jang, J. H. Wang, and B. J. Lee, "Classification of the deviated nose and its treatment," *Archives of Otolaryngology*, vol. 134, no. 3, pp. 311–315, 2008.

[7] H. R. Jin, J. Y. Lee, and W. J. Jung, "New description method and classification system for septal deviation," *Journal of Rhinology*, vol. 14, no. 1, pp. 27–31, 2007.

[8] I. Baumann and H. Baumann, "A new classification of septal deviations," *Rhinology*, vol. 45, no. 3, pp. 220–223, 2007.

[9] R. J. Rohrich, J. P. Gunter, M. A. Deuber, and W. P. Adams, "The deviated nose: optimizing results using a simplified classification and algorithmic approach," *Plastic and Reconstructive Surgery*, vol. 110, no. 6, pp. 1509–1523, 2002.

[10] P. A. A. Laura, V. H. Cortinez, L. Ercoli, and R. E. Rossi, "A simple method for the determination of the fundamental frequency of vibration of bones," *Medical Engineering and Physics*, vol. 16, no. 5, pp. 422–424, 1994.

[11] T. Mau, S. T. Mau, and D. W. Kim, "Cadaveric and engineering analysis of the septal L-strut," *Laryngoscope*, vol. 117, no. 11, pp. 1902–1906, 2007.

[12] K. Hwang, F. Huan, and D. J. Kim, "Mapping thickness of nasal septal cartilage," *Journal of Craniofacial Surgery*, vol. 21, no. 1, pp. 243–244, 2010.

[13] A. Mowlavi, S. Masouem, J. Kalkanis, and B. Guyuron, "Septal cartilage defined: implications for nasal dynamics and rhinoplasty," *Plastic and Reconstructive Surgery*, vol. 117, no. 7, pp. 2171–2174, 2006.

[14] G. S. Vicente, C. Buchart, D. Borro, and J. T. Celigüeta, "Maxillofacial surgery simulation using a mass-spring model derived from continuum and the scaled displacement method," *International Journal of Computer Assisted Radiology and Surgery*, vol. 4, no. 1, pp. 89–98, 2009.

[15] E. Peña, B. Calvo, M. A. Martínez, and M. Doblaré, "A three-dimensional finite element analysis of the combined behavior of ligaments and menisci in the healthy human knee joint," *Journal of Biomechanics*, vol. 39, no. 9, pp. 1686–1701, 2006.

[16] S. J. Lee, K. Liong, K. M. Tse, and H. P. Lee, "Biomechanics of the deformity of septal L-struts," *Laryngoscope*, vol. 120, no. 8, pp. 1508–1515, 2010.

[17] D. E. Protsenko and B. J. F. Wong, "Laser-assisted straightening of deformed cartilage: numerical model," *Lasers in Surgery and Medicine*, vol. 39, no. 3, pp. 245–255, 2007.

[18] M. Lee, J. Inman, S. Callahan, and Y. Ducic, "Fracture patterns of the nasal septum," *Otolaryngology*, vol. 143, no. 6, pp. 784–788, 2010.

[19] C. H. Kim, D. H. Jung, M. N. Park, and J. H. Yoon, "Surgical anatomy of cartilaginous structures of the asian nose: clinical implications in rhinoplasty," *Laryngoscope*, vol. 120, no. 5, pp. 914–919, 2010.

[20] R. W. Westreich, H. W. Courtland, P. Nasser, K. Jepsen, and W. Lawson, "Defining nasal cartilage elasticity: biomechanical testing of the tripod theory based on a cantilevered model," *Archives of Facial Plastic Surgery*, vol. 9, no. 4, pp. 264–270, 2007.

[21] S. J. Lee, K. Liong, and H. P. Lee, "Deformation of nasal septum during nasal trauma," *Laryngoscope*, vol. 120, no. 10, pp. 1931–1939, 2010.

[22] University of Hildesheim, "Identification of eigenmodes in vibration data," 2012, http://videolectures.net/mla09_preisach_ioeivd/.

[23] R. D. Blevins, *Formulas for Natural Frequency and Mode Shape*, Van Nostrand Reinhold, New York, NY, USA, 2nd edition, 1979.

[24] S. C. Rhee, Y. K. Kim, J. H. Cha, S. R. Kang, and H. S. Park, "Septal fracture in simple nasal bone fracture," *Plastic and Reconstructive Surgery*, vol. 113, no. 1, pp. 45–52, 2004.

[25] A. S. Lopatin, "Do laws of biomechanics work in reconstruction of the cartilaginous nasal septum?" *European Archives of Oto-Rhino-Laryngology*, vol. 253, no. 4-5, pp. 309–312, 1996.

[26] H. Fry, "The importance of the septal cartilage in nasal trauma," *British Journal of Plastic Surgery*, vol. 20, pp. 392–402, 1967.

[27] C. S. Farrow, "Chapter 14: nasal cavity disease," in *Veterinary Diagnostic Imaging: The Dog and Cat*, pp. 204–211, Mosby, Saint Louis, Mo, USA, 2003.

[28] R. J. Rohrich and W. P. Adams, "Nasal fracture management: minimizing secondary nasal deformities," *Plastic and Reconstructive Surgery*, vol. 106, no. 2, pp. 266–273, 2000.

[29] S. Shah, H. Bougherara, E. H. Schemitsch, and R. Zdero, "Biomechanical stress maps of an artificial femur obtained using a new infrared thermography technique validated by strain gages," *Medical Engineering & Physics*, vol. 34, pp. 1496–1502, 2012.

[30] R. Zdero and H. Bougherara, "Orthopaedic biomechanics: a practical approach to combining mechanical testing and finite element analysis," in *Finite Element Analysis*, D. Moratal, Ed., Intech Education and Publishing, Vienna, Austria, 2010.

6

Micromotion of Dental Implants: Basic Mechanical Considerations

Werner Winter,[1] **Daniel Klein,**[1] **and Matthias Karl**[2]

[1] *Department of Mechanical Engineering, University of Erlangen-Nuremberg, Egerlandstraße 5, 91058 Erlangen, Germany*
[2] *Department of Prosthodontics, University of Erlangen-Nuremberg, Glueckstraße 11, 91054 Erlangen, Germany*

Correspondence should be addressed to Matthias Karl; matthias.karl@uk-erlangen.de

Academic Editor: Raju Adhikari

Micromotion of dental implants may interfere with the process of osseointegration. Using three different types of virtual biomechanical models, varying contact types between implant and bone were simulated, and implant deformation, bone deformation, and stress at the implant-bone interface were recorded under an axial load of 200 N, which reflects a common biting force. Without friction between implant and bone, a symmetric loading situation of the bone with maximum loading and displacement at the apex of the implant was recorded. The addition of threads led to a decrease in loading and displacement at the apical part, but loading and displacement were also observed at the vertical walls of the implants. Introducing friction between implant and bone decreased global displacement. In a force fit situation, load transfer predominantly occurred in the cervical area of the implant. For freshly inserted implants, micromotion was constant along the vertical walls of the implant, whereas, for osseointegrated implants, the distribution of micromotion depended on the location. In the cervical aspect some minor micromotion in the range of 0.75 μm could be found, while at the most apical part almost no relative displacement between implant and bone occurred.

1. Introduction

Micromotion of dental implants has been defined as minimal displacement of an implant body relative to the surrounding tissue which cannot be recognized with the naked eye [1] (Figure 1). Various authors have shown that excessive micromotion may interfere with the process of osseointegration of dental implants [2, 3]. Although exact data are missing, it has been postulated that micromotion between implant and bone must not surpass a threshold value of 150 micrometer (μm) for successful implant healing [4–6].

In traditional loading protocols, where implants are allowed to heal undisturbed for periods of several months, the issue of implant micromotion is of limited importance. With the advent of modern treatment concepts including early and immediate loading of dental implants [7, 8], with implants being restored early in the healing phase, the issue of implant micromotion has gained significant importance [4, 5].

Numerous reports trying to relate clinical parameters to the phenomenon of implant micromotion can be found in the dental literature [8–11]. The nonuniform nomenclature, the varying experimental settings, and the partially contradicting results presented on the one hand indicate the complexity of the topic but on the other hand emphasize the need for clarifying basic engineering principles.

From a biomechanical perspective, successful osseointegration of dental implants depends on the way mechanical stresses and strains are transferred to the surrounding bone and tissues. The multiple factors hereby affecting stress and strain transfer include the type of loading that occurs, the type of implant-bone interface being present, the length and diameter of the implant, implant geometry and its surface texture, and the quality and quantity of the surrounding bone [13–19]. Only by understanding the most critical of these variables, strategies for optimizing implant stabilization can be developed. For determining how implant mobility, often referred to as micromotion, relative motion, micromovement, and so forth, or implant loading affects bone response, a closer look at implant deformation, bone deformation, and stress or strain at the implant-bone interface is required [20].

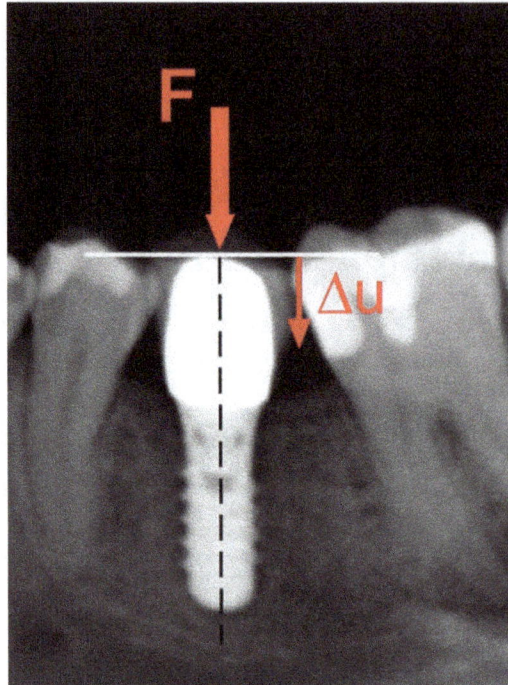

FIGURE 1: Single tooth implant used for replacing the first molar in the lower left mandible. An axial force acting on the occlusal surface of the restorations may displace the implant relative to the surrounding bone.

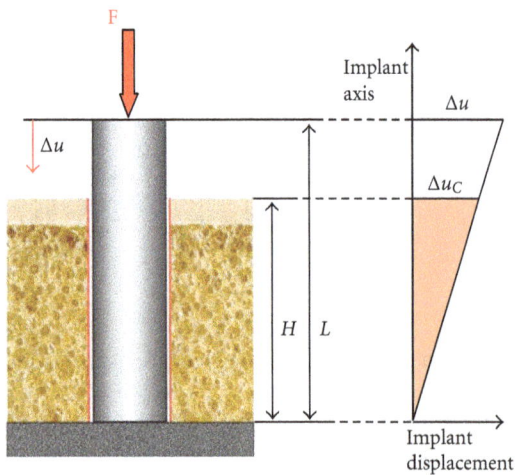

FIGURE 2: Description of scenario 1 with a dental implant resting on a fixed apical surface, with no contact existing between the vertical implant walls and the walls of the bony socket (left). When the implant is loaded vertically, deformation of the implant occurs mainly in the coronal part and decreases towards the apex. Similarly, relative displacement between implant and bone diminishes towards the apex (right).

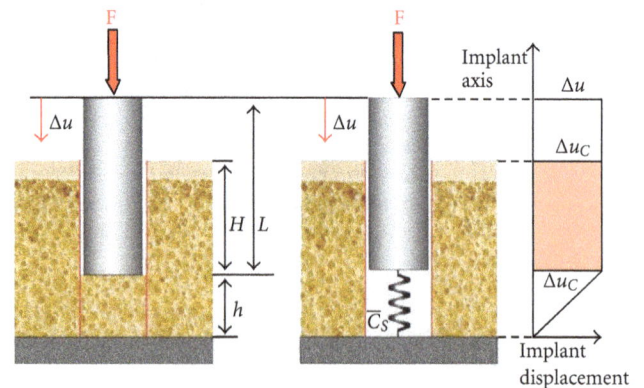

FIGURE 3: Description of scenario 2, where the implant rests on a layer of elastic trabecular bone with no contact existing between the vertical implant walls and the walls of the bony socket (left). The apically located layer of bone may be substituted by a spring which is compressed when an axial load is applied on the implant (center). Due to the great difference in elastic modulus between implant and trabecular bone, relative implant displacement is independent from the region of the implant considered (right).

In this context, it was the purpose of this paper to mechanically describe the phenomenon of micromotion occurring between implant and alveolar bone using simple spring models, continuum mechanics models, and 3D-Finite-Element models simulating varying contact types between implant and bone [21–23].

2. Material and Methods

2.1. Basic Considerations. Three basic scenarios reflecting different anchoring situations of dental implants were considered. In scenario 1, the implant rests on an apically located fixed surface but neither has contact to cortical bone nor to trabecular bone at the vertical walls of the implant (Figure 2). In this situation, maximum implant deformation under

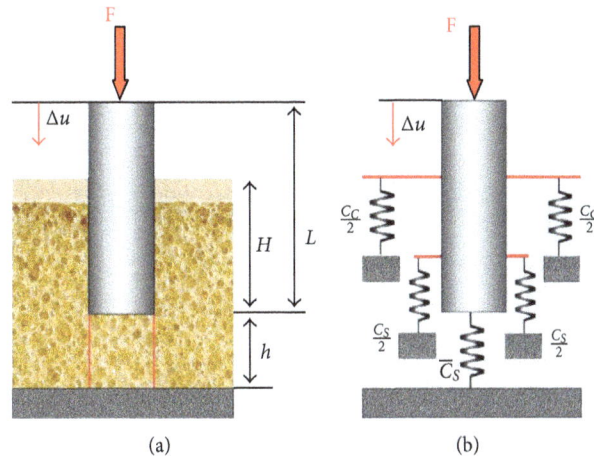

FIGURE 4: Scenario 3 showing an implant elastically supported by cortical and trabecular bone (a). The elastic support in the different regions can be replaced by a system of springs (b).

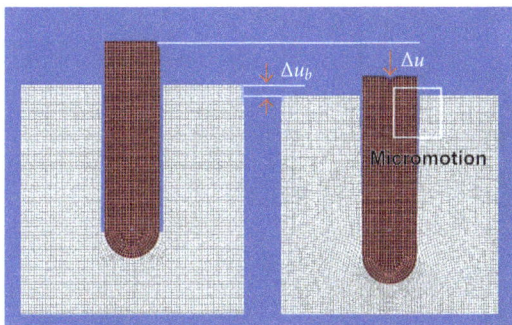

FIGURE 5: Without contact between implant and bone, an axial force acting on the implant causes implant dislocation as a result of elastic deformation of bone predominantly in the periapical region of the implant. Left: unloaded implant; right: loaded implant with implant displacement Δu and displacement of cortical bone Δu_b (displacement of a reference mark on bone).

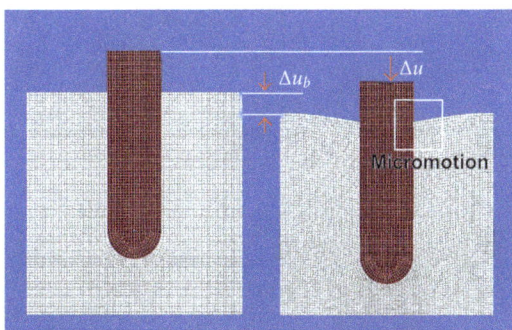

FIGURE 6: Considering an osseointegrated implant with contact between the implant surfaces and bone, axial implant loading causes elastic deformation of bone in all areas but no relative displacement between implant and bony socket, that is, no micromotion, occurs. Left: unloaded implant; right: loaded implant with implant displacement Δu and displacement of cortical bone Δu_b (displacement of a reference mark on bone).

vertical loading occurs in the coronal part and diminishes gradually towards the apex. As a result, micromotion between the implant and the vertical walls of the socket also decreases towards the apical part of the implant. Axial deformation of the implant as a consequence of vertical loading can be calculated according to

$$\Delta u = \frac{F}{EA/L} = \frac{F}{c} \qquad (1)$$

with F standing for the vertical force applied, E being the Young's modulus of the implant, A being the cross section of the implant, L being the length of the implant, and c being the stiffness of the implant. Maximum micromotion Δu_c at the cortical area can then be calculated according to

$$\Delta u_c = \frac{\Delta u}{L} H \qquad (2)$$

with H reflecting the height of cortical and trabecular bone around the implant.

For scenario 2, the fixed apical rest of the implant was altered by adding a layer of elastic trabecular bone apically to the implant. Here, an axial force acting on the implant predominantly causes compression of the elastic material the implant is resting on. Due to the drastically smaller elastic modulus of trabecular bone as compared to titanium, the deformation of the implant can be neglected and the relative movement between implant and bone is independent from the region of the implant considered (Figure 3). In this situation, implant displacement may be calculated according to

$$\Delta u = \frac{F}{E_s A/h} \left[1 + \frac{L}{h} \frac{E_s}{E} \right] = \frac{F}{C_s} [1 + \beta], \qquad (3)$$

where E_s is the Young's modulus of the trabecular bone and h the height of the bone underneath the implant.

Taking into account that the Young's modulus of the implant is much greater than the Young's modulus of the

(a)

(b)

(c)

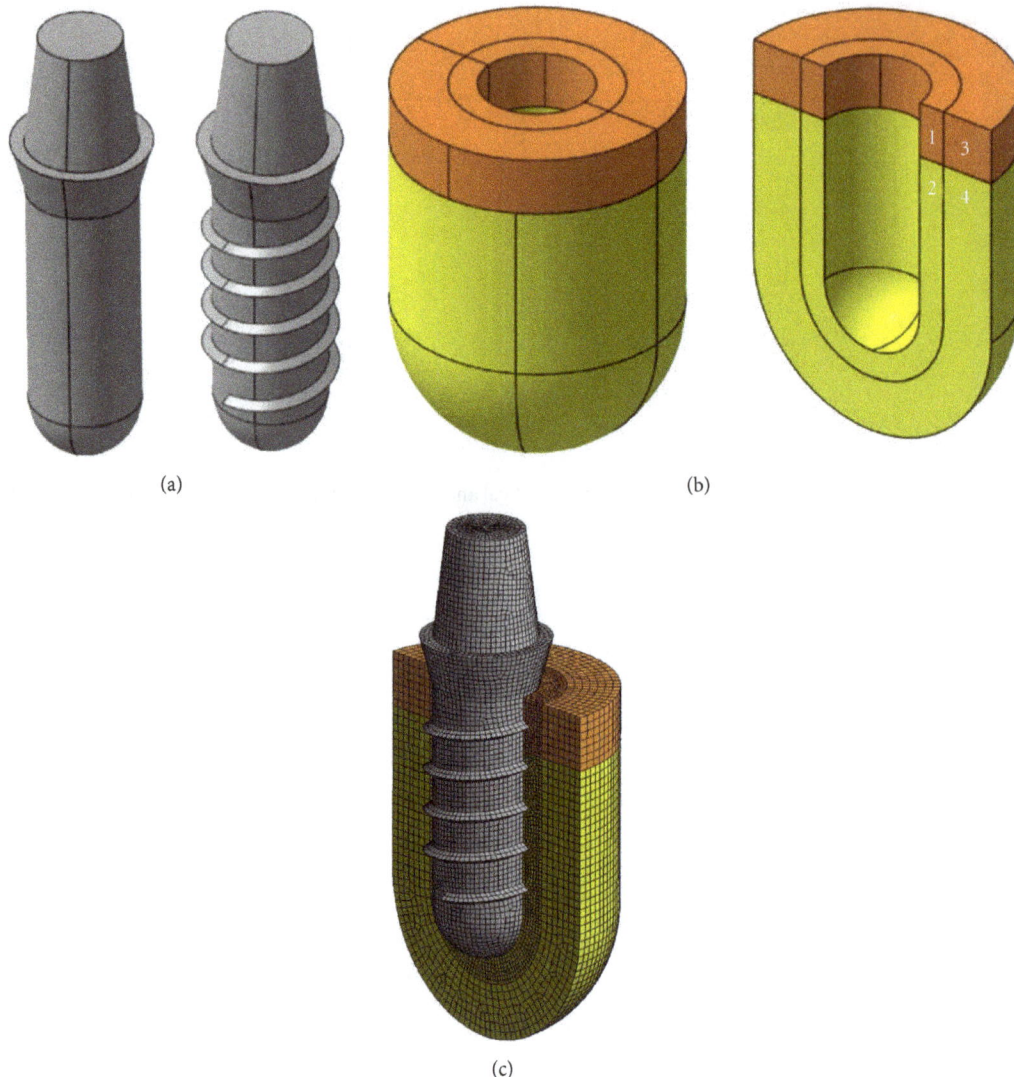

FIGURE 7: (a) Three-dimensional finite element models of dental implants with and without threads [12]. (b) Three-dimensional finite element model of a bony implant socket with cortical and trabecular bone. Areas (1) and (2) surrounding the implant are designed as an intermediate layer allowing the elastic modulus to be set independently from areas (3) and (4) representing native bone which is not affected by healing processes occurring during osseointegration [12]. (c) Three-dimensional finite element model of a single implant embedded in a bone segment consisting of cortical and trabecular bone (calculations were done on a complete model; for illustration purposes the model is cut in half) [12].

trabecular bone ($E \gg E_s$) it is accepted $\beta \ll 1$ and furthermore the approximation for the micromotion

$$\Delta u_c = \Delta u \approx \frac{F}{\overline{C_s}}. \tag{4}$$

Consequently, micromotion at the cortical area of the implant and the relative micromotion between implant and bone are identical, both being related to the stiffness $\overline{C_s}$ of the trabecular bone underneath the implant.

Further approximating the clinical situation of an osseointegrated implant, in scenario 3 the implant is elastically supported by surrounding cortical and trabecular bone. Due to the fixed contact between implant and bone, micromotion at this interface does not occur, when the implant is

axially loaded (Figure 4). Neglecting the deformation of the implant ($E \gg E_s$) the implant displacement may be calculated according to

$$\Delta u = \frac{F}{\overline{C_s}} \frac{1}{\left(1 + \left(C_s / \overline{C_s}\right) + \left(C_c / \overline{C_s}\right)\right)}. \tag{5}$$

Under the circumstances of scenario 3, no relative micromotion between implant and bone exists. The displacement of the implant equals the micromotion depending on the stiffness of both cortical and trabecular bone surrounding the implant.

Further approximating clinical reality, continuum mechanics models were considered revealing implant displacement due to elastic deformation of bone when no

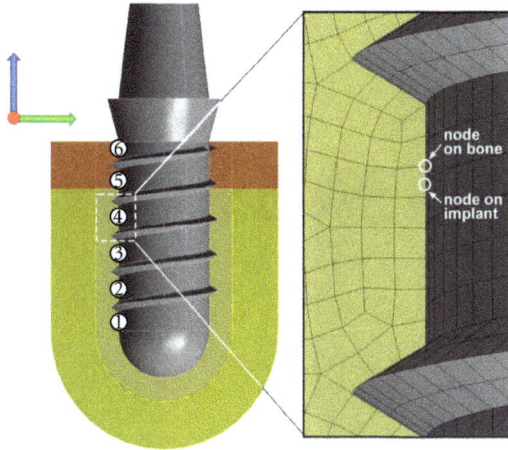

FIGURE 8: Definition of micromotion at the implant bone interface. Six corresponding nodes on the implant and on the bone were used as reference marks. For determining the relative displacement of two corresponding nodes on bone and implant, the displacement of a specific reference mark on the bone was subtracted from the displacement of the corresponding reference mark on the implant.

TABLE 1: Material properties (Young's moduli in MPa) chosen in the different models. Poisson's ratio is 0.3 for all materials.

Structure	Osseointegrated implant	Healing state
Cortical bone	14000	14000
Trabecular bone	3000	3000
Implant	110000	110000
Intermediate layer—cortical area	14000	1000
Intermediate layer—trabecular area	3000	1000

contact between implant and bone was modelled in a plane strain FE-model (Figure 5). This model correlates with scenario 2 described above (Figure 3). Simulating contact between implant and bone, bone is also elastically deformed in the cervical portion of the implant when an axial load is exerted; however, no relative micromotion between implant and bone occurs at the interface (Figure 6), and the implant displacement (micromotion) is related to the elastic properties of the cortical and trabecular bone.

2.2. Finite Element Analysis. For a more realistic representation of clinical conditions, three-dimensional FE models [12] of dental implants with and without threads were generated (Figure 7(a)) which were subsequently embedded in a bony socket consisting of cortical and trabecular bone and an intermediate layer surrounding the implant (Figure 7(b)). The geometry of the models was generated with a CAD Program (SolidWorks 2011, SolidWorks Deutschland GmbH, Haar, Germany) and imported in a FE program (ANSYS Workbench 12, ANSYS Inc., Canonsburg, PA, USA).

Combining both components, three-dimensional FE models (Figure 7(c)) were obtained for evaluating micromotion between implant and bone when an axial vertical force of 200 N was exerted which reflects an average biting force [24, 25]. Different stages of osseointegration were simulated by altering the elastic modulus of the intermediate bone layer [21–23]. The contact type between implant and bone could be modified as friction free, only transferring compressive forces and allowing for sliding and gap formation, to friction (friction coefficient 0.3) and force fit, respectively [21].

In general, isotropic linear model parameters were applied, defining the contact type between the different layers of bone as "bond." Out of the large number of possible solutions for solving contact problems, the augmented Lagrange method was chosen as accompanying optimization

method. This method was applied for defining all contacts not allowing contacting components to penetrate each other. Poisson's ratio was set at 0.3 for all materials. Based on the results of previous investigations [22, 23] indicating that the size of the models was sufficient for evaluating micromotion, model dimensions were reduced to a minimum and the borders of the models were fixed. Depending on model type, 160000 hexaeder elements and 600000 to 650000 nodes were used to set up the models using the elastic modules given in Table 1. Based on the fact that the elastic values and the strength limits of biologic materials *in vivo*—such as the bone-implant interface—are highly complex [24], only two states of osseointegration were considered (starting point and end point of osseointegration). These different states were modelled by different elastic values in the areas (1) and (2) in Figure 7(b).

Results of all simulations were recorded as von Mises equivalent stress in addition to contour plots of global displacement.

For calculating relative displacement between implant and bone (relative micromotion), a total of six corresponding nodes at the implant bone interface were established as reference marks. As the displacement of a specific reference mark on the implant represents both displacement of bone and implant, the displacement of the corresponding reference mark on the bone (Figure 8) was subtracted.

3. Results

Simulating 200 N axial force acting on an osseointegrated cylindrical implant with no friction between implant and bone caused a symmetric loading situation of the bone surrounding the implant with maximum loading and maximum displacement occurring at the apical part of the implant (Figures 9(a) and 10(a)).

Adding threads to the implant led to a decrease both in loading and displacement occurring at the apical part of the implant. Simultaneously, greater distribution of loading and displacement was observed at the vertical walls of the implants (Figures 9(b) and 10(b)).

Introducing friction between implant and bone (Figures 9(c) and 10(c)) further decreased global displacement and resulted in a more homogeneous distribution of loads as compared to the force fit situation (Figures 9(d) and 10(d)), where load transfer predominantly occurred in the cervical area of the implant, where cortical bone was modelled.

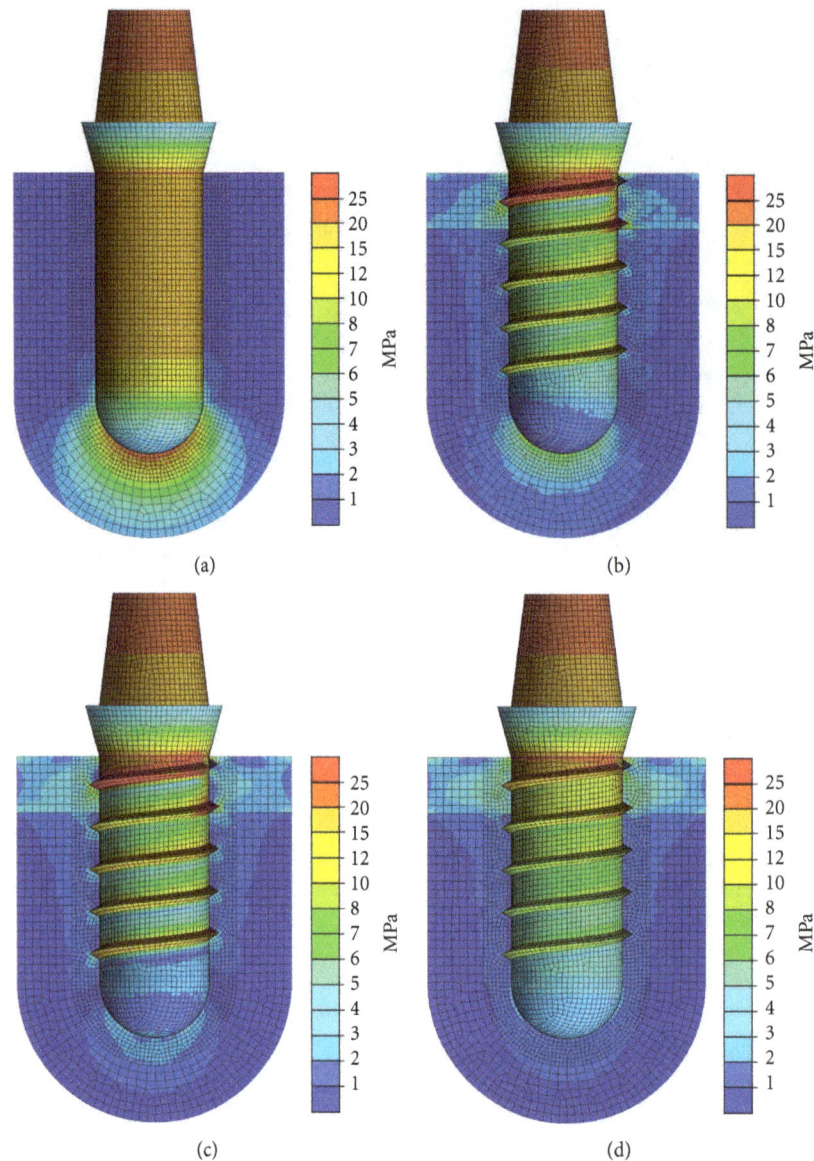

(a)

(b)

(c)

(d)

FIGURE 9: Distribution of von Mises equivalent stress around implants loaded with 200 N axial vertical force [12]: cylindrical implant without friction between implant and bone (a), threaded implant without friction between implant and bone (b), threaded implant with friction between implant and bone (coefficient of friction: 0.3) (c), and threaded implant with force fit between implant and bone (d).

For freshly inserted implants with a soft intermediate layer of bone modelled around the implants, the introduction of a friction coefficient led to a considerable reduction in micromotion between implant and bone as well as to reduced displacement of all reference marks on the implant. Displacement of the reference marks on the bone remained on a constant level. Overall, comparable values for micromotion were recorded at all corresponding reference marks (Figure 11).

Simulating an osseointegrated implant in general reduced all displacement values by about 50% compared to the situation of a freshly inserted implant. Again the introduction of a friction coefficient led to a considerable reduction in micromotion between implant and bone as well as to reduced displacement of all reference marks on the implant.

Displacement of the reference marks on the bone remained on a constant level. In contrast to a freshly inserted implant, the distribution of micromotion depended on the location of the reference mark. Whereas in the cervical aspect some minor micromotion in the range of 0.75 μm could be found, at the most apical reference almost no relative displacement between implant and bone occurred (Figure 12).

4. Discussion

Within the limitations of this investigation, the effect of friction phenomena and implant design (cylindrical versus threaded) on stress distribution and implant displacement could be demonstrated. Both the introduction of friction between implant and bone as well as the addition of threads

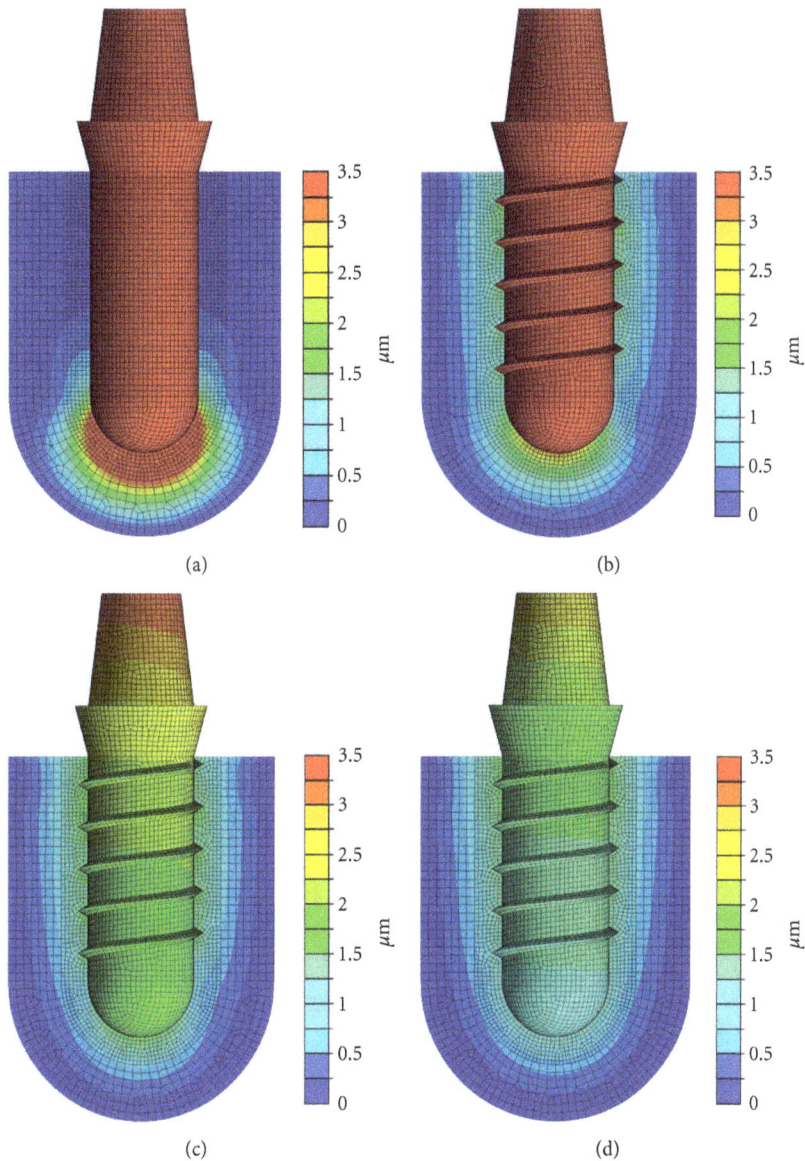

FIGURE 10: Distribution of global displacement around implants loaded with 200 N axial vertical force [12]: cylindrical implant without friction between implant and bone (a), threaded implant without friction between implant and bone (b), threaded implant with friction between implant and bone (coefficient of friction: 0.3) (c), and threaded implant with force fit between implant and bone (d).

to a cylindrically shaped implant resulted in the reduction of implant displacement under an axial load of 200 N. Simultaneously, a more homogeneously distributed loading situation at the implant bone interface could be observed. Changing the contact type between implant and bone to force fit resulted in load transfer predominantly occurring in the cervical part of the implant surrounded by stiffer cortical bone. This is in strict contrast to a situation with no friction modelled resulting in maximum loading of bone surrounding the periapical region of the implant. From a clinical perspective, these findings indicate that screw-shaped implants are advantageous while bone quality probably plays the most important role in achieving sufficient primary implant stability for immediate loading. All these factors

should be taken into account when choosing a specific loading protocol.

Based on a comparison of freshly inserted and osseointegrated implants it could be shown that the healing status affects the occurrence of micromotion phenomena along the implant bone interface. For a soft implant bone interface, reflecting early stages of osseointegration, micromotion remained on a constant level regardless of the location considered. Simulating mature bone reflecting an osseointegrated implant, the introduction of a friction coefficient between implant and bone dramatically changed the distribution of micromotion along the implant bone interface. In addition to generally reduced levels of micromotion as

(a)

(b)

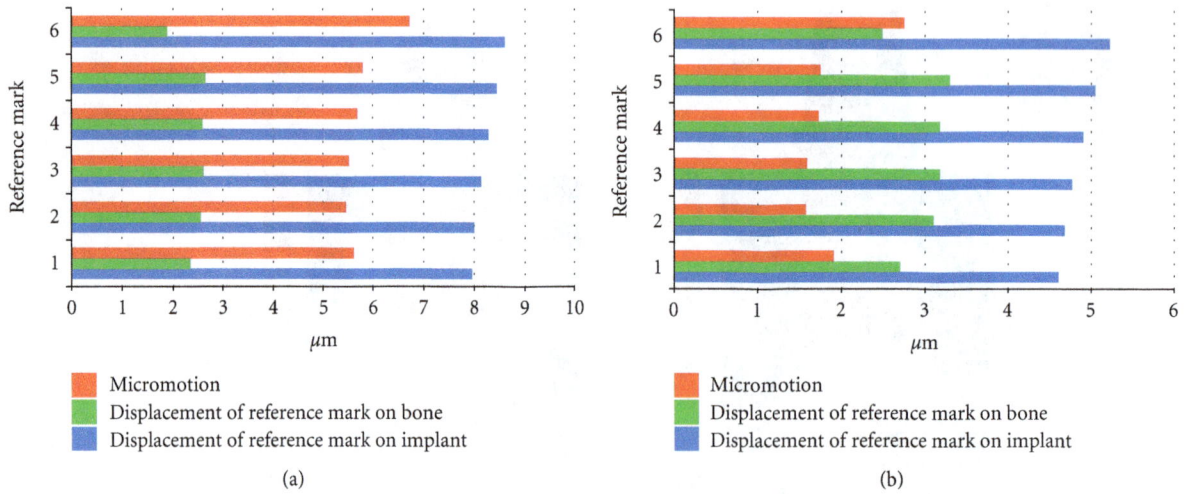

FIGURE 11: Displacement of corresponding reference marks on bone and implant for freshly inserted implants and resulting micromotion: data recorded from model without friction between bone and implant (a), data recorded from model with friction between bone and implant (b) (note the different scales).

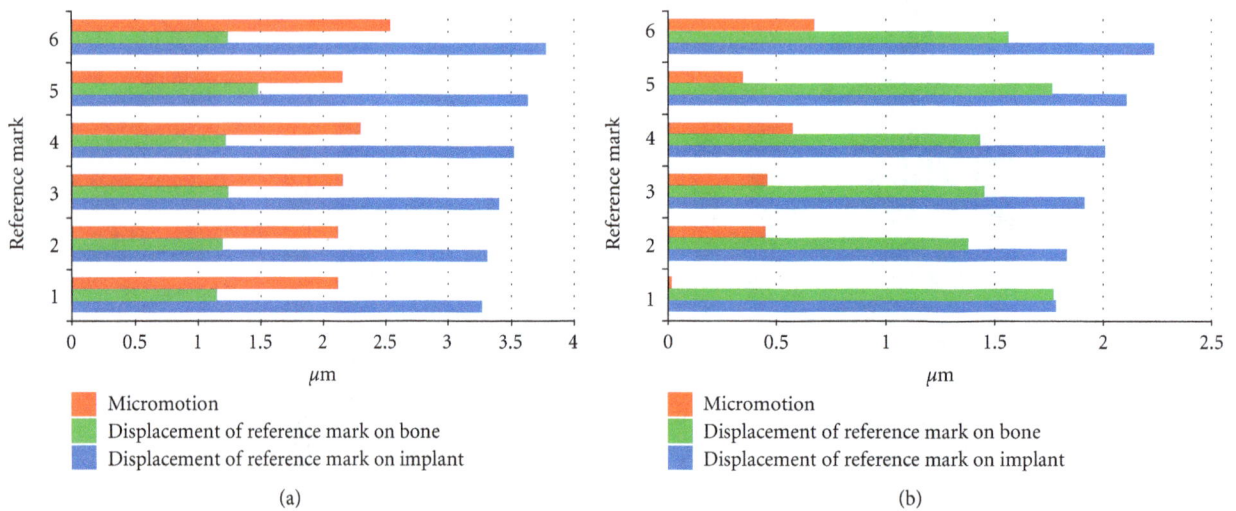

(a)

(b)

FIGURE 12: Displacement of corresponding reference marks on bone and implant for osseointegrated implants and resulting micromotion: data recorded from model without friction between bone and implant (a), data recorded from model with friction between bone and implant (b), note the different scales!

compared to a freshly inserted implant, a decrease in micromotion was noted. The amount of micromotion decreased towards the apex of the implant.

It may be seen as a limitation of this study that only one specific value for axial loading of the implants was chosen. Based on studies by Brunski and coworkers [20], axial components of biting forces can range from 100 to 2400 N, while the exact values depend on factors such as location in the mouth and nature of food. For patients having implant-supported dentures, axial closure forces ranging from 45 to 255 N have been reported [25]. It thus appears that the value chosen reflects clinical loading magnitudes.

Furthermore, besides the pure mechanical aspects addressed in this paper, also biologic factors play an important role in the process of osseointegration of dental implants. Following implant placement, the healing period starts with the adherence of serum proteins, followed by the attachment and proliferation of mesenchymal cells. Consequently, osteoid is formed in what is then mineralized. From then onwards, bone remodeling occurs as an adaptation to the implants environment [26]. With these processes occurring simultaneously to mechanical loading in an immediate loading situation, the interaction of both mechanical and biologic factors seems to be critical to the integration of the implant.

5. Conclusions

Given the nonuniform distribution of micromotion between implant and bone, it appears questionable whether currently available methods for experimentally determining this phenomenon provide meaningful data. The only valid approach for evaluating micromotion phenomena at the implant-bone interface appears to be finite element analysis. However, care has to be taken to set proper materials and interface characteristics as these parameters may greatly influence the outcome.

Acknowledgment

Figures 7(a), 7(b), 7(c); 9(a)-9(d): 10(a)-10(d) have been previously published in [12] and are reused with permission from Allen Press Publishing Services.

References

[1] W. R. Laney, *Glossary of Oral and Maxillofacial Implants, German edition*, Quintessenz, Berlin, Germany, 2008.

[2] J. B. Brunski, "The influence of force, motion and related quantities on the response of bone to implant," in *Non-Cemented Total Hip Arthroplasty*, R. Fitzgerald, Ed., pp. 7–21, Ravens Press, New York, NY, USA, 1988.

[3] J. B. Brunski, "*In vivo* bone response to biomechanical loading at the bone/dental-implant interface," *Advances in Dental Research*, vol. 13, pp. 99–119, 1999.

[4] S. Szmukler-Moncler, H. Salama, Y. Reingewirtz, and J. H. Dubruille, "Timing of loading and effect of micromotion on bone-dental implant interface: review of experimental literature," *Journal of Biomedical Materials Research*, vol. 43, pp. 192–203, 1998.

[5] S. Szmukler-Moncler, A. Piattelli, G. A. Favero, and J. H. Dubruille, "Considerations preliminary to the application of early and immediate loading protocols in dental implantology," *Clinical Oral Implants Research*, vol. 11, no. 1, pp. 12–25, 2000.

[6] H. Kawahara, D. Kawahara, M. Hayakawa, Y. Tamai, T. Kuremoto, and S. Matsuda, "Osseointegration under immediate loading: biomechanical stress-strain and bone formation—resorption," *Implant Dentistry*, vol. 12, no. 1, pp. 61–68, 2003.

[7] G. E. Romanos, "Present status of immediate loading of oral implants," *The Journal of Oral Implantology*, vol. 30, no. 3, pp. 189–197, 2004.

[8] P. Trisi, G. Perfetti, E. Baldoni, D. Berardi, M. Colagiovanni, and G. Scogna, "Implant micromotion is related to peak insertion torque and bone density," *Clinical Oral Implants Research*, vol. 20, no. 5, pp. 467–471, 2009.

[9] W. Engelke, O. A. Decco, M. J. Rau, M. C. A. Massoni, and W. Schwarzwäller, "*In vitro* evaluation of horizontal implant micromovement in bone specimen with contact endoscopy," *Implant Dentistry*, vol. 13, no. 1, pp. 88–94, 2004.

[10] S. Holst, H. Geiselhoeringer, M. Wichmann, and A. I. Holst, "The effect of provisional restoration type on micromovement of implants," *Journal of Prosthetic Dentistry*, vol. 100, no. 3, pp. 173–182, 2008.

[11] M. Goellner, J. Schmitt, M. Karl, M. Wichmann, and S. Holst, "The effect of axial and oblique loading on the micromovement of dental implants," *International Journal of Oral & Maxillofacial Implants*, vol. 26, pp. 257–264, 2011.

[12] W. Winter, D. Klein, and M. Karl, "Effect of model parameters on finite element analysis of micromotions in implant dentistry," *The Journal of Oral Implantology*. In press.

[13] J. E. Bechtold, O. Mouzin, L. Kidder, and K. Søballe, "A controlled experimental model of revision implants: Part II. Implementation with loaded titanium implants and bone graft," *Acta Orthopaedica Scandinavica*, vol. 72, no. 6, pp. 650–656, 2001.

[14] L. C. Jones, C. Frondoza, and D. S. Hungerford, "Effect of PMMA particles and movement on an implant interface in a canine model," *Journal of Bone and Joint Surgery B*, vol. 83, no. 3, pp. 448–458, 2001.

[15] O. Mouzin, K. Soballe, and J. E. Bechtold, "Loading improves anchorage of hydroxyapatite implants more than titanium implants," *Journal of Biomedical Materials Research*, vol. 58, pp. 61–68, 2001.

[16] R. Skripitz and P. Aspenberg, "Early effect of parathyroid hormone (1–34) on implant fixation," *Clinical Orthopaedics and Related Research*, no. 392, pp. 427–432, 2001.

[17] S. Toksvig-Larsen, L. P. Jorn, L. Ryd, and A. Lindstrand, "Hydroxyapatite-enhanced tibial prosthetic fixation," *Clinical Orthopaedics and Related Research*, no. 370, pp. 192–200, 2000.

[18] H. M. Van Der Vis, P. Aspenberg, R. K. Marti, W. Tigchelaar, and C. J. F. Van Noorden, "Fluid pressure causes bone resorption in a rabbit model of prosthetic loosening," *Clinical Orthopaedics and Related Research*, no. 350, pp. 201–208, 1998.

[19] H. G. Willert and G. H. Buchhorn, "Osseointegration of cemented and noncemented implants in artificial hip replacement: long-term findings in man," *Journal of Long-Term Effects of Medical Implants*, vol. 9, no. 1-2, pp. 113–130, 1999.

[20] J. Brunski, J. Currey, J. A. Helms, P. Leucht, A. Nanci, and R. Wazen, "Implant geometry, interfacial strain and mechanobiology of oral implants revisited," in *Proceedings of the first P-I Branemark Scientific Symposium, Gothenburg 2009*, R. Gottlander and D. van Steenberghe, Eds., pp. 45–59, Quintessence Publishing, London, UK, 2011.

[21] D. C. Holmes and J. T. Loftus, "Influence of bone quality on stress distribution for endosseous implants," *The Journal of Oral Implantology*, vol. 23, no. 3, pp. 104–111, 1997.

[22] W. Winter, S. Möhrle, S. Holst, and M. Karl, "Parameters of implant stability measurements based on resonance frequency and damping capacity: a comparative finite element analysis," *The International Journal of Oral & Maxillofacial Implants*, vol. 25, no. 3, pp. 532–539, 2010.

[23] W. Winter, P. Steinmann, S. Holst, and M. Karl, "Effect of geometric parameters on finite element analysis of bone loading caused by nonpassively fitting implant-supported dental restorations," *Quintessence International*, vol. 42, pp. 471–478, 2011.

[24] J. B. Brunski, "Biomechanical factors affecting the bone-dental implant interface," *Clinical Materials*, vol. 10, no. 3, pp. 153–201, 1992.

[25] A. B. Carr and W. R. Laney, "Maximum occlusal force levels in patients with osseointegrated oral implant prostheses and patients with complete dentures," *The International Journal of Oral & Maxillofacial Implants*, vol. 2, no. 2, pp. 101–108, 1987.

[26] S. Raghavendra, M. C. Wood, and T. D. Taylor, "Early wound healing around endosseous implants: a review of the literature," *International Journal of Oral and Maxillofacial Implants*, vol. 20, no. 3, pp. 425–431, 2005.

GPS and GPRS Based Telemonitoring System for Emergency Patient Transportation

K. Satyanarayana,[1] **A. D. Sarma,**[2] **J. Sravan,**[1] **M. Malini,**[1] **and G. Venkateswarlu**[3]

[1] Department of Biomedical Engineering, Osmania University, Hyderabad 500 007, India
[2] Research and Training Unit for Navigational Electronics, Osmania University, Hyderabad 500 007, India
[3] Department of ECE, Vasavi College of Engineering, Hyderabad 500 031, India

Correspondence should be addressed to A. D. Sarma; ad_sarma@yahoo.com

Academic Editor: Michel Labrosse

Telemonitoring during the golden hour of patient transportation helps to improve medical care. Presently there are different physiological data acquisition and transmission systems using cellular network and radio communication links. Location monitoring systems and video transmission systems are also commercially available. The emergency patient transportation systems uniquely require transmission of data pertaining to the patient, vehicle, time of the call, physiological signals (like ECG, blood pressure, a body temperature, and blood oxygen saturation), location information, a snap shot of the patient, and voice. These requirements are presently met by using separate communication systems for voice, physiological data, and location that result in a lot of inconvenience to the technicians, maintenance related issues, in addition to being expensive. This paper presents design, development, and implementation of such a telemonitoring system for emergency patient transportation employing ARM 9 processor module. This system is found to be very useful for the emergency patient transportation being undertaken by organizations like the Emergency Management Research Institute (EMRI).

1. Introduction

Immediate medical attention to critically ill patients and accident victims followed by transportation to a well-equipped medical facility within the golden hour saves many lives. Numbers of road accidents in India are the highest across the world. According to the National Transportation Planning and Research Center (NTPRC) the number of road accidents for 1000 vehicles in India is about 35 while the figure is between 4 to 10 in developed countries. About 1,05,000 accidents take place every year [1]. There are several governmental and nongovernmental agencies like the Emergency Management Research Institute (EMRI), located across the country, which have been dedicated to the cause of transporting critically ill patients and accident victims. About 2,87,000 lives have been saved by EMRI in the past six years. The ambulances are specially designed to carry emergency drugs and instruments. Inner area of ambulance is fabricated in such a way that it houses emergency medicines,

sterilizer, stretcher, and so forth. A typical inside layout of an ambulance employed for emergency patient transportation is shown in Figure 1. The paramedics that accompany the ambulances are specially trained to be emergency technicians. There exists a need to augment the skill set of such paramedics with the expert doctor's advice from the central monitoring station (CMS). CMS helps in identifying the nearest and appropriate hospital and coordinating with the medical personnel of that hospital.

Hence there is a need for communication between the staff of the ambulance and the central monitoring station. The doctor needs to understand the physical and physiological condition of the patient so that the right decision regarding administration of drugs and transport destination can be appropriately taken. Administrative requirements include the location of the vehicle and personnel attendance details. Deploying these instruments would mean that separate communication mechanism, namely, the usage of separate SIM cards and separate GSM/GPRS modems, would be

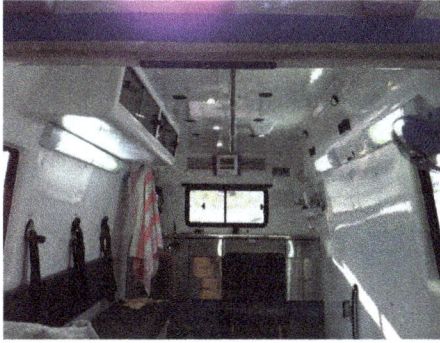

FIGURE 1: Interior view of a typical ambulance.

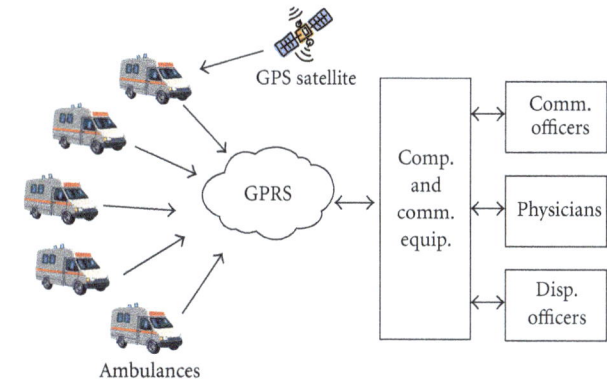

FIGURE 2: Architecture of the proposed system.

required resulting in increased recurring expenditure. Hence a comprehensive, cost effective system that can acquire physiological data from the patient keyed-in data from a keyboard and location data from a GPS receiver as well as voice signals and send them using cellular network is needed.

Several systems related to telemonitoring are reported in the literature [2]. Liszka et al. [3] and Plesnik et al. [4] have developed a real-time remote monitoring system for combining ECG data and GPS data before transmitting to a Central Monitoring Station (CMS). Similar systems are independently developed by researchers at the Glenn Research Center (NASA), University of Akron and Case Western University. Khan and Mishra [5] reported that position and velocity of vehicle can be estimated using GPS receiver fitted in the vehicle and sent to the central monitoring station using GSM. Zhang and Lu [6, 7] developed an ECG telemonitoring system with GPS and GPRS to continuously monitor ECG of the cardiac patient along with the position and posture. Fang and Lai [8] have developed a system to monitor the ECG of cardiac patient who is away from the hospital. A mobile ECG telemonitoring system along with an accelerometer, to sense sudden postural changes that reflect sudden cardiac failure, has been developed [9]. Philips Company has developed systems for remote monitoring of cardiac patients from their research centers worldwide in 2009. Exact and continuous blood pressure monitoring system along with location information and time synchronization facility to work with other measurement modules has also been proposed [10]. Recent advances in telemonitoring systems resulted in early diagnosis and management of chronic and degenerative conditions, significantly prevalent in elderly people. The number of recurring visits to the hospital can be reduced. Physicians are provided with better insight into the patient's health. Telemonitoring can also be applied on a long-term basis to elderly persons to detect gradual deterioration in their health status resulting in inability to live independently. Monitoring the activity of upper limbs can help in assessing the health status [11]. Several digital signal processing techniques have also been employed to denoise and study GPS and ionospheric studies [12, 13]. Companies including QRS diagnostic, VivoMetrics, Human Network Technology, AMD Telemedicine, Health Frontier,

and CardioNet have been engaged in the development of telemonitoring systems. Though all these systems are intended towards remote monitoring of cardiac patients, the concept of a unified system is not satisfied by any of the existing systems. They do not cater to the specific requirement of acquiring the physiological data, keyed-in data, patient snapshot, and voice signals and transmitting them using cellular network. Hence, it is proposed to develop an indigenous system to meet these requirements (Figure 2). The system developed by us is a comprehensive, cost-effective system (approximately USD 4500), which can acquire the physiological data, GPS data, vehicle parameters, patient information, patient snap shot, and SOS messages and transmit them as a single data packet using a cellular network.

This system has been built around ARM 9/11 microcontroller based module exclusively designed for this purpose and integrated with commercially available GPS and GSM modules and physiological signal acquisition systems like Medicaid and GSM phone with Blue Tooth (BT) facility. The necessary software for the embedded system has been developed using KEIL compiler for ARM family of microcontrollers. The front end application at the central monitoring system is developed using .net technologies. The system has been deployed in an ambulance for initial evaluation. This system is found to be very useful for emergency patient transportation being undertaken by organizations like EMRI.

2. Design Aspects of the System

The designed and developed ambulance system consists of GPS receiver, signal and image acquisition system, GSM/GPRS modem, and microcontroller unit. The central monitoring system has a bank of GPS/GPRS modems, central server and application software. Hardware design challenges include combining different commercial off-the-shelf (COTS) devices and specifically built subsystems, design, and implementation of interprocessor communication interfaces. Software design challenges include different sampling rates on separate channels to meet the Nyquist sampling rate of signals with varying bandwidths and reducing the data

TABLE 1: Typical specifications of the proposed system.

Operating voltage	12 V DC
Operating current	400 mA
GPS receiver update	Every one second
GPRS modem frequency	900/1800 MHz
LCD display	800 × 600 pixel 7″
Camera	1 M pixel
Serial interfaces	3
USB interfaces	3
Programming interfaces	1 USB + 1 serial
Parameters monitored	ECG, respiration, temp.

packet size. Implementation of interprocessor communication protocols is also complex. The typical specifications of the proposed system are as shown in Table 1.

3. Ambulance Electronics Unit

The electronics system present in the ambulance consists of a powerful low cost 32 bit RISC based microprocessor based system. It acquires location, speed, and time data from GPS receiver and physiological signals from an exclusively developed and built medical data acquisition module. Image from a camera, administrative data from a keyboard, and vehicle parameters from a vehicle data module are also acquired (Figure 3). All these data are combined to form a data packet, in predefined format, and transmitted using GPRS module to a central monitoring station. The central monitoring station houses call center personnel as well as trained medical personnel and expert doctors, who analyze the signals and data coming from field ambulances.

3.1. Microprocessor Module. The heart of the ambulance unit is built around ARM 9 Samsung processor S3C2440A operating at 400 MHz. The unit consists of two major building blocks, namely, processor module and base module. The processor module and base module are separately designed to provide flexibility and reduction in cost. Processor module consists of 6 layers whereas base module is a double layered one. The S3C2440A has ARM920T core. It adopts a new bus architecture known as Advanced Microcontroller Bus Architecture (AMBA). The integrated on-chip functions include SDRAM Controller, LCD Controller, 4-channel DMA Controllers, 3-channel UARTs, 2-channel SPI, IIC, IIS, Audio AC′97, 2-channel USB Host Controller, Camera Interface, and RTC Calendar function. Processor module includes nonvolatile NOR Flash of 2 MB (SST39VF1601), nonvolatile NAND Flash of 256 MB (K9F2G08U0B), and 32 bit SDRAM of 64 MB (2 X HY57V561626FTP-H), operating at a clock frequency of 100 MHz. NOR flash houses BIOS routines for interfaces and resources. The activities of microcontroller are indicated by three LEDs mounted on the board. NAND flash houses KERNEL, OS, and file system. SDRAM is used for temporarily storing the data for processing.

3.2. Base Board Module. Baseboard houses key board, programming interfaces, memory card, and audio devices. Alcor micromake USB hub IC (AU9254A) enhances the number of USB host ports to 4. USB host 1 connects camera and USB host 2 connects PDA system; USB host ports 3 and 4 are used for connecting external keyboard and vehicle parameter modules, respectively. USB device port helps in connecting to PC for downloading programs and OS images from the computer. There are three serial ports provided. Serial port 1 is connected to PC for programming purpose. Serial port 2 is connected to GSM/GPRS/3G modem for transmitting data, voice, and images. Serial port 3 is connected to GPS receiver to collect Position, velocity, and time (PVT) information at a speed of 4800/9600 bps. The data is updated every second. Transcend make Micro-SD card with necessary SD/MMC interface (clock, command, and 4 data lines) with a storage capacity of 4 GB is used for local storage of data and housing related file system. 10/100 MBPS LAN connectivity is provided with Ethernet IC DM9000. Audio CODEC Philips make UDA1341TS provides sound functionality of the system. The core module is connected to the base board using three 50 pin connectors. This board also connects LCD module through 40 pin flat cable. General purpose I/O pins are terminated in a connector for future expansion activity. The assembled microcontroller module along with the base board is shown in Figure 4.

3.2.1. SD/MMC Card Interface. The SD interface consists of 4 data lines (SDDATA0-SDDATA3) connected to the core module (PA24, PA25, PA22, and PA23) which are in turn connected to the processor through R8, M8, P8, and J9, respectively. Apart from data lines the SD interface has two control signals SDCMD and SDCLK which are connected to K8 and N8. The interface supports DMA transfer. Serial clock line synchronizes shifting and sampling of the information on data lines. The transmission frequency is controlled by making the appropriate bit settings to the SDIPRE register. The frequency can be changed to adjust the baud rate of the connected peripheral.

3.2.2. Graphic LCD Module. The graphic LCD and touch screen controllers are incorporated in the Samsung microcontroller itself. A 7-inch LC display (Innolux make model AT070TN83) having a resolution of 800X3 (RGB) X480 along with 4 wire resistive touch screen interface is included. The LCD is connected with a forty pin flat ribbon cable. The touch panel is connected using a 4-pin connector. In order to achieve energy savings, a power supply control switch to LCD is incorporated. There is a provision to adjust the brightness also. It displays several messages related to the DAS, GPS/GPRS modem, GPS receiver, and patient details and forms the human machine interface.

3.2.3. Programming Interfaces. Programming of the embedded system is carried out using JTAG and device USB interfaces. A 10-pin JTAG programming interface is provided on the processor module to program NOR flash using the programming software H flasher. USB device is used for

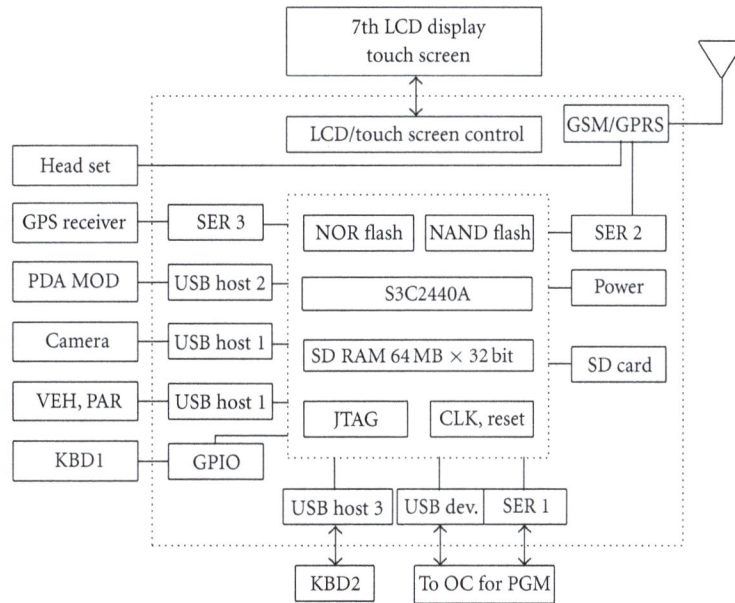

FIGURE 3: Architecture of ambulance electronics system.

FIGURE 4: Photograph of the base board with core module.

programming the NAND flash that houses the OS and application programs.

3.2.4. Power Supplies. Power supply circuits generate 5 VDC, 3.3 VDC, and 1.8 VDC required for peripherals and other interfaces that include camera, GPS receiver, and USB hub, and so forth. Switched mode regulator 2576 family ICs (LM2576-5 V, LM2576-3.3 V, and LM1117-1.8 for 1.8 V) are employed. RTC battery backup is provided with a 3 V Lithium battery. Battery backup to the unit is provided using a 7.5 V 4000 mAH nickel cadmium battery. A fuel gage circuit incorporating TI make BQ 2019 IC is employed.

3.2.5. Operating System and Embedded Software. Operating systems supported for this board are Linux 2.6 and WinCE 6. The Linux kernel source code used is Linux 2.6.29 and it is configured and cross-compiled using the cross-platform development tool chain arm-Linux-gcc 4.3.2. The drivers for all the peripherals are loaded statically to the kernel. The application code interacts with the hardware device through the device driver associated with the corresponding hardware and accesses the device.

3.3. Peripheral Modules. GPS receiver, camera module, vehicle parameter module, GSM/GPRS/3G module, patient information module, and SOS messaging module are connected to the system using appropriate interfaces as discussed below.

3.3.1. GPS Receiver. The GPS receiver used in the system is a 12-channel GPS receiver module EB818 from Compass Systems. It is built around SiRF-Star III (GSC3f/LP base band processor with integrated flash memory, and RF front end) chipset technology. It is easily integrated in the system being proposed. The module has the advantage of fast acquisition hardware, integrated RF filtering, and a TCXO. GPS receiver is connected through serial interface 1.

3.3.2. Camera Module. The camera captures the image of the patient and surroundings (in the case of an accident). Logitech/Emprex camera with a resolution of 5 megapixels is connected using USB host 1 interface to the ARM processor. There is a provision to connect a camera with I^2C interface also. The CMOS image sensor captures the image and forwards it to ARM processor. The CMOS image sensor used is a 1.3 megapixel IC OV9650. It provides the functionality of a single-chip camera and image processor in a small footprint package and outputs full-frame, subsampled, or windowed 8-bit/10-bit images. Timing generator, analog signal processor, A/D converters, digital signal processor, output formatter, and serial camera control bus (SCCB) interface are the main blocks present in the image sensor. Camera is controlled through the SCCB interface and can provide up to 15 frames per second (FPS). SCCB provides a programming interface

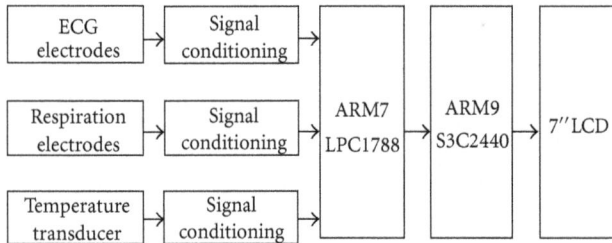

FIGURE 5: Block diagram of physiological data acquisition system.

for all required image processing functions including image quality, formatting, and output data transfer that can be controlled. The image is stored in the micro-SD card. Later the captured image is transferred to the CMS.

3.3.3. Vehicle Parameter Module. Vehicle parameter acquisition module is built around a low cost PIC family microcontroller 18F452 with a multichannel 10-bit ADC and USB interface for transferring the vehicle data to the microcontroller system. The parameters acquired include fuel level, ignition status, and engine oil level.

3.3.4. GSM/GPRS/3G Module. WAVECOM make GSM/GPRS/3G modem is chosen for transmission of voice, data, and images. Selection of exact model depends upon the service provider. This is connected to the ARM board through RS 232 with a connection speedup to 19,200 bps. In addition to the standard AT commands, the module supports an extended set of AT commands. This facilitates tasks like reading, writing, deleting SMS messages, and sending SMS messages.

3.3.5. Patient Information Module. KBD2 is a regular QWERTY keyboard connected using USB host 3 to enter administrative and patient information data, mainly alphanumeric. Paramedic staff present in the ambulance can key in any data related to the patient including condition and particulars like his/her name, age, address, and so forth through the keyboard. This message is transmitted to the Central Monitoring Station (CMS) for the medical advice from the doctors. Apart from sending patient information, the staff details like leave, present, or absent can be sent to CMS to notify the authorities.

3.3.6. SOS Messaging Module. KBD1, a 5X5 matrix keyboard, is used to initiate transmission of precanned messages like start of the ambulance from its base station, reaching the site of patient, start of ambulance after picking up the patient, and so forth. Fifteen such messages are provided presently, leaving the other 10 buttons for future expansion.

3.3.7. Physiological Signal Acquisition Module. The subsystems of this module are shown in Figure 5.
    Physiological data acquisition system acquires the patient's physiological data like ECG, respiration, and temperature through separate modules. The outputs from

these modules are connected to an LPC 1788 based data acquisition module through a 200 KSPS, 16-bit resolution, and 8-channel ADC (Linear technologies make LTC1867). They are processed in the system and are transferred to microprocessor based system which in turn transmits to CMS for further processing and action initiation.

3.3.8. Central Monitoring Station. The CMS consists of several networked computers with voice facility. An accident victim or patient in emergency or an attendant dials a toll free, predesignated number belonging to the central monitoring station being managed by an NGO. The call is answered by the call center executive, who obtains details like the nature of emergency, assesses the help needed, and then transfers the call to a medical expert, emergency response care physician (ERCP) in the call cenetr. Initial advice is given to the patient's attendant. Based on the information related to the location, an ambulance in the vicinity of the site is dispatched. Once the ambulance reaches the site, the emergency medical technician (EMT) provides prehospital care to the victim at the scene. He then connects the physiological data acquisition to the patient, which acquires required data from the victim and transfers the data (including patient's image, previous history, etc.) to the call center ERCP's computer system, where the doctor is available. The data received through internet is extracted to obtain the individual parameters. These extracted parameters include ambulance position, patient's physiological parameters, and patient's image. These are displayed on the graphical user interface of the system. The precise position of the ambulance and its movement is also displayed on the monitor with the help of mapping software on PC at the CMS. Because of this mapping interface the dispatch officer (DO) at the CMS has a chance to know the position of ambulance exactly and guide that ambulance to the accident location on receiving the accident call. Later this software helps the staff at the central monitoring station to guide the ambulance to the appropriate, nearest hospital. EMT is advised by the ERCP, who assesses the patient with the available clinical information made available to him, regarding the care and drugs to be administered to the patient. This is followed by the shifting of the emergency victim to the nearest and appropriate hospital. GPS information is useful in identifying the appropriate hospital with requisite medical care facilities. The ambulance is equipped with a variety of medical equipment, medical consumables, and disposables to ensure the victim is stabilized (prehospital care) by the time the ambulance reaches the hospital for further treatment. Once the patient is handed over to the hospital, the ambulance returns to its predesignated base location and is ready for attending next emergency.

4. Testing and Evaluation/Validation

The individual modules of the developed system are tested separately for their independent satisfactory performance. Later they are integrated into a unified system and again tested as a whole.

4.1. Power Supply. The power supply is tested for regulation, ripple, load regulation, and line regulation. The required DC voltages are 5.01 V, 3.29 V, and 1.79 V with a ripple content less than 2 mV.

4.2. Signal Conditioners. The ECG signal conditioner is evaluated for gain, frequency response, and CMRR. The values are found to be 1000, 0.03–150 Hz, and 100 db, respectively. The temperature measurement is calibrated using a standard clinical thermometer and water bath for the temperature range 25°C to 50°C. The respiratory module is tested for the impedance variation. As the respiratory activity results in a small change of transthoracic impedance, calibration is done by using a variable resistance. A voltage variation of 0.5 V is obtained for a resistance change of 0.1 Ω for a base resistance of 1500 Ω.

4.3. LPC1788 Board. Clock and reset circuits are tested using an oscilloscope. ADC is tested for a proper conversion by applying variable DC voltage to analog input pins of the board. Specifically developed program reads this data and transfers through USB port to a computer. The data is transferred to a PC through a USB interface. The application running on PC reads this data and displays values on the monitor. Then, a sinusoidal waveform with a varying frequency (0.05 Hz to 200 Hz) is applied and the values are again read and presented on PC screen. This procedure is repeated for all other analog input channels.

4.4. Vehicle Module. The parameters monitored in this module are ignition status (digital input), fuel level, and engine oil level (analog inputs). The analog voltages vary from 0 to 3.3 V for minimum to maximum levels.

4.5. GPS Receiver. As the receiver output is available through serial interface, at 4.8 Kbps, the module is tested for accuracy and update frequency by running hyperterminal program on PC.

4.6. GSM/GPRS Module. GSM functionality is tested by making a voice call to another cell phone whereas GPRS functionality is tested by connecting to a website on the net.

4.7. Camera Module. This module is tested by running the software along with the camera.

4.8. Patient Information Module. The patient details as required by the central monitoring station are entered using the QWERTY keyboard and checked for successful transmission.

4.9. SD/MMC Card. This card is tested by storing and retrieving typical text files.

4.10. Programming Interfaces. The programming interfaces are tested by connecting serial and USB cables between

FIGURE 6: Compilation of the kernel source.

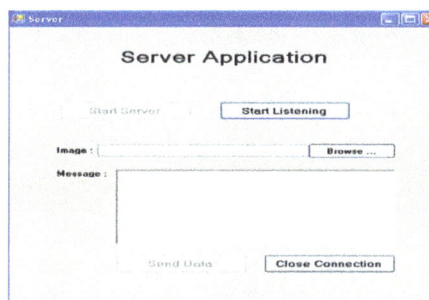

FIGURE 7: Typical CMS computer screen for testing the connectivity between ambulance unit and CMS.

ARM9 unit and a PC running flash magic and DNW software. Typical snapshots on the PC are indicated in Figure 6.

4.11. SOS Messaging Module. This is tested by pressing the keys on the matrix keyboard for successful transmission to the microcontroller ARM9 unit and subsequent transmission to CMS.

4.12. LCD Module. This is tested by transferring LCD test code from PC to ARM9 board and running the program. The module displays a preselected image.

The communication between the ambulance unit and CMS computer is tested using the developed diagnostic software. Typical screenshots are shown in Figures 7 and 8.

5. Data Acquisition

An investigation using the developed physiological data acquisition module and GPS receiver has been carried out to explore the possibility of relationship between physiological parameters and geographical location. Two locations with different GPS parameters are chosen for data collection. First location is the Department of Biomedical Engineering, University College of Engineering, Osmania University and

FIGURE 8: Sample test message transfer from CMS to ambulance unit.

TABLE 2: GPS parameters of the two locations.

Location	Latitude	Longitude	Altitude
Department of BME	$17°\,24'\,32.858$	$78°\,31'\,09.903$	$451.77\,\mathrm{m}$
Rangapur observatory	$17°\,05'\,45.024$	$78°\,43'\,04.695$	$605.74\,\mathrm{m}$

the second one is the Rangapur Observatory. The location details are shown in Table 2.

The signals recorded are ECG, body temperature, and respiration. The recording is repeated three times for the same subject and the average value of these three recordings is considered. Three parameters, namely, heart rate, body temperature, and respiration rate are computed from these recordings. The data are recorded from the subjects of different age groups ranging from nineteen years to fifty-nine years and the mean age is twenty-five years. Students and staff of the Department of Biomedical Engineering, University College of Engineering and Navigational Electronics Research and Training Unit of the Osmania University, volunteered to be subjects. 18 female subjects and 28 male subjects participated as volunteers in this study.

6. Results and Discussion

The base board PCB is designed and fabricated. The circuit is assembled and tested for its power supply functioning as per specifications. The core module is a bought out item. The core module is placed in the connector combination (Figure 8). The unit is tested for power consumption. The serial interface and USB device cables are connected to PC. H-JTAG-Dongle is connected to the J-Tag connector on the core module and the other side to PC's parallel port interface. Using the software H flasher boot loader (Super vivi128.) is loaded into the NOR flash. The boot loader (Super vivi128), kernel (zimage_A70), and file systems (root_qutopia128 M) are subsequently downloaded loaded into the 256 MB NAND flash using DNW software. The application is developed using gnu gcc compiler. Application software, after thorough testing, is also downloaded into the NAND flash.

The use of the processor S3C2440 resulted in compact design of the system. The software modules H-JTAG flasher,

FIGURE 9: System with all peripheral devices.

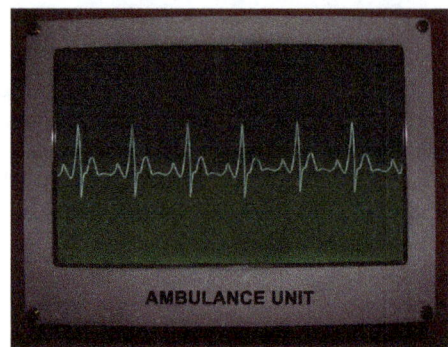

FIGURE 10: ECG display at the ambulance unit.

DNW, and gnu gcc compilers are employed in the development of software. With user friendly software development tools, the development and change management at site are relatively simpler. The prototype has been installed in an ambulance. The system facilitated acquisition of GPS data (PVT), images, physiological signals, vehicle fuel tank level, keyboard entered administrative data, and SOS messages. The acquired signals are made in to a data packet and sent to a central monitoring station using GPRS. The data are stored locally, when there is connection loss to the internet and transferred as soon as the connection is restored. The application code interacts with camera, data acquisition system, keyboards, GPS receiver, and GSM/GPRS modems through their respective drivers.

The developed ambulance electronics system with all peripherals (Figure 9) is installed inside an ambulance. Figure 10 indicates the display of ECG in the unit. The image of the surroundings as captured by the camera is also transmitted to the CMS.

6.1. Location Based Analysis of Physiological Parameters. It is clearly noted that heart rate increased at Rangapur Observatory compared to that at the Department of BME. The mean heart rate increased from 83 at the Department of BME to 90 at Rangapur Observatory. Thus the percentage increase in heart rate between these two locations is 8. Out of the 46 subjects, the heart rate increased for 33 subjects and it decreased for 13 subjects (Figure 11). The maximum

TABLE 3: Gender-wise physiological parameters at the two locations.

Gender	Department of BME			Rangapur		
	Heart rate (beats/min)	Temp. (°C)	Resp. (breaths/min)	Heart rate (beats/min)	Temp. (°C)	Resp. (breaths/min)
Female	85	37	20	95	37	20
Male	82	37	19	87	38	20

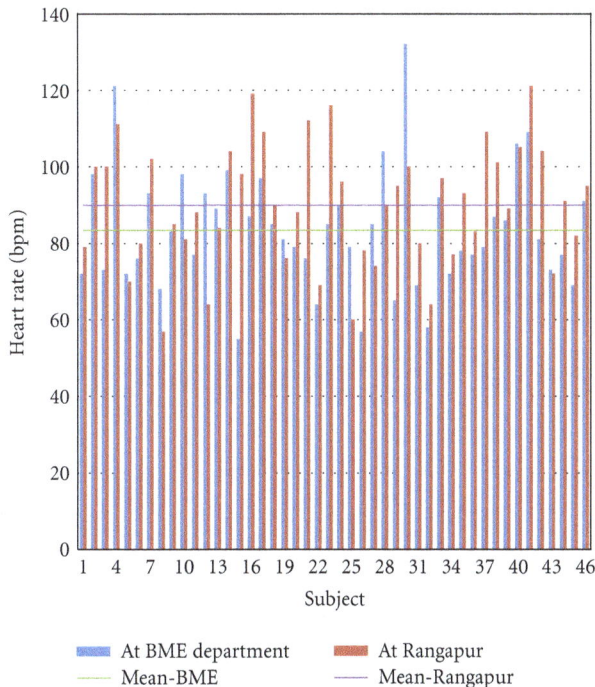

FIGURE 11: Heart rate of all subjects at the two locations.

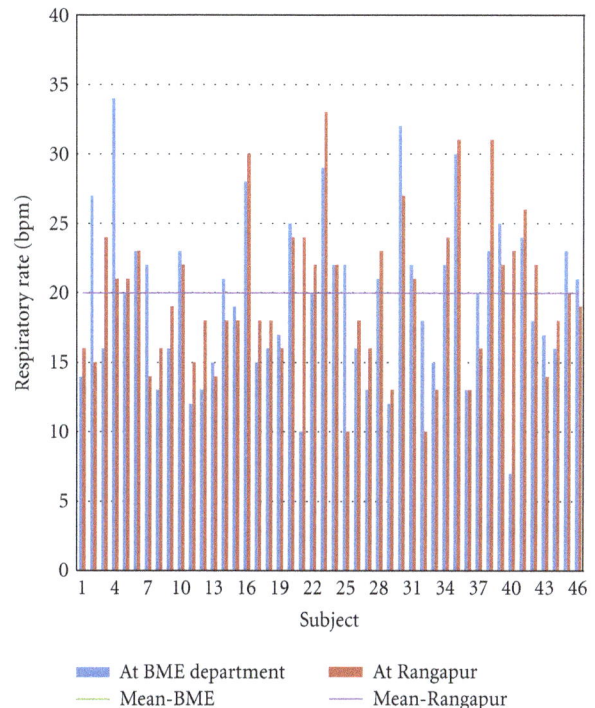

FIGURE 12: Respiratory rate of all subjects at the two locations.

increase is 43 beats per minute and the minimum increase is 2 beats per minute; while the maximum decrease is 32 beats per minute, the minimum decrease is 1 beat per minute.

The respiratory rate of 23 subjects increased and that of 23 subjects decreased. The increase in respiratory rate ranged from 1–14 breaths per minute while the decrease ranged from 1–13 breaths per minute. As seen from Figure 12, there is no change in the mean value of respiratory rate between the two locations.

A slight variation in the parameters is observed between female and male subjects. Though the mean values of heart rate, respiration rate of female subjects are slightly higher than those of male subjects at both locations, the difference between male and female subjects is different at the two locations as shown in Table 3. There is no significant change in the body temperature between the two locations.

7. Conclusion

The ambulance electronics system consisting of ARM 9 microcontroller has been designed, fabricated, and tested. The prototype has proved that it is possible to integrate and deploy an integrated medical data acquisition system, administrative data collection system, and communication system that is cost effective as well as life saving. An engineered model based on this prototype can be built and installed in the ambulances engaged for patient transportation. Engineered models, when installed in the ambulances, will enhance the capability of emergency patient transportation organizations in terms of reduction in time to reach the nearest hospital and extending timely prehospital medical care. The features of the system can be enhanced to meet the requirements from time to time based on the experience gained while using the system.

There has been a significant change in the heart rate of the subjects between the two locations. The changes in physiological parameters might be attributed to the difference in altitude (about 150 meters) between the two locations. As the data has been recorded from a small and healthy group of subjects, further work on a much larger database is required to establish the results obtained in this study regarding the effect of location on physiological parameters. Hence to derive more useful conclusions, larger group of subjects, both normal and abnormal, with different age groups have to be studied. This work can be extended by taking the recordings at different geographical locations. Diagnostically significant information could be extracted and the effect of location (as obtained by GPS receiver) on physiological parameters could be identified more successfully. Critically ill patients can be warned about moving to high altitude locations.

The system developed here is a product of extensive need-based research projected by government and nongovernment organizations engaged in emergency patient transportation. Different manufacturers provide data in different formats. One cannot employ systems from different manufacturers in the same project, as interoperability becomes difficult. They do not provide the data that can be easily read by scientific software for further processing. The system, as it provides data in text format, alleviates all such problems. Such a system will also avoid dependence on proprietary software to store and process the data.

Acknowledgments

The authors wish to thank the staff and students of NERTU and the Department of BME for their technical support in developing and evaluating the system. This work is a by-product of project sponsored by the Department of Science and Technology (DST) project vide sanction order SR/S4/AS, 53/2010, dated 12th July 2010.

References

[1] WHO Report, Decade of Action for Road Safety 2011-2010, 2011.

[2] Y. S. Yen, W. C. Chiang, S. F. Hsiao, and Y. P. Shu, "Using WiMAX network in a telemonitoring system," in *Proceedings of the 3rd International Conference on Computer Research and Development (ICCRD '11)*, pp. 313–318, March 2011.

[3] K. J. Liszka, M. A. Mackin, M. J. Lichter, D. W. York, D. Pillai, and D. S. Rosenbaum, "Keeping a beat on the heart," *IEEE Pervasive Computing*, vol. 3, no. 4, pp. 42–49, 2004.

[4] E. Plesnik, O. Malgina, J. F. Tasič, and M. Zajc, "ECG signal acquisition and analysis for telemonitoring," in *Proceedings of the 15th IEEE Mediterranean Electrotechnical Conference (MELECON '10)*, pp. 1350–1355, April 2010.

[5] A. Khan and R. Mishra, "GPS–GSM based tracking system," *International Journal of Engineering Trends and Technology*, vol. 3, no. 2, 2012.

[6] J. Zhang and Z. Lu, "The mobile ECG telemonitoring system based on GPRS and GPS," in *Proceedings of the International Conference on Networks Security, Wireless Communications and Trusted Computing (NSWCTC '09)*, pp. 454–456, Wuhan, China, April 2009.

[7] J. Zhang and Z. Lu, "The mobile ECG telemonitoring system based on GPRS and GPS," in *Proceedings of the International Conference on Networks Security, Wireless Communications and Trusted Computing (NSWCTC '09)*, vol. 2, pp. 454–456, IEEE Computer Society Ishington, Wuhan, China, April 2009.

[8] Z.-X Fang and D.-K Lai, "Uninterrupted ECG mobile monitoring," *International Journal of Bioelectromagnetism*, vol. 9, no. 1, 2007.

[9] Á. Alesanco and J. García, "Clinical assessment of wireless ECG transmission in real-time cardiac telemonitoring," *IEEE Transactions on Information Technology in Biomedicine*, vol. 14, no. 5, pp. 1144–1152, 2010.

[10] O. Krejcar, Z. Slanina, J. Stambachr, P. Silber, and R. Frischer, "Noninvasive continuous blood pressure measurement and GPS position monitoring of patients," in *Proceedings of the IEEE 70th Vehicular Technology Conference Fall (VTC '09)*, pp. 1–5, Anchorage, Alaska, USA, September 2009.

[11] N. Kiss, G. Patai, P. Hanák et al., "Vital fitness and health telemonitoring of elderly people," in *Proceedings of the 34th International Convention on Information and Communication Technology, Electronics and Microelectronics (MIPRO '11)*, pp. 279–284, Opatija, Croatia, 2011.

[12] K. Yedukondalu, A. D. Sarma, and V. SatyaSrinivas, "Estimation and Mitigation of GPS Multipath Interference using Adaptive filtering," *Journal of Progress in Electromagnetics Research M (PIER M), U.S.A.*, vol. 21, pp. 133–148, 2011.

[13] D. Venkata Ratnam, A. D. Sarma, V. Satya Srinivas, and P. Sreelatha, "Performance evaluation of selected ionospheric delay models during geomagnetic storm conditions in low-latitude region," *Radio Science*, vol. 46, no. 3, Article ID RS0D08, 6 pages, 2011.

8

An Adaptive Control Method for Ros-Drill Cellular Microinjector with Low-Resolution Encoder

Zhenyu Zhang and Nejat Olgac

Department of Mechanical Engineering, ALARM Lab, University of Connecticut, Storrs, CT 06269, USA

Correspondence should be addressed to Nejat Olgac; olgac@engr.uconn.edu

Academic Editor: Thomas Boland

A novel control methodology which uses a low-resolution encoder is presented for a cellular microinjection technology called the Ros-Drill (rotationally oscillating drill). It is developed primarily for ICSI (intracytoplasmic sperm injection) operations, with the objective of generating a desired oscillatory motion at the tip of a micro glass pipette. It is an inexpensive setup, which creates high-frequency (higher than 500 Hz) and small-amplitude (around 0.2 deg) rotational oscillations at the tip of an injection pipette. These rotational oscillations enable the pipette to drill into cell membranes with minimum biological damage. Such a motion control procedure presents no particular difficulty when it uses sufficiently precise motion sensors. However, size, costs, and accessibility of technology to the hardware components severely constrain the sensory capabilities. Consequently, the control mission and the trajectory tracking are adversely affected. This paper presents two contributions: (a) a dedicated novel adaptive feedback control method to achieve a satisfactory trajectory tracking capability. We demonstrate via experiments that the tracking of the harmonic rotational motion is achieved with desirable fidelity; (b) some important analytical features and related observations associated with the controlled harmonic motion which is created by the low-resolution feedback control structure.

1. Introduction and Motivation

We provide some background on the main task at hand and relevant motivation, before the control methodology is elaborated. ICSI (intracytoplasmic sperm injection) is a broadly utilized technique for artificial fertilization. This procedure is successfully performed in human oocytes as well as other species such as mouse and bovine. First, a holding pipette is used to immobilize an individual oocyte with a slight suction. Then an injection pipette (with outer diameter of about 8 μm), which contains the sperm head to be injected, is forced into the cell. The piercing through the zona layer and the membrane needs to be achieved with minimal biological damage to facilitate rapid healing. A significant amount of research effort has been devoted towards developing microscopic instruments for ICSI from this perspective alone. The most popular procedure at the present is the piezo-assisted ICSI [1]. However, its piercing performance is successful only by using a small mercury droplet in the pipette tip [2, 3]. Without this addition, undesirable lateral oscillations

occur at the tip and severely hamper the performance of piercing. Due to high toxicity of mercury, on the other hand, the piezo-assisted ICSI procedure is forbidden in many biological laboratories. In recent years, an improved remedial technology, called rotationally oscillating drill (Ros-Drill), is introduced [4]. This technique shows comparable results to those obtained by the piezo-assisted ICSI process, with one major difference, that Ros-Drill does not have the mercury problem [5].

A schematic of the Ros-Drill assembly is shown in Figure 1. The injection pipette is connected to a small-precision micromotor which is controlled to track a desired sinusoidal trajectory:

$$\theta_d = A_d \sin\left(2\pi f_d \cdot t\right), \tag{1}$$

where A_d is the oscillation amplitude (typically around 0.2 deg), and f_d is the frequency (in the range of 500–700 Hz). These selections are based on a simple bandwidth analysis; at this frequency range of external stimuli the cell membrane is

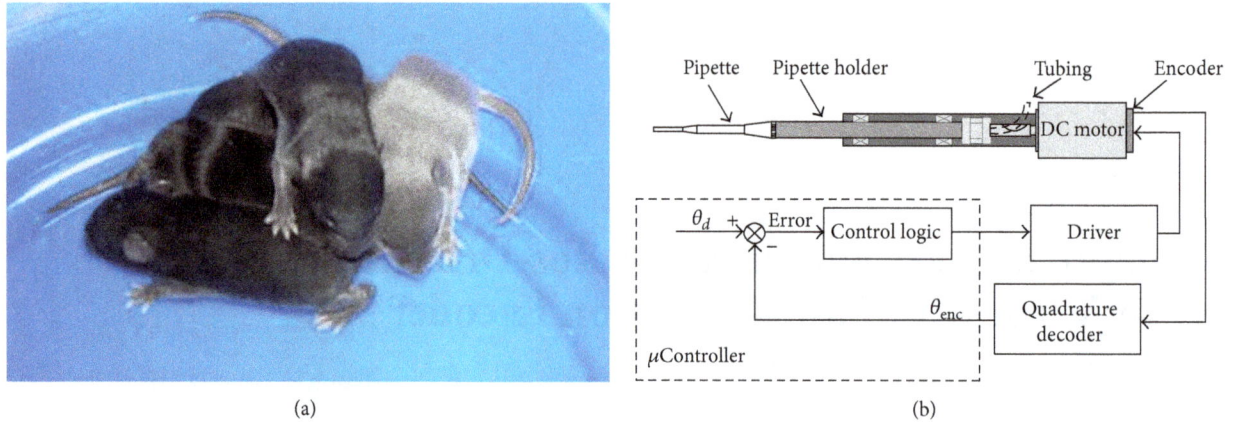

FIGURE 1: (a) Black mice reproduced using Ros-Drill technology (white mouse is the surrogate mother) [5]. (b) Ros-Drill assembly and control system.

not expected to follow the pipette tip motion. The ensuing relative rotational motion between the pipette and the cell membrane creates a clean piercing action, which facilitates the rapid healing of membrane after piercing.

Considerable amount of experimental effort has been invested to demonstrate the validity of Ros-Drill technology [4]. It is shown that the success rate in ICSI by Ros-Drill is comparable to that of piezo-assisted technology, provided that the pipette oscillations are maintained as close as possible to the desired trajectory in (1). Many healthy mouse offspring are produced using the Ros-Drill methodology, as shown in Figure 1(a). These biological tests are conducted by a group of experimentalists from the University of Connecticut and the University of California, Davis, USA. Two outstanding requirements are noted in these experiments towards an acceptable Ros-Drill performance: (i) rotational oscillation to track the harmonic trajectory very closely; (ii) the flexible pipette to be concentric to the rotational axis of Ros-Drill. The later condition is shown to satisfy because of the extreme bending compliance and whirling effects [2, 3]. The condition (i) is the topic of this paper. The smooth variation of the rotational motion in accordance with a harmonic function is the most natural desired trajectory. The contribution of this paper is a proper control law which can serve the objective under severe sensing constraints. Overarching restriction in this study comes from the pricing aspect. In order to make the Ros-Drill accepted by a wide range of IVF clinics, the cost of this automated device must remain within $1500 per copy; otherwise, it would be difficult to compete against the commonly-used but inefficient method adopted by trained ICSI specialists. These conditions set the tone of the key research issues.

Let us take a closer look at the physics of the Ros-Drill microinjection procedure. The rotational stiffness of the pipette holder and the pipette including the extremely fine tip are assumed to be high. Consequently, the angular displacement, θ, of the injection pipette is transmitted from the shaft of the micromotor to the pipette tip without a loss. The lateral (bending) vibrations are the major concern at the tip, and the present operating scheme is intended to

suppress them to ignorable levels especially when compared with those oscillations caused by the piezo-assisted ICSI [2, 3]. With these assumptions, the major objective of this study is directed to insure desirably precise harmonic motion tracking capabilities at the pipette tip [6–8] despite the very coarse sensory capabilities.

For position servoing, in general, the sensors are expected to have high resolution vis-à-vis the range of the intended motion, which can yield a desirable tracking capability. For instance, in representing a harmonic trajectory, one expects to have a minimum of 10 discrete data points per cycle. However, resolution of digital encoders is limited by the number of slots on a rotating disk, through which the encoder's light beam travels. Although advances in encoder technology have wonderfully progressed to increase the number of slots in order to improve the resolution of the encoders, the trade-off between resolution and cost is unavoidable. In such applications as Ros-Drill where the cost limitations are very stringent, sensor resolution is often compromised.

We encounter considerable past research on control methods using some low-resolution sensors. Recent model-based speed observers [9] make the velocity estimate robust using an interesting disturbance observer. In another effort, Kwon et al. [10] incorporate acceleration measurement in velocity estimation and motion control. Bautista-Quintero and Pont [11] propose an H-infinity control algorithm for sensor-constrained mechatronic systems using the position sensors with relatively low resolutions. They demonstrate how this procedure allows a faithful reproduction of observed motion starting from a limited sensing ability using relatively common (and inexpensive) microcontrollers. Furthermore, the methods which deal with control applications with low-resolution sensors usually have high computational demands to compensate the sensory shortfall. This aspect quickly makes the method prohibitive in a cost-sensitive design such as Ros-Drill.

We wish to familiarize the reader with the components of the first-generation Ros-Drill setup and the current improvements. The first-generation design having a 512 lines/revolution-encoder with quadrature signature is the finest selection we could find within the cost and spatial confines. Its resolution is 0.17 deg (including the quadrature signature feature). It allows a maximum of 2-step reading over the desired peak-to-peak stroke (note that the desired amplitude is 0.2 deg) [4, 12]. This sensor makes our best observation of a Ros-Drill harmonic cycle with a 2-step representation, which is a colossal handicap to perform the control. The spirit of the proposed control scheme and the focus of this paper are primarily on this crucial feature.

Moving on to another component, the first trial generation of Ros-Drill employs a PLC (programmable logic controller) as its digital controller, which has a maximum sampling speed of 1 KHz. This constraint clearly limits the maximum frequency of the controlled trajectory to 500 Hz. Most recently reported ICSI tests use injection pipette oscillations up to 0.3 deg amplitude and maximum frequency of 500 Hz. These oscillations last a certain length of time which we name the *duration of oscillation, D*. Typically, *D* varies within 250–500 msec. The preliminary reports claim that Ros-Drill-assisted ICSI results in embryo survival, embryo development, birth, and weaning rates comparable to those of piezo-assisted ICSI using mercury [5]. Although these biological results are very promising, the trajectory tracking performance of the existing prototype is not satisfactory mainly because of the low resolution of position sensor and low control sampling rate of PLC. This work represents an effort to further improve this performance.

In the second-generation (current) Ros-Drill microinjector prototypes, the low-sampling-frequency matter is considerably improved by replacing the PLC with a microcontroller (which brings the sampling speed from 1000 Hz to 10000 Hz) and the encoder resolution issue by selecting one-level higher-capacity sensor (1000 lines and 0.09 degree resolution after quadrature [12]). However, despite this upgrade, the position sensor with the resolution of 0.09 degrees still presents the biggest hurdle in this control system design because the required harmonic motion of 0.2 degree amplitude displays only 4-step peak-to-peak encoder recording, which is still low. In [12], a look-up-table-based adaptively tuned PID control law is used to cope with the limitations in the hardware. The most appropriate PID (proportional, integral, and derivative) control gains are a priori selected corresponding to the given operating frequency using dynamic simulations. For a given desired trajectory, θ_d, with A_d and f_d attributes, an ad hoc 3D search routine is performed using the representative SIMULINK simulation program. This yields a set of feedback gains, which are then utilized through a look-up adaptation structure. They are regarded as nominal control gains as they are tuned based on the nominal model of Ros-Drill, and the control design is considered as the first adaptation stage towards the second-generation Ros-Drill.

An important point to discuss for this first adaptation stage is that the look-up-table-based control law can handle the potential parameter uncertainties, disturbances, and the unmodeled mechanical properties of the system only via the supervisory interference from the user. These uncertainties are important and unavoidable. For instance, the external load originating from the membrane resistance torque may vary with the type of species and it may affect the control performance of the system. Still worse is the extremely stringent requirement of concentricity on the shaft of motor and pipette holder via a coupling and ensuing resistance to the servosystem. In order to achieve better updates of the control gains, we introduce a second adaptation scheme in this paper. This procedure constitutes the primary contribution of the paper. Furthermore, the low-resolution encoder is a range indicator (as discussed later) rather than a measurement device. We offer some novel observations on the actual angular motion.

The text is organized as follows. In the first part of the paper, a dedicated novel adaptive feedback control method is presented to achieve a satisfactory trajectory tracking capability with poor-quality sensors. Some important definitions about encoder signal are listed in Section 2. The first adaptation scheme is revisited in Section 3. In Section 4, the second-stage adaptive law is developed to update the control gains in situ, which is validated by experiments. In the second part of the paper, a stochastic perspective is presented which correlates the encoder readings with the actual angular motion. Section 5 presents an intriguing observation over the trajectory tracking process. Finally, conclusions are given in Section 6.

2. Some Descriptors of the Encoder Signal

The highlighted feature in this study is the unusually low-resolution sensor (encoder). Figure 2 gives a description of the operating principle of an optical encoder which is our main sensor. For simplicity, we overlook the quadrature effect and depict an encoder disk as the combination of transparent and dark lines. Δ is the angular resolution of the encoder. The small circle indicates the position of light beam. The angular motion θ denotes the rotational motion of the encoder. At the start, the encoder is declared at its "zero position" and $\theta = 0$.

We now present a critical argument on the position detection ability of the optical encoder. It is about an encoder offset angle, which we will denote by a, $0 < a < \Delta$. It is the angular motion needed for the first encoder pulse to register under a counterclockwise rotation (see Figure 2). "a" can be taken as a dead zone during which the encoder does not respond. Notice that this offset angle is not really an "offset position" from a baseline configuration. On the contrary, it is a measurement where we observe the first pulse. This value presents no importance at all, when the monitored motion is a few orders of magnitude larger than the resolution, Δ. However, for the particular application here the complete range of motion is composed of only a few Δ's (e.g., 4Δ). Thus, the offset angle a plays a very critical role and we present a set of novel observations on this issue, later in the paper.

Clearly, a is an unknown quantity which is random and uniformly distributed within $0 < a < \Delta$. That is, from the starting position $\theta = 0$ the encoder does not register any reading until $-a$ degree of rotation (in counterclockwise sense) or $-a + \Delta$ degree (in clockwise sense) is completed (see Figure 2). In Figure 3, we further depict the sensing ability of the encoder on a hypothetical oscillation θ (shown as the red line). The encoder can only register when θ reaches angular displacements of $-a + \Delta$, $-a + 2\Delta, \ldots, -a + m\Delta$ clockwise

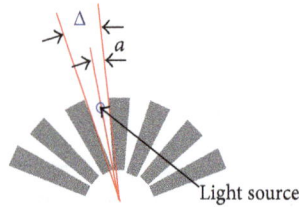

FIGURE 2: Encoder schematic and functionalities.

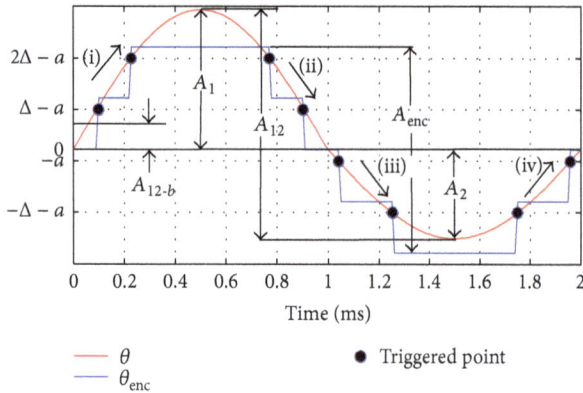

FIGURE 3: Oscillatory motion θ read by encoder (blue discrete lines represent the encoder signal); $0 < a < \Delta$ is the random offset angle.

and $-a, -a - \Delta, \ldots, -a - m\Delta$ counterclockwise. Therefore, the responses to the same angular motion may vary; that is, the encoder may sense the same θ differently, depending on encoder offset angle, a.

Of course, if the resolution of encoder is high compared to the total stroke, such an offset angle would not cause a noticeable difference in detecting the motion. We represent this behavior using a quantizer block in SIMULINK if the resolution is high. However, for low-resolution encoders (i.e., relative to the stroke), such a quantizer block cannot reveal the true response of the encoder, as depicted in Figure 3. For such cases, the correct model will be introduced in Section 3.2.

Some definitions are provided next, in Figure 3; A_1 and A_2 are the upper and lower amplitudes of θ, respectively. Without loss of generality, we assume $(A_1 > A_2 > 0)$ and define the complete actual peak-to-peak stroke as $A_{12} = A_1 + A_2$. The blue line represents the rotational motion being sensed by the encoder and the encoder reading is denoted as θ_{enc}. The peak-to-peak angular stroke which is recorded by the encoder is named A_{enc}, which is an integer multiple of Δ (e.g., 4Δ in Figure 3). We denote the average of A_{12} and A_{enc} over a certain number of oscillations by $\overline{A_{12}}$ and $\overline{A_{\text{enc}}}$, respectively. The bias of actual angular motion is defined as the distance of center of A_{12} from "zero" and it is expressed as

$$A_{12\text{-}b} = \frac{A_1 - A_2}{2}. \qquad (2)$$

For a certain frequency, the hypothetical actual oscillations of the pipette can be described by the bias and the amplitude of actual angular motion. Let us express θ as

$$\theta = A_{12\text{-}b} + A \sin \omega t, \qquad (3)$$

where $\omega = 2\pi f$ and A and f are amplitude and frequency of harmonic wave, respectively, and we have $A_{12} = 2A$. To summarize these definitions, we list them again as follows:

(i) resolution of position sensor = Δ;

(ii) encoder offset angle = a (deg);

(iii) upper stroke of actual angular position from rest = $A_1 > 0$ (deg);

(iv) lower stroke of actual angular position from rest = $A_2 > 0$ (deg);

(v) actual peak-to-peak angular stroke = $A_{12} = A_1 + A_2$ (deg);

(vi) bias of actual angular stroke = $A_{12\text{-}b}$;

(vii) peak-to-peak angular stroke detected by the encoder = A_{enc} (deg);

(viii) average actual angular stroke over a certain number of oscillations = $\overline{A_{12}}$ (deg);

(ix) average encoder stroke over a certain number of oscillations = $\overline{A_{\text{enc}}}$ (deg).

3. Review of Earlier Work: The First-Stage Adaptation Scheme

This section presents a brief review of the earlier work [12], which establishes the departure points of the present effort.

3.1. Control Objective and the Sensitivity Analysis. In Ros-Drill application, the desired angular trajectory requires high control sampling frequency, f_s, in the microcontroller (such as 10 KHz) in order to perform a meaningful tracking. This imposes some further constraints on the limited computational capabilities of selected microcontroller. Keeping these restrictions in mind, a proportional integral and derivative (PID) control logic is adopted here.

The Ros-Drill microinjector can be considered as a simple rotational mass, which is attached to a torque-generating DC servomotor (Figure 1). This yields a transfer function of Ros-Drill system and the corresponding frequency response creates the magnification factor as a function of ω, K_p, K_i, and K_d:

$$M = |G(s = j\omega)| = M(f_d, K_p, K_i, K_d), \qquad (4)$$

where $\omega = 2\pi \cdot f_d$ and K_p, K_i, and K_d are the control gains. The objective of the control is to achieve a flat response over a given range of operating frequencies. That is, $M(f_d, K_p, K_i, K_d) \approx 1$. We wish to emphasize that the phase angle between the desired harmonic input and the resulting output is not considered as a part of the performance description. Another important point to stress is that since we have limited access to the actual rotationally oscillating motion via the coarse measurements of the encoder, this flat response characteristic can be achieved, at best, by enforcing the peak-to-peak strokes of the encoder readings to be equal to $A_{\text{enc}} = \text{floor}(2A_d/\Delta) \cdot \Delta$ deg.

It is shown in [12] that M is much more sensitive to the K_d variations rather than K_p and K_i. We compare the effects of

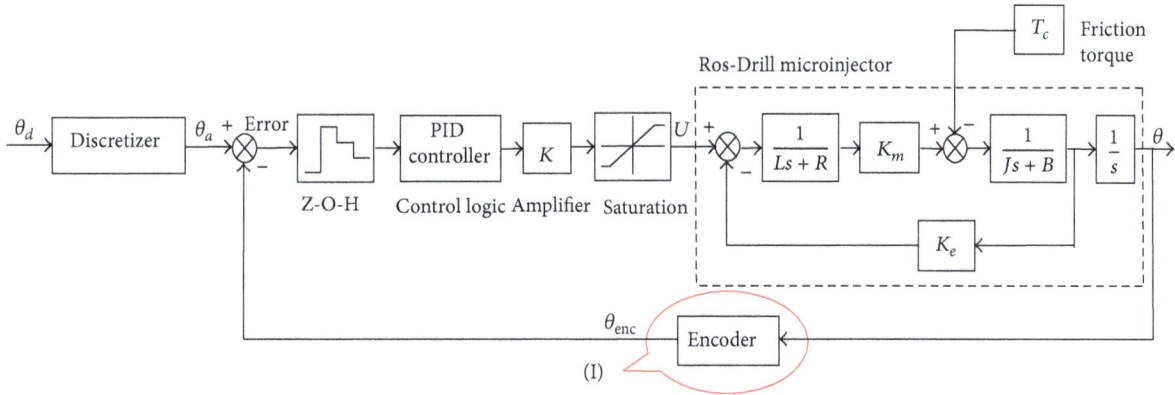

FIGURE 4: SIMULINK model of the spatially and temporally discrete control system.

variation of these three control gains around their nominal values ($\Delta K/K$). This results in some percentage variation of magnification factor ($\Delta M/M$). Among the three gains (K_d, K_p, and K_i), the most effective one is K_d, from this perspective. Thus, the tuning of the magnification factor can be achieved more directly by utilizing only K_d variations. This tuning methodology needs to be further adapted to the actual sample data system when using a low-resolution encoders. The logic steps of this tuning methodology, or the adaptive update laws of K_d, form the main contributions of this study, and they will be detailed in the following sections.

3.2. SIMULINK Model. In this section, we present a realistic dynamic model of the system using MATLAB-SIMULINK platform. Figure 4 shows this model for the sample data PID control system. The encoder block is created to reflect the limitations of the position sensor as closely as possible to the reality. Two pieces of information are used from the quad signature encoder outputs: one is the angular step counter, and the other is the velocity of the angular motion. These data divide the angular movement space into four sections:

$$\text{(i) } \theta > 0, \qquad \dot{\theta} > 0$$

$$\text{(ii) } \theta > 0, \qquad \dot{\theta} < 0$$

$$\text{(iii) } \theta < 0, \qquad \dot{\theta} < 0 \qquad\qquad (5)$$

$$\text{(iv) } \theta < 0, \qquad \dot{\theta} > 0.$$

Notice that only one of these four cases will be active at any sampling period; thus, the summation of the four outputs will actually declare the one-step angular accumulation. The angular measurements, A_d, in zero-order-hold (Z-O-H) mode is created with sampling time T_s just to simulate the sample data procedures of the microcontroller, see Figures 5(a) and 5(b). The integration step size for the simulation routine is chosen to be one-tenth of the sampling period of microcontroller (10 μs), which is sufficient to represent the transient behavior within a sampling period with a desirable fidelity.

The amplifier gain, K, in Figure 4 is chosen by the user based on experimental knowledge such that the saturation

in the D/A converter is prevented. The section enclosed by dotted lines in Figure 4 is the dynamic model for Ros-Drill microinjector itself. Using this SIMULINK tool, we first determine some starting values of the control gains K_p, K_i, and K_d using a pole-placement-based method as described in [12]. Then, we systematically vary K_d. The objective in this tuning procedure is to obtain $M = 1$ as close as possible for the particular interval of frequencies. For a desired motion of $\theta_d = A_d \sin(2\pi f_d \cdot t)$, $M = 1$ implies a peak-to-peak stroke of $2A_d$. This is the continuous rotational angle and the corresponding peak-to-peak encoder recording should be $A_{\text{enc}} = \text{floor}(2A_d/\Delta) \cdot \Delta$.

We continue the systematic variations of K_d until the frequency response amplitude condition, $M = 1$, is achieved for the given operating frequency. We then repeat the same operation for other f_d values on a list of potential (i.e., biologically required) operating frequencies. This set of K_d gains will form a look-up table which can be used by the real-time control program on the microcontroller. Once the user identifies a preferred harmonic frequency, the program selects a K_d feedback gain from a given list and adaptively sets the new control logic. This completes the first adaptation scheme, which follows the steps below:

(a) declare the desired amplitude and frequency via GUI (A_d, f_d);

(b) select the corresponding control gain (K_d) using the look-up table;

(c) observe the encoder registrations of peak-to-peak strokes, A_{enc};

(d) evaluate the average of A_{enc} over a predetermined duration D, $\overline{A_{\text{enc}}}$;

(e) signal the operator the direction of the error $e = \text{floor}(2A_d/\Delta) \cdot \Delta - \overline{A_{\text{enc}}}$ when it is outside a tolerance range, $|e| > \text{tolerance}$. Notice that A_d is typically not an integer multiple of Δ. Therefore, we deploy $2A_d/\Delta$ to create a comparable basis with $\overline{A_{\text{enc}}}$;

(f) manually adjust the gain, K_d, as supervisory fine tuning and repeat the steps (c)–(f) until no violation of tolerance in (e) remains.

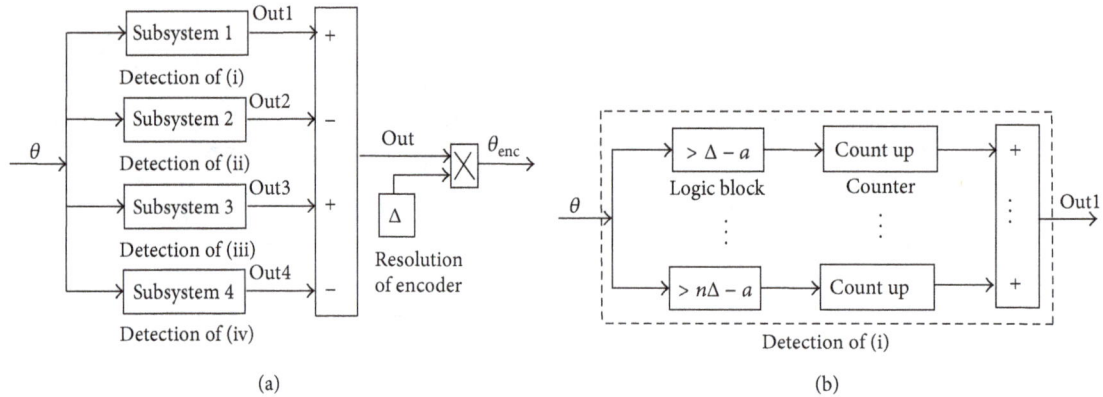

Figure 5: (a) Encoder reading scheme. (b) Model of the spatially discrete θ.

Figure 6: Experimental setup: (a) microinjector. (b) Control box.

This procedure is shown to perform the task so that the peak-to-peak strokes of Ros-Drill are guaranteed to stay desirably close to $A_{enc} = floor\, (2A_d/\Delta) \cdot \Delta$ [12].

3.3. Experimental Setup Used for All the Tests. One common setting was utilized during the different phases of progress. Figure 6 shows that experimental setup, the mechanical device along with the controller box. A FAULHABER series 2342 micromotor is the actuator driving the pipette holder. An optical encoder with 1000 lines is utilized as position sensor, enabling the angular resolution at 0.09 deg (with quad-signature characteristics). It is important to clarify that this is the best option for an encoder in terms of size, resolution, and costs demanded by the application. A control box contains the components that handle the logical operations between the sensor (encoder) and the actuator (micromotor). Its main unit (the CPU and I/O device) is a Silicon Lab's C8051F121 μ controller. This box also contains the necessary peripheral circuits such as the converter of encoder's quadrature-signatures into the rotational pulse counter and direction determination of the rotation, optoisolators, and power output chip which feeds directly into the DC micromotor.

We wish to give an idea to the reader about the microcontroller's time management structure. For the aimed Ros-Drill application, 10 KHz SISO (single input-single output) control

sampling rate is selected, and the C8051F121 micro-controller can easily accommodate this speed. Out of this $100\,\mu$s total loop time, $71.8\,\mu$s is used for sensing, control logic evaluation, and D/A conversion; the remaining $28.9\,\mu$s is the idle time which can be devoted to other applications (such as filtering) to be added later.

In order to crossvalidate the capabilities of the tracking control, we use an independent angular motion monitoring tool. It is a dSPACE 1104 DSP card which simply performs encoder decoding duties. The motion validations provided in the entire text are from this sensing channel instead of the unit which handled the same task in the loop.

4. New Procedure: Second-Stage Adaptation Scheme

We present the main novelty of this paper in this section. The look-up-table-based control law offered in [12], as we summarized above, can handle the potential parameter uncertainties, disturbances, and the unmodeled mechanical properties of the system only via the supervisory interference from the user, see step (f) above. For a typical application such as Ros-Drill, these uncertainties originate mainly from friction-related sources, which are impossible to model accurately. We include them in our SIMULINK model (Figure 4) as some arbitrary combinations of viscous and coulomb friction effects, just to get a feel.

The control objective is again to achieve a sustained peak-to-peak stroke of $A_{enc} = \text{floor}(2A_d/\Delta) \cdot \Delta$ as the encoder registers it (as an integer multiple of Δ). Remember that this is the only real-time measuring ability we are given. The average deviation of the A_{enc} (over a predetermined number of cycles) from this desired $A_{enc} = \text{floor}(2A_d/\Delta) \cdot \Delta$ is used as the error for tuning of K_d. One must pay attention, however, to another angle of these encoder readings. During the experiments, it is commonly observed that A_{enc} fluctuates typically between two successive steps (say 3Δ and 4Δ). This feature occurs due to the random offset angle a as mentioned above, but also stems from the transient regimes where the oscillations are not yet set into periodically repeated format. To discriminate the two causes of the same effect is an impossible task. Throughout this study, we consider the motion to have reached a steady regime for simplicity, therefore disqualifying the latter cause. The effects of the transient behavior are left for a future document.

In this paper, we confine the execution of supervisory gain adjustments: (a) through a windowing and averaging procedure; (b) with sufficient update-and-wait period. In short, before we adaptively select K_d we allow sufficient length of oscillations to be recorded. Therefore, K_d can change only after a certain number of oscillations. In the ICSI experiment, typically entire duration of oscillations, D, lasts less than 500 ms. The tuning of K_d is expected to be completed within a small fraction of D and in the very early stages of the period D. After extensive tests using SIMULINK with additional disturbances imposed, it is observed that the angular position of the servo system (θ) reaches the steady state no later than 5 oscillations for all the operating frequencies (from 400 to 1000 Hz). Therefore, we use the average of peak-to-peak strokes within 15 cycles of oscillations, which we denote by $\overline{A_{enc}}$, to assess the performance of current control gains and to update them. Obviously, if $\overline{A_{enc}}$ is equal to $A_{enc} = \text{floor}(2A_d/\Delta) \cdot \Delta$ that is considered to be satisfactory for the objectives of the control.

In summary, this adaptation scheme follows the steps listed below:

(a,b,c) are identical to the first adaptation scheme above;

(d) evaluate the average of A_{enc} over 15 cycles, $\overline{A_{enc}}$;

(e) utilize an update law for K_d adjustments if the error, e, is outside a tolerance range, that is, $|e| >$ tolerance. If the error is within the tolerance, no K_d update is needed.

(f) repeat step (d).

These steps will again guarantee the execution of the desired A_{enc} using an adaptive control gain update law. The details of this adaptation are given below. For the sake of simplicity, from this point onwards we will take $A_d = 0.2$ degrees and the corresponding $A_{enc} = 4\Delta = 0.36$ degrees.

4.1. The Adaptive Update Law. Let us define $K_d(n)$ as the nth update of the derivative gain and $e(n)$ as ensuing amplitude error which is obtained over 15 oscillatory cycles (an informed selection based on observation of experimental data), after $K_d(n)$ is applied. That is,

$$e(n) = 4\Delta - \overline{A_{enc}}. \qquad (6)$$

For the next 15-cycle period, we use a new feedback gain with the following update law:

$$K_d(n) = K_d(n-1) + \Delta K_d(n), \qquad (7a)$$

where

$$\Delta K_d(n) = C \cdot e(n-1), \qquad (7b)$$

and C is an update constant. This update process is performed in the following sequences:

(i) start with $K_d(0)$ which is taken from a look-up table as explained in Section 3.

(ii) Determine the resulting $e(0)$ at the end of the following 15 cycles.

(iii) For $n = 1$, use $\Delta K_d(1) = C \cdot e(0)$ and evaluate $K_d(1)$ from (7a).

(iv) Again after 15 cycles of using this control gain, determine $e(1)$.

(v)

   (A) If $|e(1)| \leq |e(0)|$, evaluate $\Delta K_d(2) = C \cdot e(1)$, $K_d(2) = K_d(1) + \Delta K_d(2)$, assign $K_{d\text{-ref}} = K_d(1)$ and $e_{ref} = e(1)$, and go to (vi).

   (B) If $|e(1)| > |e(0)|$, evaluate $\Delta K_d(2) = 0.5 C \cdot e(0)$, $K_d(2) = K_d(0) + \Delta K_d(2)$ and determine $e(2)$. If $|e(2)| \leq |e(0)|$, assign $K_{d\text{-ref}} = K_d(2)$ and $e_{ref} = e(2)$, go to (vi). If $|e(2)| > |e(0)|$, repeat (vB) but this time with $\Delta K_d(3) = 0.5 * 0.5C \cdot e(0)$ and continue until the absolute error falls below $e(0)$. Assign the current $K_d(n)$ value to $K_{d\text{-ref}}$, current $e(n)$ value to e_{ref} and move to (vi).

(vi) After 15 cycles, determine the new $e(n)$ and go to (vA). Use comparison of $|e(n)| \leq |e_{ref}|$.

Some nuances on this gain adaptation process are discussed later in this section over an example data set along with the role of $K_{d\text{-ref}}$ and e_{ref}.

The following portion is devoted to the selection of the update constant C. Uncertain friction term is modeled as some combination (in SIMULINK model of Figure 4) of viscous and coulomb frictions:

$$T_f(t) = T_c \cdot \text{sgn}(\dot{\theta}) + B \cdot \dot{\theta}, \qquad (8)$$

where T_c is the coulomb friction torque and B is the viscous friction coefficient. Note that different friction scenarios refer to the model in Figure 4 with the same system parameters but with combination of different T_c and B. Scenarios with various combinations of frictions are artificially created by varying T_c and B components in (8). The selection of C is performed offline via the following steps:

(a) estimate the bounds of T_c and B based on experimental studies.

(b) Deploy $K_d(0)$ in SIMULINK and find the corresponding $e(0)$ over a certain number of cycles (say 15).

(c) Determine $\Delta K_d(1) = C \cdot e(0)$, using systematically increasing C values (starting from zero), so that the resulting error $e(1)$ becomes desirably small (e.g., 75% as we used in our tests) compared with $e(0)$ under most adverse friction conditions. At the same time, we note that K_d should be upper-bounded as it multiplies the encoder-based angular speed and tends to saturate the actuator input (U in Figure 4). Note that the determination of C becomes critical from the concern of minimizing the number of gain updates before the error converges to zero. Using the proposed method, determination of constant C is done offline.

The following portion is devoted to the convergence of e to zero and the determination of K_d over some experimental data. We give an interpretation of the earlier assigned variable $K_{d\text{-ref}}$. "Reference K_d, $K_{d\text{-ref}}$" is the updated derivative gain which leads to smaller amplitude error $|e_{\text{ref}}|$ than that which evolves under the previous $K_{d\text{-ref}}$. And the corresponding new error is denoted by e_{ref}. To ensure that e converges to zero, the update law of K_d always moves the gain in the direction of smaller $|e_{\text{ref}}|$. The update of $K_{d\text{-ref}}$ is based on e_{ref} (i.e., $\Delta K_d = C \cdot e_{\text{ref}}$). If the resultant $|e| < |e_{\text{ref}}|$, this update results in a new $K_{d\text{-ref}}$ and e_{ref}. Otherwise, we can continue reducing this ΔK_d by 50% until a new $K_{d\text{-ref}}$ as described in the updating sequence earlier.

This update process of K_d is shown over an example experimental case study. In this case, experiment is done on our Ros-Drill setup for the desired amplitude of 0.2 degrees and frequency of 500 Hz. The update of K_d is performed every 15 oscillations and throughout D (i.e., 500 ms). Figure 7 shows the experimental result. Following the logic above, $C = 0.0256$ is selected for the present operating conditions. The microcontroller is programmed to monitor the average peak-to-peak value of θ_{enc} over 15 oscillations (i.e., $\overline{A_{\text{enc}}}$). Table 1 shows the update process of K_d for this experiment. In the first round of tuning (marked as $k = 0$ in Table 1 and Figure 7), $K_d(0) = 0.0036$ is selected from the look-up table for operating frequency of 500 Hz. However, $e(0) = 1.9\Delta$ is large (see the inset in Figure 7), meaning that the peak-to-peak strokes are much smaller than 4Δ. In the second round of tuning ($k = 1$ in Table 1 and Figure 7), $C \cdot e(0)$ is used to update K_d and $K_d(1) = K_{d\text{-ref}} + C \cdot e(0) = 0.008$. Because $K_d(1)$ results in $|e(1)| > |e_{\text{ref}}|$, $\Delta K_d(1)$ is bisected for the third round of tuning ($k = 2$ in Table 1 and Figure 7), $K_d(2) = K_{d\text{-ref}} + \Delta K_d(1)/2$. Because $|e(2)| < |e_{\text{ref}}|$, $K_d(2)$ and $e(2)$ are taken as new $K_{d\text{-ref}}$ and e_{ref}, respectively. Tuning updating process terminates here. The controlled oscillation of pipette lasts for a predetermined duration and then the pipette is returned to starting "zero position" in order to prevent the wrap-around effect (on the tubing attachments). In Figure 7, at 0.98 sec, oscillation of pipette ends and begins to return to "zero position." Figure 8 shows the flowchart of this monitoring and on-line tuning mechanism.

TABLE 1: Update process of K_d for Figure 7.

k	$K_d(k)$	$e(k)$	e_{ref}	$K_{d\text{-ref}}$	$\Delta K(k+1)$
0	0.0036	1.9Δ	1.9Δ	0.0036	$C \cdot e(0)$
1	0.0080	-2.66Δ	1.9Δ	0.0036	$\Delta K_d(1)/2$
2	0.0058	0	0	0.0058	0

The first part of the paper, which is related to the new, second-stage adaptation scheme, ends here. In what follows, we will demonstrate the influence mechanism for the actual angular motion, if the encoder reading of the stroke is $A_{\text{enc}} = 4\Delta$ just to provide a better insight to the reader.

5. Comprehension of Actual Angular Motion from Coarse Encoder Readings

The above control scheme is designed to perform an adaptively updating logic to assure a given A_{enc} peak-to-peak stroke. Let us take $A_{\text{enc}} = 4\Delta$ for the sake of simplified arguments, without loss of generality. We query which operational conditions are satisfied if $A_{\text{enc}} = 4\Delta$ reading is guaranteed, next. This quest results in some interesting observations which are stated in this section. Note that the derivations pertaining to $A_{\text{enc}} = 4\Delta$ in what follows can be generalized with ease to cases when desired control performance is $A_{\text{enc}} = 2m\Delta$, for $m = 1, 2, \ldots$.

To clarify some definitions about encoder reading, we list them as follows:

(i) a duration of oscillations $= D$;

(ii) absolute deviation of A_{12} from its nearest odd integer multiple of $\Delta = \varepsilon$.

A very important observation is stated next. One can see from Figure 3 that for the encoder to register 4-step stroke (i.e., $A_{\text{enc}} = 4\Delta$), the following are the necessary and sufficient conditions:

$$-a + 2\Delta < A_1 < -a + 3\Delta, \tag{9a}$$

$$a + \Delta < A_2 < a + 2\Delta. \tag{9b}$$

By summing (9a) and (9b), we obtain

$$3\Delta < A_{12} < 5\Delta. \tag{10}$$

Here, we define ε as the absolute deviation of A_{12} from the nearest odd integer multiple of Δ and $0 < \varepsilon < \Delta$. For $3\Delta < A_{12} < 5\Delta$, we have

$$\text{either } 3\Delta < A_{12} < 4\Delta, \quad \varepsilon = A_{12} - 3\Delta \tag{11a}$$

$$\text{or } 4\Delta \leq A_{12} < 5\Delta, \quad \varepsilon = 5\Delta - A_{12}. \tag{11b}$$

For the simplicity of expressions, let us denote

$$\overline{A_{12\text{-}b}} = \frac{\Delta - 2a}{2}. \tag{12}$$

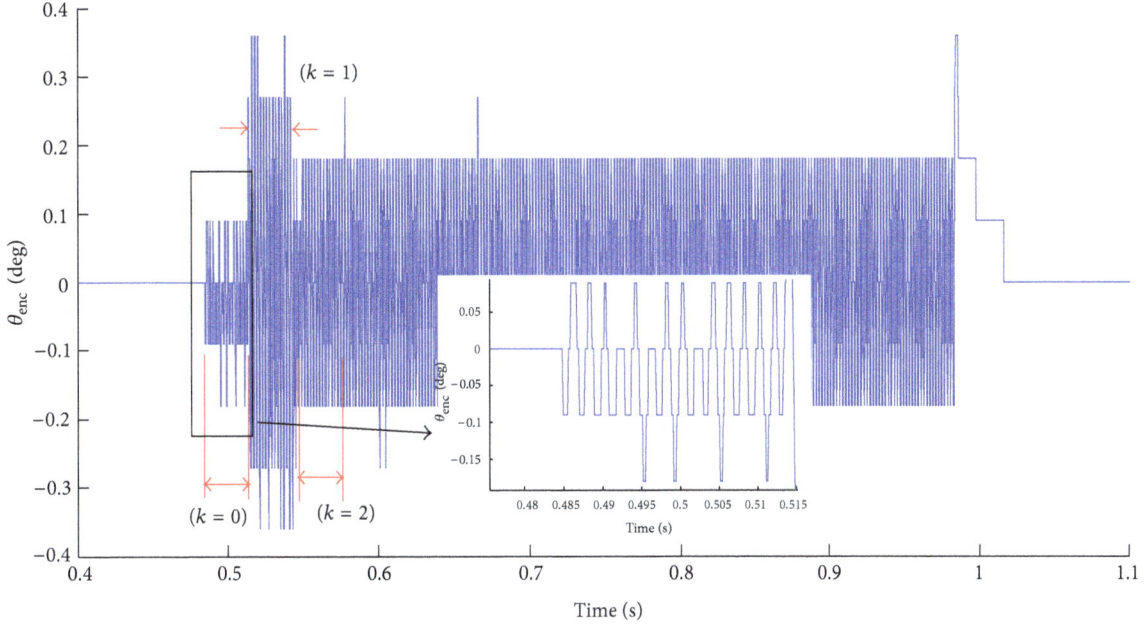

FIGURE 7: Experimental results with second adaptation scheme ($k = 0, 1, 2$ refer to the stages in Table 1).

Note that the offset angle is an uniformly distributed random variable between zero and Δ. Hence, $\overline{A_{12\text{-}b}}$ is also an uniformly distributed random variable but between $-\Delta/2$ and $\Delta/2$.

Proposition 1. *For any harmonic motion with $(2m - 1)\Delta < A_{12} < (2m+1)\Delta$, peak-to-peak encoder reading will always be $A_{enc} = 2m\Delta$ if and only if the bias of stroke ($A_{12\text{-}b}$) satisfies*

$$-\frac{\varepsilon}{2} < A_{12\text{-}b} - \overline{A_{12\text{-}b}} < \frac{\varepsilon}{2}. \qquad (13)$$

Furthermore if $A_{12\text{-}b} = \overline{A_{12\text{-}b}}$, for such a harmonic motion the encoder reading will always be $2m\Delta$.

Proof. Without loss of generality, let us take $m = 2$ and prove the sufficiency of (13) for $A_{enc} = 4\Delta$. By substituting (2) and (12) into (13), we obtain

$$\Delta - 2a - \varepsilon < A_1 - A_2 < \Delta - 2a + \varepsilon. \qquad (14a)$$

(a) First consider the interval $3\Delta < A_{12} < 4\Delta$. Note from (11a):

$$A_{12} = A_1 + A_2 = 3\Delta + \varepsilon. \qquad (14b)$$

Using (14a) and (14b) once for A_1 and again for A_2, we obtain

$$-a + 2\Delta < A_1 < -a + 2\Delta + \varepsilon, \qquad (15a)$$

$$a + \Delta < A_2 < a + \Delta + \varepsilon. \qquad (15b)$$

Keeping in mind that $0 < \varepsilon < \Delta$, one can see that satisfying (15a) and (15b) automatically satisfies (9a) and (9b), which are the necessary and sufficient conditions for $A_{enc} = 4\Delta$ measurement (as mentioned at the beginning of Section 5).

(b) One should next consider the interval $4\Delta < A_{12} < 5\Delta$. Note from (11b):

$$A_{12} = A_1 + A_2 = 5\Delta - \varepsilon. \qquad (16)$$

Using (14a) and (16) once for A_1 and again for A_2, we obtain

$$-a + 3\Delta - \varepsilon < A_1 < -a + 3\Delta, \qquad (17a)$$

$$a + 2\Delta - \varepsilon < A_2 < a + 2\Delta. \qquad (17b)$$

With the condition $0 < \varepsilon < \Delta$, (17a) and (17b) guarantee the fulfillment of (9a) and (9b), respectively. Again they are the necessary and sufficient conditions for $A_{enc} = 4\Delta$ measurement. This completes the proof of sufficiency of Proposition 1. □

Next we handle the necessity clause of the proposition. We take into account that $A_{enc} = 4\Delta$ and, therefore, ((9a) and (9b)) conditions hold. We wish to show that this necessitates (13). Let us again consider $3\Delta < A_{12} < 4\Delta$ first. Under this assumption, the right inequalities of ((9a) and (9b)) are satisfied automatically. The left inequality of (9a) is $-a + 2\Delta < A_1$ and it can be expressed as

$$-(A_1 + A_2 - 3\Delta) < A_1 - A_2 - (\Delta - 2a). \qquad (18a)$$

Similarly, the left inequality of (9b) is $a + \Delta < A_2$ and it can be rewritten as

$$A_1 - A_2 - (\Delta - 2a) < A_1 + A_2 - 3\Delta. \qquad (18b)$$

Using the definitions in (2) and (12) and the constraint (14b), the combined inequalities of ((18a) and (18b)) render exactly the conditions given in (13).

Let us now focus on the interval $4\Delta < A_{12} < 5\Delta$. We also claim that this condition, this time, together with the right

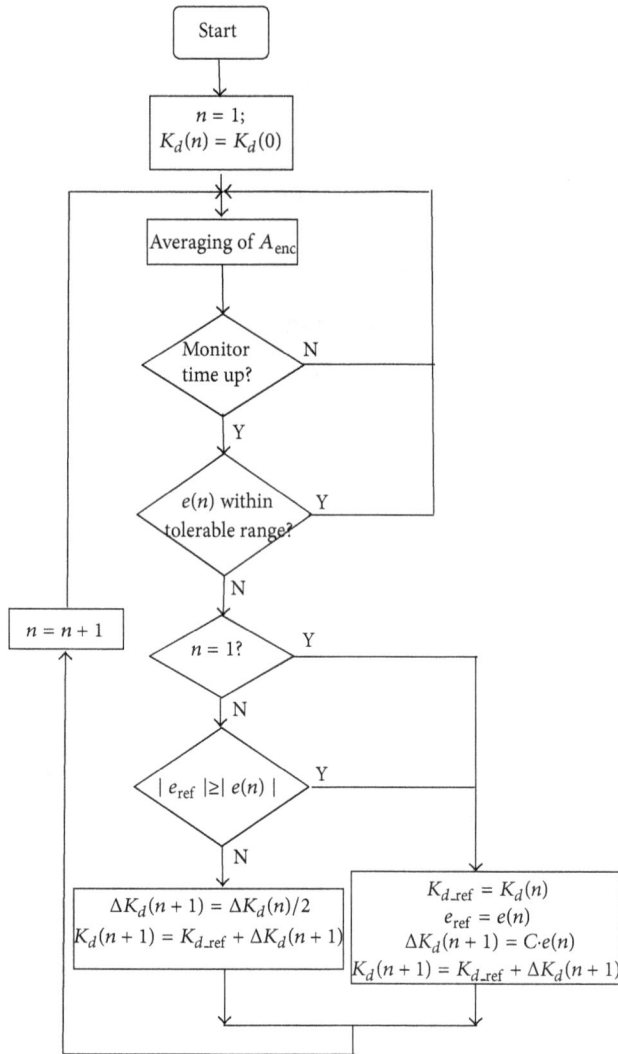

FIGURE 8: Flowchart of monitoring and on-line tuning mechanism.

(i) $\overline{A_{12\text{-}b}}$ is a random number (due to the term "a"). However, "a" remains unchanged during the oscillatory period D; thus, $\overline{A_{12\text{-}b}}$ remains constant in the same period.

(ii) Bounds expressed by (13) suggest that the bias, $A_{12\text{-}b}$, of each cycle during D stays within a bounded distance from $\overline{A_{12\text{-}b}}$ (i.e., $\varepsilon/2$) if and only if we maintain the encoder reading of $A_{\text{enc}} = 2m\Delta$.

(iii) In essence, if a control mechanism which is described earlier assures the $A_{\text{enc}} = 2m\Delta$, this also brings a guarantee for the boundedness of $|A_{12\text{-}b} - \overline{A_{12\text{-}b}}| < \varepsilon/2$, where ε is defined as (11a) and (11b). This implies that the bias $A_{12\text{-}b}$ has an attractive bonding to $\overline{A_{12\text{-}b}}$. It can fluctuate around this value but cannot run away from it.

6. Conclusions

A novel control system with low-resolution encoder for the desired harmonic trajectory is studied on a cellular microinjection technology called the Ros-Drill (rotationally oscillating drill). In the first part of paper, a novel adaptive control logic is developed to facilitate the tracking of the harmonic rotational motion under uncertainties, especially frictions. We demonstrate via dynamics simulations first, followed by experiments that the tracking of the harmonic rotational motion is achieved with desirable fidelity. In the second part, a stochastic analysis connecting the actual motions and their low-resolution sensory recordings is presented. It is observed that when the control structure guarantees a fixed peak-to-peak stroke, the bias of actual angular motion is bounded in a range and it is attracted to a certain predefined value.

Nomenclature

θ :   Angular position of pipette holder (deg)
θ_{enc} :   Angular position sensed by encoder (deg)
θ_d :   Desired harmonic trajectory (deg)
A_d :   Amplitude of θ_d (deg)
f_d :   Frequency of θ_d (Hz)
f_s :   Control sampling frequency (Hz)
K_p :   Proportional control gain
K_i :   Integral control gain
K_d :   Derivative control gain
Δ :   Resolution of position sensor
a :   Encoder offset angle (deg)
A_1 :   Upper amplitude of angular position (deg)
A_2 :   Lower amplitude of angular position (deg)
A_{12} :   Peak-to-peak angular stroke of θ (deg)
$A_{12\text{-}b}$:   Bias of angular stroke (deg)
A_{enc} :   Peak-to-peak angular stroke of θ_{enc} (deg)
$\overline{A_{12}}$:   Average angular stroke over a certain number of oscillations (deg)
$\overline{A_{\text{enc}}}$:   Average encoder angular stroke over a certain number of oscillations (deg)
D :   A duration of oscillations

inequalities of ((9a) and (9b)), will satisfy the left inequalities of ((9a) and (9b)). Following the similar procedure above, with the right inequality of (9a), $A_1 < -a + 3\Delta$, we arrive at

$$A_1 - A_2 - (\Delta - 2a) < 5\Delta - A_1 - A_2. \tag{19a}$$

The right inequality of (9b), $A_2 < a + 2\Delta$, creates

$$-\left(5\Delta - A_1 - A_2\right) < A_1 - A_2 - (\Delta - 2a). \tag{19b}$$

Once again, using the definitions in (2) and (12) and the constraint (16), the combined inequalities of ((19a) and (19b)) render exactly the conditions given in (13). This completes the necessity clause of the proposition.

Furthermore, if $A_{12\text{-}b} = \overline{A_{12\text{-}b}}$, (13) automatically holds for any A_{12} within $3\Delta < A_{12} < 5\Delta$; therefore, the causality follows. This proof for $m = 2$ can be extended for cases when $A_{\text{enc}} = 2m\Delta$, for $m = 1, 2, \ldots$.

   QED

The important implications of Proposition 1 can be summarized in the following logical sequences:

$\varepsilon :$ Absolute deviation of A_{12} from its nearest odd integer multiple of Δ

$U :$ Voltage input to the micromotor (V)

$J :$ Moment of inertia of the microinjector $(\text{g} \cdot \text{cm}^2)$

$L :$ Armature inductance (μH)

$R :$ Armature resistance (Ω)

$K_e :$ Motor back EMF constant (mV/rpm)

$K_m :$ Motor torque constant (mNm/A)

$T_c :$ Coulomb friction torque (Nm)

$B:$ Viscous friction coefficient (Nms/rad).

Acknowledgments

This work is partly sponsored by NIH R24RR018934-01 and NSF CBET-0828733.

References

[1] Y. Kawase, T. Iwata, O. Ueda et al., "Effect of partial incision of the zona pellucida by piezo-micromanipulator for in vitro fertilization using frozen-thawed mouse spermatozoa on the developmental rate of embryos transferred at the 2-cell stage," *Biology of Reproduction*, vol. 66, no. 2, pp. 381–385, 2002.

[2] K. Ediz and N. Olgac, "Microdynamics of the piezo-driven pipettes in ICSI," *IEEE Transactions on Biomedical Engineering*, vol. 51, no. 7, pp. 1262–1268, 2004.

[3] K. Ediz and N. Olgac, "Effect of mercury column on the microdynamics of the piezo-driven pipettes," *Journal of Biomechanical Engineering*, vol. 127, no. 3, pp. 531–535, 2005.

[4] A. F. Ergenc and N. Olgac, "New technology for cellular piercing: rotationally oscillating μ-injector, description and validation tests," *Biomedical Microdevices*, vol. 9, no. 6, pp. 885–891, 2007.

[5] A. F. Ergenc, M. W. Li, M. Toner, J. D. Biggers, K. C. K. Lloyd, and N. Olgac, "Rotationally oscillating drill (Ros-Drill) for mouse ICSI without using mercury," *Molecular Reproduction and Development*, vol. 75, no. 12, pp. 1744–1751, 2008.

[6] P. Schellekens, N. Rosielle, H. Vermeulen, M. Vermeulen, S. Wetzels, and W. Pril, "Design for precision: current status and trends," *CIRP Annals*, vol. 47, no. 2, pp. 557–586, 1998.

[7] X.-D. Lu and D. L. Trumper, "Ultrafast tool servos for diamond turning," *CIRP Annals*, vol. 54, no. 1, pp. 383–388, 2005.

[8] K. J. Astrom, *PID Controllers: Theory, Design, and Tuning*, ISA, 2nd edition, 2002.

[9] H.-W. Kim and S.-K. Sul, "A new motor speed estimator using kalman filter in low-speed range," *IEEE Transactions on Industrial Electronics*, vol. 43, no. 4, pp. 498–504, 1996.

[10] S.-J Kwon, W. K. Chung, and Y. Youm, "A combined observer for robust state estimation and kalman filtering," in *Proceedings of the American Control Conference*, pp. 2459–2464, Denver, Colorado, June 2003.

[11] R. Bautista-Quintero and M. J. Pont, "Implementation of H-infinity control algorithms for sensor-constrained mechatronic systems using low-cost microcontrollers," *IEEE Transactions on Industrial Informatics*, vol. 4, no. 3, pp. 175–184, 2008.

[12] Z. Zhang and N. Olgac, "Adaptive hybrid control for rotationally oscillating drill (Ros-Drill), using a low-resolution sensor," in *Proceedings of the ASME Dynamic and System Control*, pp. 564–569, 2011.

9

Enabling 3D-Liver Perfusion Mapping from MR-DCE Imaging Using Distributed Computing

Benjamin Leporq,[1] Sorina Camarasu-Pop,[1] Eduardo E. Davila-Serrano,[1] Frank Pilleul,[1,2] and Olivier Beuf[1]

[1] Université de Lyon, CREATIS, CNRS UMR 5220, Inserm U1044, INSA-Lyon, Université Lyon 1, 69622 Villeurbanne Cedex, France
[2] Departement d'imagerie Digestive, Hospices Civils de Lyon, CHU Edouard Herriot, 69008 Lyon, France

Correspondence should be addressed to Olivier Beuf; olivier.beuf@univ-lyon1.fr

Academic Editor: Nicusor Iftimia

An MR acquisition protocol and a processing method using distributed computing on the European Grid Infrastructure (EGI) to allow 3D liver perfusion parametric mapping after Magnetic Resonance Dynamic Contrast Enhanced (MR-DCE) imaging are presented. Seven patients (one healthy control and six with chronic liver diseases) were prospectively enrolled after liver biopsy. MR-dynamic acquisition was continuously performed in free-breathing during two minutes after simultaneous intravascular contrast agent (MS-325 blood pool agent) injection. Hepatic capillary system was modeled by a 3-parameters one-compartment pharmacokinetic model. The processing step was parallelized and executed on the EGI. It was modeled and implemented as a grid workflow using the Gwendia language and the MOTEUR workflow engine. Results showed good reproducibility in repeated processing on the grid. The results obtained from the grid were well correlated with ROI-based reference method ran locally on a personal computer. The speed-up range was 71 to 242 with an average value of 126. In conclusion, distributed computing applied to perfusion mapping brings significant speed-up to quantification step to be used for further clinical studies in a research context. Accuracy would be improved with higher image SNR accessible on the latest 3T MR systems available today.

1. Introduction

Liver fibrosis is an important cause of mortality and morbidity and contributes substantially to increase health care costs in patient with chronic liver diseases [1]. Fibrosis can lead to cirrhosis, for which the complications such as hepatic decompensation, hepatocellular carcinoma, and portal hypertension involve growing public health concerns. Cirrhosis and chronic liver disease were the 10th leading cause of death for men and the 12th for women in the United States in 2001, leading to the death of about 27,000 people each year [2]. Cirrhosis was first considered as an irreversible process, but, with the growing understanding of hepatic fibrogenesis mechanisms, more effective treatments have been developed [3, 4]. However, the latter must be initiated at a specific and early stage in fibrous development, and their administration requires regular clinical followup. While histological analysis after liver biopsy is the gold standard for the diagnosis,

inherent risk of a recognized morbidity and mortality renders this method unsuitable for clinical monitoring [5, 6]. Furthermore liver biopsies have other limitations such as interobserver variability and sampling errors [7]. It has been demonstrated that perfusion imaging has the potential to detect and assess vascular modifications [8] associated with liver fibrosis [9]. Several studies, using magnetic resonance dynamic contrast-enhanced imaging (MR-DCE) to quantify liver perfusion, have shown that some perfusion parameters were relevant indicators for liver fibrosis assessment [10–12]. In a previous work, an MRI protocol associated to a dedicated processing step to quantify liver perfusion was developed [12, 13]. Several parameters showed significant correlations between hepatic perfusion modifications and fibrosis stage. Results demonstrated that MR perfusion imaging could be a noninvasive method for the clinical followup in patient with chronic liver diseases. Nevertheless, the evaluation was restricted to an ROI, and regional variations often met in diffuse liver diseases could not be observed. ROI-based perfusion quantification already requires heavy processing methods such as image registration, denoising, and data fitting. Processing time drastically increases and becomes really prohibitive for clinical application for 2D or 3D mapping. In this context, parallel computing on distributed infrastructures such as clusters, grids, or clouds proves to be an interesting solution. Such infrastructures can bring significant speedup for a large spectrum of applications from various scientific domains. They have already been used for medical imaging as described in [14, 15] but never before for 3D-liver perfusion mapping. Significant effort has been put in rendering distributed infrastructures as user friendly as possible. Nevertheless, new applications still require extra work for adapting (porting) them on the considered infrastructure. This work describes an MR acquisition protocol and a processing method using distributed computing on the European Grid Infrastructure (EGI) to allow 3D liver perfusion parametric mapping after MR-DCE imaging with the MS-325 blood pool agent. Processing speed, reproducibility, and accuracy were assessed and adequate acquisition requirements were defined.

2. Materials and Methods

2.1. Subjects. The study protocol was approved by the local experimentation ethics committee, and informed consent was obtained from each patient. Privacy rights of subjects have always been observed. Seven subjects (4 women, 3 men; average age, 40 ± 12 years; mean weight, 75 ± 8 kg) were enrolled. Among this group, one healthy subject was used as control and six patients with chronic liver diseases were prospectively enrolled (maximum prospective period of one month) after having had a liver biopsy. Biopsies were performed by percutaneous sampling of the right lobe with a 15-gauge needle. All biopsies were 1.5 cm or more in length. Tissue samples were fixed in buffered formalin and embedded in paraffin. 4μm-thick sections were stained with hematoxylin-eosin-saffron, iron stain, and Masson trichrome reagents and evaluated by two pathologists. The histopathological evaluation was performed masked from any clinical information. Fibrosis was evaluated on trichrome-stained slides according to the METAVIR classification [16] (*score F0*: absence of fibrosis; *score F1*: portal fibrosis; *score F2*: portal fibrosis with isolated bridges *score F3*: fibrosis with numerous bridges without cirrhosis; *score F4*: cirrhosis).

2.2. 3D MR Dynamic Acquisition. Acquisitions were performed using a Siemens Magnetom Symphony Maestro Class 1.5T imaging system (Siemens Medical Solutions, Erlangen, Germany). A T_1-weighted VIBE 3D sequence with a parallel imaging technique was used (GRAPPA, *R*-factor = 2). The sequence parameters were as follows: TE/TR/α, 1.22/2.87 ms/12°; K-space partial filling, 6/8th according to slice and phase direction; reduction of the slice and phase encoding step, 63 and 50%, respectively. The plane was coronal oblique with a rectangular FOV (400×300 mm^2) for a rebuild matrix of 256×192 pixels with right/left phase-encoding direction. The rationale behind the use of coronal

imaging was to minimize the flow-related enhancement of the aorta. Moreover, it allowed covering a larger liver volume. Volume angulation was not systematic and aorta orientation independent. The exploratory volume was acquired with a 1-sec temporal resolution with 6.4 cm slab thickness (16 slices of 4 mm). The signal was collected using two circularly polarized phased array coils (CP Body Array and CP Spine Array) with a bandwidth of 650 Hz·pixel^{-1}. Acquisition has begun at the time of injection of the contrast medium (MS-325;Epix Pharmaceutical, Inc., Lexington, MA, USA), and it continued for 2 minutes [12, 13]. Patients were instructed to breathe calmly. All subjects were asked to undergo fast before MR acquisition. Injection was performed with an injection rate of 1 mL·s^{-1}, a posology of 0.03 mM·Kg^{-1}, and flushed with 25 mL of physiologic saline injected at the same rate. Finally, sixteen 2D + t volumes with $t = 120$ were acquired leading to 1920 images per examination.

2.3. Images Preprocessing. Images were first imported on a personal computer running an in-house developed application written in Matlab r2010a (The MathWorks, Natick, MA, USA).

Due to free-breathing acquisition, spatial shifts linked to motion had to be corrected. Hence, each volume was automatically registered. The registration method ignored nonrigid aspects of liver transformation during breathing. Only translations and rotations (rigid transformations) were taken into account. This method consisted in the estimation of the transformation vector needed to register each moving images in relation to a static reference image. For each image from 2D + t volumes, a pixel-based method was used (iconic approach) to control the transformation of an input image. An error measure was used to measure the registration error between the moving and static image. The reduced-memory Broyden-Fletcher-Goldfarb-Shanno (BFGS) quasi-Newton algorithm was used to move the control points to achieve affine rigid registration between both images with a minimal registration error. Then pixels were interpolated with a bicubic method.

Secondly, native 2D + t volumes were converted in to 2D + t MS-325 mass concentration maps from a pixel-by-pixel operation based on the relationship between signal intensity and MS-325 concentration. The latter was established in a previous work using a calibration phantom [13].

Thirdly, native arterial and portal input functions, $C_A(t)$ and $C_P(t)$, were measured using squared ROIs of 25 pixels placed by an experienced radiologist (*F.P.* 12 years of postgraduate experience in digestive imaging) at the level of the abdominal aorta close to the cœliac trunk and the main portal vein. Finally, definitive arterial and portal input functions were converted into continuous form (function of the time) instead of vectorial form (discrete form), by an interpolation using spline curves. Measurements were previously filtered using a moving average filter to reduce noise effect.

2.4. Image Modeling. Hepatic capillary system was modeled by a 3-parameter one-compartment pharmacokinetic model adapted to hepatic dual supply (portal and arterial). The

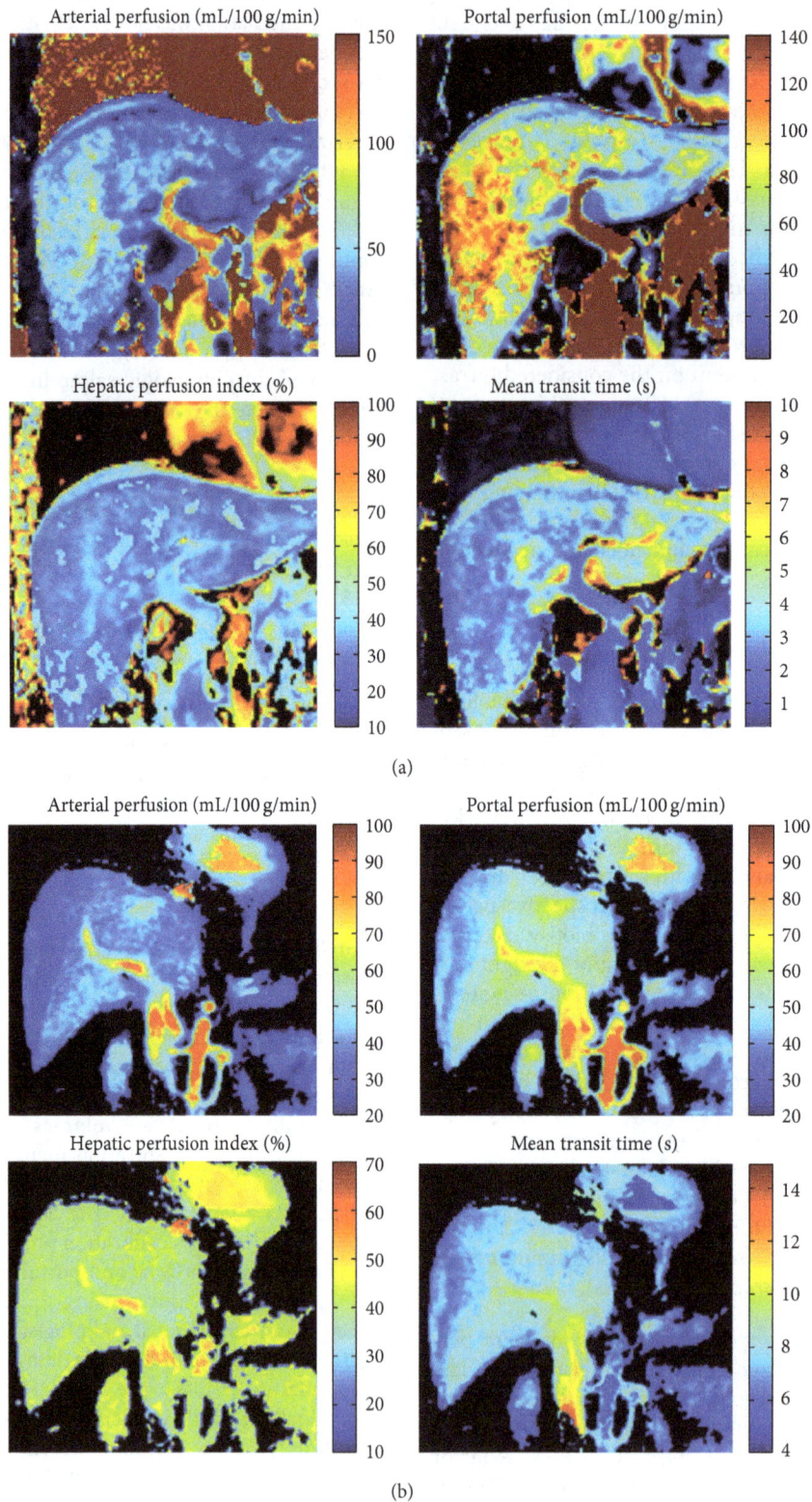

FIGURE 1: Representative liver perfusion parametric maps: arterial perfusion, portal perfusion, mean transit Time (MTT), and Hepatic perfusion Index (HPI) computed on a healthy subject (a) and on a patient with chronic liver diseases classified $F2$ according to METAVIR classification (b).

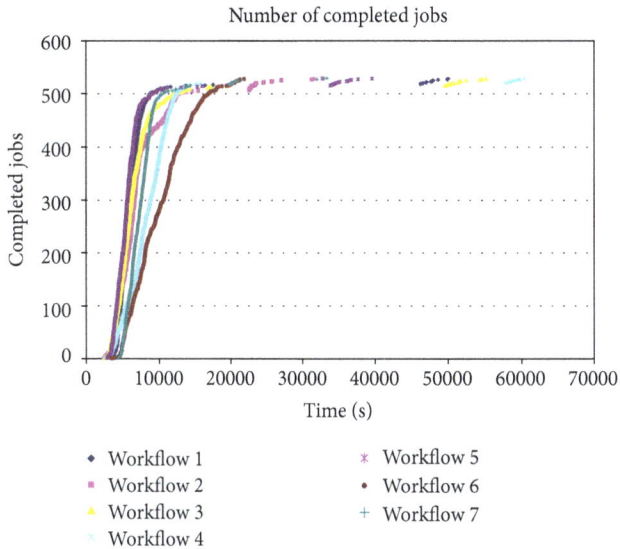

FIGURE 2: Number of completed jobs over time. Each workflow needs to wait for the completion of all its jobs in order to produce the final result. While most of the jobs finish within 10000 seconds, the last ones need much more time to complete. These last jobs almost triple the makespan (from roughly five to fifteen hours).

central compartment includes the hepatic sinusoids and the space of Disse. The leakage of tracer through venous washout is carried out exponentially over time and the inverse of the constant of elimination is the mean transit time (MTT). The equation describing this model is as follows:

$$C(t) = \rho \left[C_A (t - \tau_A) \times \phi_A + C_p (t - \tau_P) \times \phi_P \right] \otimes e^{-t/\text{MTT}},$$
(1)

where $\otimes$ designates the convolution product and ρ the volemic mass considered to be equal to $1 \, \text{g·mL}^{-1}$. The parameters, ϕ_A and ϕ_P, are the arterial and portal perfusion, respectively, expressed as $\text{mL·100 g}^{-1} \cdot \text{min}^{-1}$. C_A and C_P, are respectively, the arterial and portal input functions. The two delays, τ_A and τ_P, take into account the temporal offset between central compartment input and measured input from arterial and portal ROIs. While ϕ_A, ϕ_P, and MTT are model parameters, delays are independent of the fit procedure. The hepatic perfusion index (HPI), defined as the arterial perfusion to total perfusion (arterial + portal perfusion) ratio, was also calculated. For each part of the image, pixel-by-pixel tissular time activity curves were obtained and a nonlinear least-square fit was performed according to the model previously described (1) using the Levenberg-Marquard algorithm. Because some coefficients are closely connected, in particular portal perfusion and arterial perfusion with arterial and portal delay, the results of optimization were strongly influenced by the choice of starting coefficients, and algorithm may converge to local minima. In order to improve the robustness and reliability of optimization, but also to avoid any convergence to local minima, the algorithm needed to be started with a grid of pseudorandom starting points generated within two bounds (multistart technique). So, each fit procedure was done two-hundred-fold, with two-hundred different initializations. For

each fit procedure, delays were determined as the time between the beginning of tissular enhancement and the beginning of arterial enhancement in celiac trunk. These starting points are chosen as the maximum of second-order derivative of tissue time activity curve and arterial input function. From this step, three perfusion parametric maps were obtained, one for each parameter of the model used.

2.5. *Distributed Processing.* The processing step was parallelized and executed on EGI within the biomed virtual organization (VO). The parallelization was handled at the input data level, by splitting each volume into several pieces. Each piece was processed by independent jobs running in parallel on multiple grid resources and eventually merged. The whole processing operation was modeled and implemented as a grid workflow using the Gwendia language [17] and the MOTEUR workflow engine [18]. The splitting and merging algorithms were developed in C++, while the processing algorithm was developed in Matlab. All three programs were compiled on a grid compliant operating system (CentOS) and deployed on the fly on the grid nodes. For the Matlab code we used the Matlab Compiler and the Matlab Compiler Runtime (MCR).

The interface with the grid resources was provided by the VIP web platform (https://vip.creatis.insa-lyon.fr/) [19]. A specific cartography workflow was developed for this application and integrated into the VIP platform.

The user uploaded the input volumes on the grid and launched the processing workflow from a web portal. In order to evaluate the speedup provided by our parallel approach, the total CPU time to make span ratio was determined. The makespan was defined as the time elapsed between the launch and the completion of the workflow, and the total CPU time as the sum of CPU times of all jobs in a workflow.

2.6. *Statistical Analysis.* In order to evaluate the reproducibility of our distributed computing algorithm, the 3D mapping procedures (workflow) were repeated three times for each subject. Relative standard variation (coefficient of variations) was then mapped for each parametric map for all patients and defined as the standard deviation to arithmetic mean ratio.

Next, to evaluate the accuracy of our method, results between ROI-based quantification method described in [12] and the method presented in this paper were compared. Quantitative perfusion parameters from three ROIs were calculated and averaged. The difference between methods was evaluated using the Bland-Altman representation for each perfusion parameters, the Spearman's coefficient calculation, and the nonparametric Wilcoxon test.

3. Results

3.1. *Subjects.* Among the 6 biopsied patients, histological results were as follows: 2 patients were scored $F0$, 3 patients scored $F2$, and 1 patient scored $F4$.

3.2. *Quantification Results.* A representative set of 2D parametric maps extracted from 3D volumes on the healthy patient (METAVIR $F0$) is shown in Figure 1. Perfusion

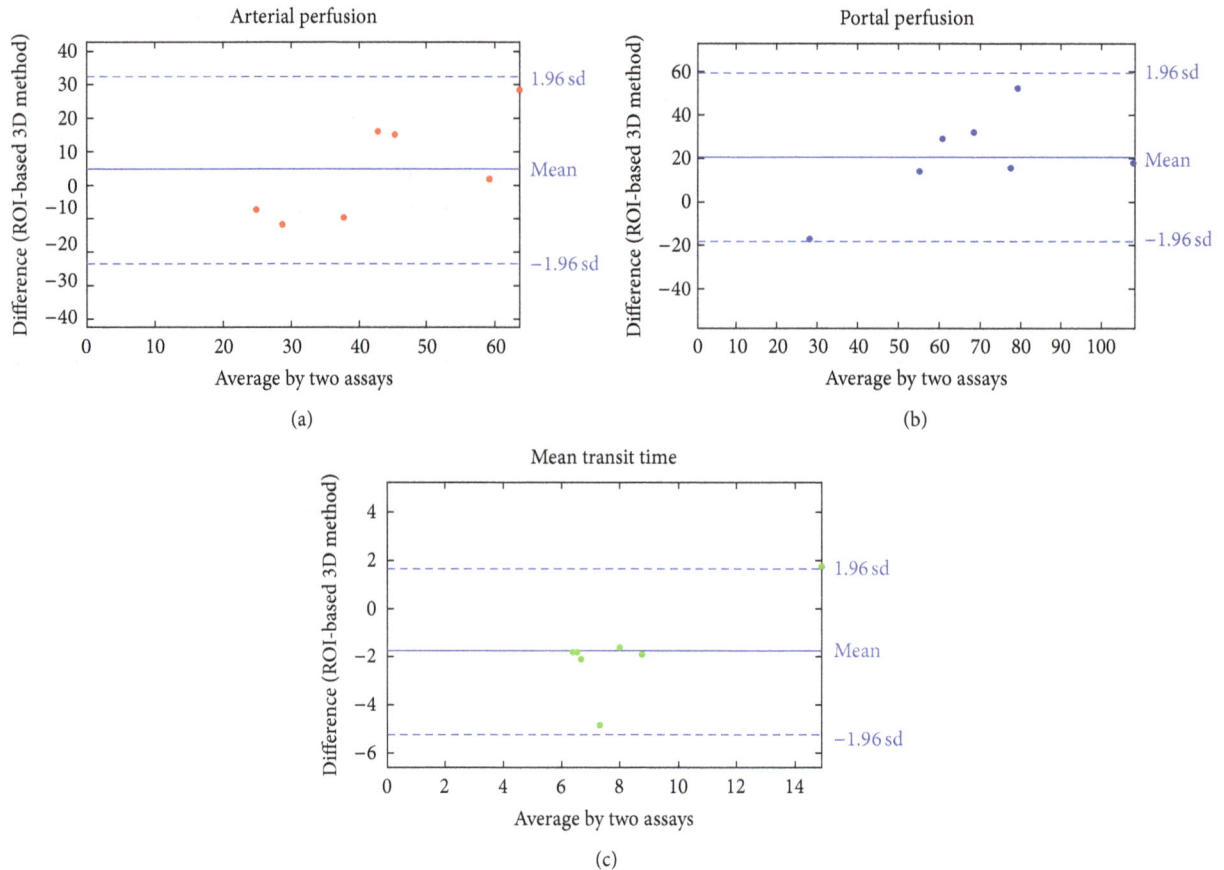

FIGURE 3: Bland-Altman representations computed for each perfusion parameters: (a) arterial perfusion, (b) portal perfusion, and (c) mean transit time quantified with the ROI-based method and the distributed method.

TABLE 1: Quantified mean values of perfusion parameters for all subject obtained with the two compared methods.

Method	Arterial perfusion $(mL \cdot min^{-1} \cdot 100\,g^{-1})$	Portal perfusion $(mL \cdot min^{-1} \cdot 100\,g^{-1})$	MTT (s)
ROI-based (reference)	32.9 ± 20.8	78.7 ± 31.7	7.5 ± 3.8
3D (presented)	40.7 ± 10.2	57.9 ± 20.6	8.9 ± 1.6

parameters values computed from the two compared methods are summarized in Table 1. Then, parameter values were stratified according to the fibrosis severity. Results corresponding to the advanced stage *(METAVIR stage ≥ F2)* and early stage *(METAVIR stage < F2)* are presented in Table 2.

3.3. Statistical Analysis. A significant correlation was observed between ROI-based method and distributed method for each parameter. Spearman's coefficients (ρ) were 0.86, 0.92, and 0.80 ($P < 0.01$) for arterial perfusion, portal perfusion and MTT, respectively. Regarding the Wilcoxon test and the Bland-Altman representations (Figure 3), significant difference was shown between compared methods. However, Bland-Altman representations showed a systematic decrease of MTTs values calculated

with distributed method compared to ROI-based reference method.

About method reproducibility, all computed relative standard variation maps were null or negligible.

3.4. Distributed Processing Performance. The major drawback of the perfusion-based method, its prohibitive computing time, has been overcome with the help of the EGI. By using the resources of distributed European infrastructure, 1 CPU year (corresponding to twenty-one 3D mapping procedures) was computed in only 9.5 days. The speedup varied among the 21 workflows from 20 to 94 with an average value of 48. The average error ratio for the experiments presented here was of 18%, with a maximum of 43% for one of the workflows. As shown in Figure 2, the late completion of the last jobs significantly increases the makespan. Most of the jobs finish within 10000 seconds. Nevertheless, a small percentage of the jobs need much more time to complete. These are typically failed jobs that need to be resubmitted one or multiple times. The workflow needs to wait for the completion of all jobs in order to produce the final result. Thus, for the experiments presented here, the makespan was almost tripled (from roughly five to fifteen hours) because of these late jobs. Similar results have been reported in other studies such as [20], where the authors propose a dynamic

TABLE 2: Mean values of perfusion parameters stratified according to fibrosis severity (advanced, METAVIR stage $\geq F2$ and not advanced, METAVIR stage $< F2$) obtained with the presented method.

Fibrosis stage		Arterial perfusion (mL·min^{-1}·100 g^{-1})	Portal Perfusion (mL·min^{-1}·100 g^{-1})	MTT (s)
stage $< F2$	3D	35.0 ± 7.1	72.9 ± 26.3	8.7 ± 1.2
	Ref	25.6 ± 6.3	92.6 ± 26.3	6.9 ± 1.1
stage $\geq F2$	3D	44.9 ± 10.9	47.6 ± 7.5	9.1 ± 2.1
	Ref	60.4 ± 12.3	68.2 ± 7.5	8.0 ± 5.2

load balancing approach in order to improve performance. Nevertheless, the dynamic load balancing approach proposed in [20] only works for Monte-Carlo-based simulations.

4. Discussion

The presented method was reproducible and results were correlated with ROI-based reference method run locally on a personal computer. The quantified parameters were found to be in the same range as those obtained with the reference ROI-based method and those related in the literature [9, 10, 12]. Nevertheless, patient-wise, quantified values were slightly modified even if the shift was not found significant. Indeed, compared to the ROI-based method, blood flow quantified with the presented 3D method is overestimated whereas; on the contrary, MTTs are underestimated. Additionally, for each parameter, standard deviation observed with 3D methods run on EGI was found lower compared to the ROI-based method. When results are globally stratified according to fibrosis severity, the difference between mean values for each parameter computed with presented method is systematically lower than with ROI-based method. These findings confirm the smoothing effect induced by 3D quantification algorithm. Indeed, in ROI-based estimation method, arterial and portal delays are optimally set by user. However, these delays depend on spatial location and take into account the time shift between the measured input functions and the position where modeling takes place in parenchyma. Hence, manual setting is not possible in the 3D case, and an automatic estimate of both delays was mandatory. Due to relatively low image signal-to-noise ratio (SNR) of about eighteen, this step requires hard smoothing filtering, affecting quantification results with the acquisition data currently available. Another limitation is the restricted exploration volume. Indeed, to keep an acceptable SNR acquired with a high temporal resolution of 1 sec, the number of encoding steps in the slice-encoding direction was limited and the whole liver volume was not covered. These restrictions (SNR and coverage) can be overcome with the latest imaging MR systems with improved acquisition capabilities using 32 receiver channels with multiple element array coils. An SNR value of 80 was measured based on preliminary test performed at our institution with a 3T GEHC MR 750 (GEHC, Milwaukee, WI, USA) with 32 ch body coil. The parallelization of the method brings significant speedup and renders it feasible despite its prohibitive computing time. Nevertheless, performance can be still significantly improved. Currently the poor scheduling of the last tasks is largely due to platform heterogeneity and multiple task resubmissions

caused by high error ratios. Data transfers account for most of the errors, while the rest are mostly application failures due to improper grid node configuration.

As future work, scheduling will be improved by taking into account these considerations.

To conclude, this preliminary study demonstrated that the described method allows 3D liver perfusion quantification within a reasonable processing time. It is now suitable to be used for similar clinical studies in a research context. While the distributed processing method was validated compared to the ROI-based quantification, such fully automatic processing requires high-quality images. The required SNR, together with a high temporal resolution and large volume exploration, can now be achieved on the latest 3T MRI systems available. Further work will have to demonstrate the interest of parametric 3D perfusion maps for fibrosis assessment on a larger number of subjects with chronic liver disease.

Acknowledgments

This work was conducted in the framework of the LabEX PRIMES. The authors thank France-Grilles and EGI for the grid infrastructure and technical support. They also thank T. Glatard, R. Silva, F. Bellet, J-P. Roux, V. Romanello-Perez, and M. Orkisz for technical support and fruitful discussions.

References

[1] N. H. Afdhal and D. Nunes, "Evaluation of liver fibrosis: a concise review," *American Journal of Gastroenterology*, vol. 99, no. 6, pp. 1160–1174, 2004.

[2] R. N. Anderson and B. L. Smith, "Deaths: leading causes for 2001," *National Vital Statistics Reports*, vol. 52, no. 9, pp. 1–85, 2003.

[3] K. Kazemi, B. Geramizadeh, S. Nikeghbalian et al., "Effect of D-penicillamine on liver fibrosis and inflammation in Wilson disease," *Experimental and Clinical Transplantation*, vol. 6, no. 4, pp. 261–263, 2008.

[4] E. J. Heathcote, M. L. Shiffman, W. G. E. Cooksley et al., "Peginterferon alfa-2a in patients with chronic hepatitis C and cirrhosis," *The New England Journal of Medicine*, vol. 343, no. 23, pp. 1673–1680, 2000.

[5] F. Piccinino, E. Sagnelli, and G. Pasquale, "Complications following percutaneous liver biopsy. A multicentre retrospective study on 68,276 biopsies," *Journal of Hepatology*, vol. 2, no. 2, pp. 165–173, 1986.

[6] J. F. Cadranel, P. Rufat, and F. Degos, "Practices of liver biopsy in France: results of a prospective nationwide survey," *Hepatology*, vol. 32, no. 3, pp. 477–481, 2000.

[7] J. Poniachik, D. E. Bernstein, K. R. Reddy et al., "The role of laparoscopy in the diagnosis of cirrhosis," *Gastrointestinal Endoscopy*, vol. 43, no. 6, pp. 568–571, 1996.

[8] Y. Billaud, O. Beuf, G. Desjeux, P. J. Valette, and F. Pilleul, "3D contrast-enhanced MR angiography of the abdominal aorta and its distal branches: interobserver agreement of radiologists in a routine examination," *Academic Radiology*, vol. 12, no. 2, pp. 155–163, 2005.

[9] R. Materne, A. M. Smith, F. Peeters et al., "Assessment of hepatic perfusion parameters with dynamic MRI," *Magnetic Resonance in Medicine*, vol. 47, no. 1, pp. 135–142, 2002.

[10] M. Hagiwara, H. Rusinek, V. S. Lee et al., "Advanced liver fibrosis: diagnosis with 3D whole-liver perfusion MR imaging— initial experience," *Radiology*, vol. 246, no. 3, pp. 926–934, 2008.

[11] B. E. Van Beers, R. Materne, L. Annet et al., "Capillarization of the sinusoids in liver fibrosis: noninvasive assessment with contrast-enhanced MRI in the rabbit," *Magnetic Resonance in Medicine*, vol. 49, no. 4, pp. 692–699, 2003.

[12] B. Leporq, J. Dumortier, F. Pilleul, and O. Beuf, "3D-liver perfusion MR imaging with the MS-325 blood pool agent: a non-invasive protocol to assess liver fibrosis," *Journal of Magnetic Resonance Imaging*, vol. 35, pp. 1380–1387, 2012.

[13] B. Leporq, O. Beuf, and F. Pilleul, "Perfusion MR imaging of the liver with a vascular contrast agent," *Journal de Radiologie*, vol. 92, no. 3, pp. 257–261, 2011.

[14] R. Lützkendorf, J. Bernarding, F. Hertel, F. Viezens, A. Thiel, and D. Krefting, "Enabling of grid based diffusion tensor imaging using a workflow implementation of FSL," *Studies in Health Technology and Informatics*, vol. 147, pp. 72–81, 2009.

[15] T. Glatard, R. S. Soleman, D. J. Veltman, A. J. Nederveen, and S. D. Olabarriaga, "Large-scale functional MRI study on a production grid," *Future Generation Computer Systems*, vol. 26, no. 4, pp. 685–692, 2010.

[16] French METAVIR Cooperative Study Group, "Intraobserver and interobserver variations in liver biopsies in patients with chronic hepatitis C," *Hepatology*, vol. 20, pp. 15–20, 1994.

[17] J. Montagnat, B. Isnard, T. Glatard, K. Maheshwari, and M. B. Fornarino, "A data-driven workflow language for grids based on array programming principles," in *Proceedings of the 4th Workshop on Workflows in Support of Large-Scale Science (WORKS '09)*, pp. 1–10, November 2009.

[18] T. Glatard, J. Montagnat, D. Lingrand, and X. Pennec, "Flexible and efficient workflow deployment of data-intensive applications on grids with moteur," *International Journal of High Performance Computing Applications*, vol. 22, no. 3, pp. 347–360, 2008.

[19] T. Glatard, C. Lartizien, B. Gibaud et al., "A virtual imaging platform for multi-modality medical image simulation," *IEEE Transactions on Medical Imaging*, vol. 32, no. 1, pp. 110–118, 2013.

[20] S. Camarasu-Pop, T. Glatard, J. T. Mościcki, H. Benoit-Cattin, and D. Sarrut, "Dynamic partitioning of GATE Monte-Carlo simulations on EGEE," *Journal of Grid Computing*, vol. 8, no. 2, pp. 241–259, 2010.

10

Use of Molecular Dynamics for the Refinement of an Electrostatic Model for the *In Silico* Design of a Polymer Antidote for the Anticoagulant Fondaparinux

Adriana Cajiao,[1] **Ezra Kwok,**[1] **Bhushan Gopaluni,**[1] **and Jayachandran N. Kizhakkedathu**[2]

[1] *University of British Columbia, Chemical and Biological Engineering, 2360 East Mall, Vancouver, BC, Canada V6T 1Z3*
[2] *University of British Columbia, Centre for Blood Research, Department of Pathology and Laboratory Medicine and Department of Chemistry, 2350 Health Sciences Mall, Vancouver, BC, Canada V6T 1Z3*

Correspondence should be addressed to Ezra Kwok; ezra@chbe.ubc.ca

Academic Editor: Jun Liao

Molecular dynamics (MD) simulations results are herein incorporated into an electrostatic model used to determine the structure of an effective polymer-based antidote to the anticoagulant fondaparinux. *In silico* data for the polymer or its cationic binding groups has not, up to now, been available, and experimental data on the structure of the polymer-fondaparinux complex is extremely limited. Consequently, the task of optimizing the polymer structure is a daunting challenge. MD simulations provided a means to gain microscopic information on the interactions of the binding groups and fondaparinux that would have otherwise been inaccessible. This was used to refine the electrostatic model and improve the quantitative model predictions of binding affinity. Once refined, the model provided guidelines to improve electrostatic forces between candidate polymers and fondaparinux in order to increase association rate constants.

1. Introduction

While anticoagulation therapy is widely used, it has certain undesirable side effects such as the potential to cause life-threatening hemorrhages. Such bleeding complications can be mitigated, in the event of an overdose of anticoagulants, by the administration of antidotes which neutralize the anticoagulants while still avoiding thrombosis [1, 2]. The most commonly used anticoagulants are heparin-derived drugs [3], which include unfractionated heparin (UFH), low molecular weight heparins (LMWHs), and the synthetic pentasaccharide derivatives fondaparinux and idraparinux [4–7]. Because of its predictable dose response, almost complete bioavailability [4, 7], increased half-life [1], and no occurrence of heparin-induced thrombocytopenia [5], fondaparinux is becoming increasingly important in clinical medicine; however, its widespread use is limited by a lack of a specific antidote. Administration of protamine, the antidote for UFH and LMWHs, does not reverse the anticoagulant effect of fondaparinux, and hemodialysis only reduces fondaparinux plasma levels by 20% [1]. Hence, the development of a clinically safe antidote for this anticoagulant has become critical [8].

Currently, only limited experimental work has been reported for the development of an antidote to fondaparinux. It has been shown that heparinase I and the recombinant factor VII (rVIIa) can partially reverse fondaparinux *in vitro*; however, these studies were limited in scope: there is no clinical data for heparinase I, and there is only one volunteer study and one clinical case for rVIIa [1]. More recently, Borgel et al. have experimentally developed antithrombin (AT) variants as potential antidotes for heparin derivatives, including fondaparinux [9, 10]. Although the first of these was shown to neutralize fondaparinux *in vitro* and *in vivo*, its production was severely limited [9]. To overcome this problem, a new chemically modified AT variant has been produced but so far it lacks critical clinical data such as pharmacokinetics, safety, and immunogenicity [10].

The experimental design of a polymeric antidote for fondaparinux is a daunting challenge due to the multitude of structures that need to be synthesized to arrive at the molecule with optimum binding. Given the large number of possible structural configurations for a polymer antidote, a traditional trial-and-error approach to the development of a novel antidote would be a very expensive, labour-intensive, and time-consuming process. Moreover, obtaining experimental data on the interactions between these polymer structures and fondaparinux is also very difficult. Computer simulations therefore provide the only feasible method to screen putative polymer structures that show promise to be effective antidotes for fondaparinux, even though a rigorous experimental validation of these *in silico* predictions is not viable due to the arduousness of polymer synthesis and characterization.

While fondaparinux has been studied *in silico* to some extent [11], the lack of experimental data for the polymer presents a challenge to the application of computational modeling techniques to antidote polymer design. Computer programs for structure-based design strategies require the use of 3D structures, which are typically generated by X-ray crystallography [12]. However, producing an X-ray crystallographic structure of the polymer candidates is difficult because they do not crystallize under normal conditions.

The aim of this work is therefore to use molecular dynamic (MD) simulations to gain a deeper insight—at a microscopic level—into the interactions between fondaparinux and individual polymer's cationic binding groups. This information will guide the selection of favourable binding groups that will promote improved binding between the polymer antidote and fondaparinux. Furthermore, the knowledge gained from these MD simulations will allow for the improvement of the electrostatic model that we have previously reported to characterize the polymer-fondaparinux complex formation but which, due to the lack of interaction data, contained binding simplifications that consequently overpredicted k_a values [13].

In the next sections, the MD simulations and the calculation of the free energy as well as the main equations of the electrostatic model is explained. Then, the selection of the most promising binding groups based on the results obtained from MD simulations and free-energy calculations is discussed. This is followed by a description of the modifications made to the previously published electrostatic model [13] and a discussion of the impact of these changes on the model's predictions.

2. Molecular Dynamics Simulations

All MD simulations in this work were performed using the commercially available software package Materials Studio 5.5 (MS). Each MD simulation system consisted of one individual fondaparinux molecule interacting with one individual cationic binding group surrounded by water molecules. For each MD simulation, the system under study was first prepared, and then its free energies were calculated. The preparation of each model system was performed on an Intel i5 2400 quad-core, 3.1 GHz computer and took approximately 44 h. The MD calculations of the free energies of the prepared

systems were then run using 48 2.66 GHz processors from the Bugaboo cluster maintained by WestGrid and Compute/Calcul Canada; these calculations took, on average, 90 h to complete for each system studied.

2.1. Binding Group and Fondaparinux 3D System Preparation. Because electrostatic interactions drive the binding between fondaparinux and the cationic binding groups, five different binding groups (Figure 1) were chosen in order to determine the effect of valency on complex formation. Based on our previous calculations [13], the R4-1 binding group structure was chosen as the basis for modification. Therefore, the range of theoretically predicted cationic charges was chosen to have values lower and higher than that calculated for R4-1 (i.e., +3). Specifically, binding groups with one, three, four, and six nitrogen, N, atoms connected by $-CH_2-CH_2-$ linkages were selected. To observe the structural impact on the binding to fondaparinux, an additional binding group was constructed based on the structure of the binding group R4-1 but consisting of $-CH_2-CH_2-CH_2$ linkages between the four N atoms.

The initial 3D atomistic structure of fondaparinux was obtained from the DrugBank database [14]. The Na^+ atoms were deleted from the original 3D atomistic structure in the Visualizer module of MS in order to assign a net charge of -10 to fondaparinux as determined at physiological conditions with the ChemAxon pKa Calculator Plugin (Marvin 5.5.5, 2011) [15]. The partial charges and force field types were assigned with the force field COMPASS using the Discover module. COMPASS is an *ab initio* force field designed for use with a broad range of organic and inorganic molecules and polymers [16] and is optimized for the simulation of condensed phases [17]. As with all current force fields, COMPASS does not properly describe heparin [18, 19], in particular some of the sulfonamide functional groups found within heparin derivatives such as fondaparinux [20]. Therefore, the force field types and charges assigned by COMPASS to the three sulfonamide oxy anions were modified to match those of the undissociated analog and the sulfonylmethoxy oxy anion, respectively. The charges were then adjusted based on a desired net charge of -10 for fondaparinux. Lastly, the structure of fondaparinux was minimized after 575 iterations in the Forcite module of MS using the force field COMPASS. Similarly, the 3D atomistic structures of the different binding groups were sketched using MarvinSketch from ChemAxon (Marvin 5.5.5, 2011) [15]; their protonation state was calculated with ChemAxon pKa Calculator Plugin (Marvin 5.5.5, 2011) [15], and their force field types and partial charges were assigned with the Discover module using the force field COMPASS. The structures were then minimized in the Forcite module of MS using COMPASS.

Each model system was constructed using the Amorphous Cell module of MS and consisted of one deprotonated fondaparinux molecule and one protonated binding group randomly dispersed in 2,600 water molecules. Appropriate amounts of sodium counterions (Na^+) were added to achieve charge neutrality. A cubic simulation box was constructed with periodic boundary conditions in all directions to avoid

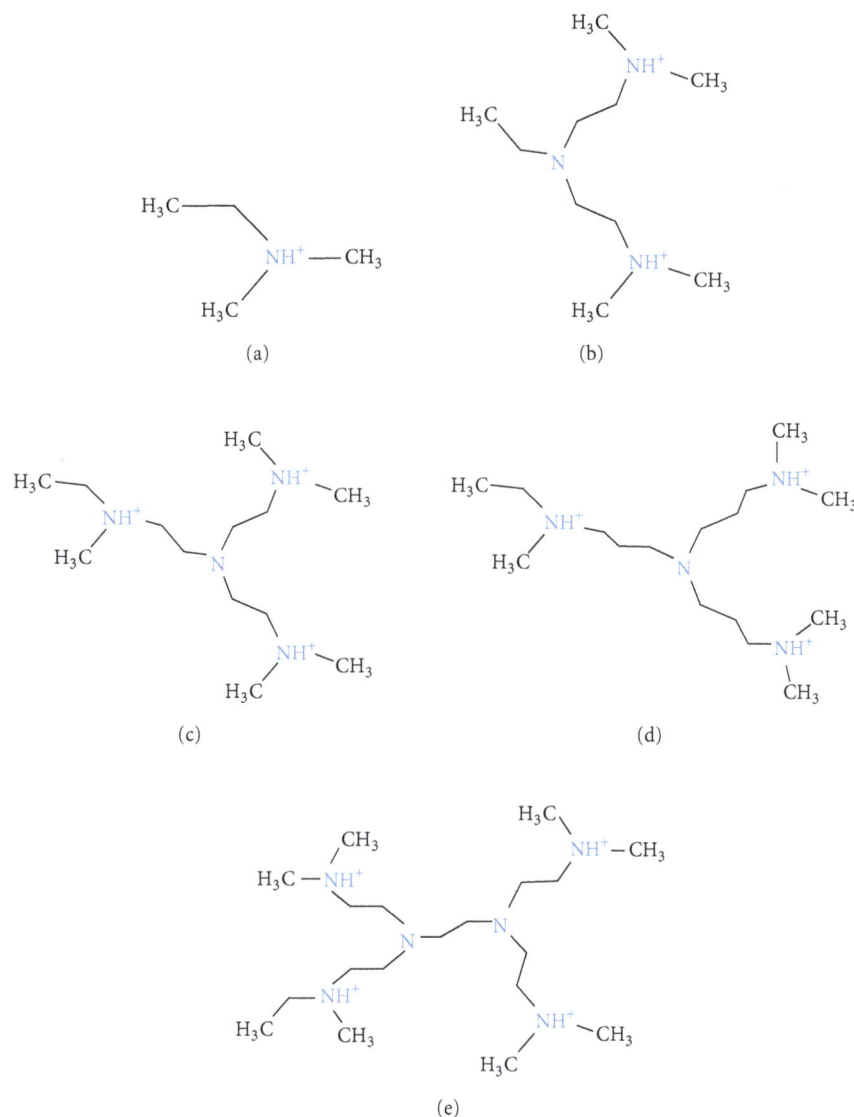

FIGURE 1: Structures of the various amines used as binding groups: (a) R1-1; (b) R3-1; (c) R4-1; (d) R4-2; and, (e) R6-1. The nitrogen atoms are connected by $-CH_2-CH_2-$ linkages in (b), (c), and (e) and by $-CH_2-CH_2-CH_2$ linkage in (d). The protonation state is for physiological pH of 7.4 and was calculated with ChemAxon pKa Calculator Plugin (Marvin 5.5.5, 2011) [15].

surface effects [21, 22], a density of $1\,g\,cm^{-3}$ since the system consists mainly of water molecules, and a side length of approximately 43 Å. A distance object was created between the center of mass (centroid) of fondaparinux and the centroid of the binding group. Then, a harmonic restraint with a harmonic force constant of $100\,kcal\,mol^{-1}\,Å^2$ and a harmonic minimum of 21.65 Å was applied to this distance. Energy minimization and MD simulations were performed to equilibrate the system using the force field COMPASS in the Forcite module of MS. The MD simulations were carried out under NVT conditions, with temperature held at 298 K by the Nose-Hoover thermostat [16, 22–24]. The time step was 1 fs, and the simulation time was 200 ps. This simulation time proved to be sufficient to obtain equilibrium conditions, namely, the potential energy and temperature of the system as shown in Figure 2 for the representative system of fondaparinux and an R4-1 molecule. Once the model

system was relaxed, the restraint on the centroid-centroid distance was removed from the system.

2.2. Free-Energy Calculation. Using restraint forces, the intermolecular separation of fondaparinux and a binding group was sampled at equal intervals of 0.5 Å [25] from an initial separation of 21.5 Å to a final separation of 1.5 Å. At each interval, a commercially available code distributed by Accelrys was used to solve the following equation [26]:

$$F_R\left(R'\right) = -\int_{R'_o}^{R'} \left\langle K^r\left(R\left(r\right) - R_o\right)\right\rangle_{r,R''} dR''$$
$$- 2k_B T \ln\left(\frac{R'}{R'_o}\right) + F_R\left(R'_o\right),$$

(1)

where K^r is the harmonic force constant that enhances the restraint of the system to the R-coordinate value R_o. The

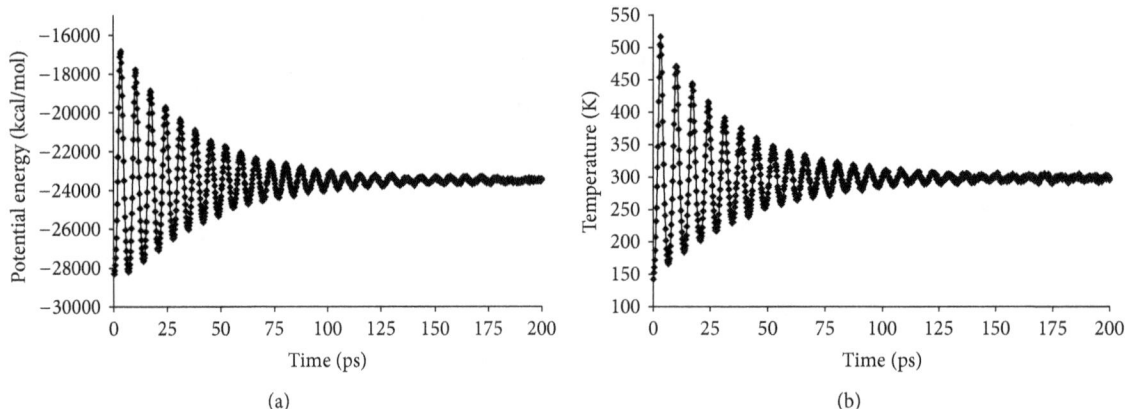

FIGURE 2: Time evolution of (a) the potential energy (kcal mol^{-1}) and (b) the temperature (K) for a model system consisting of 1 R4-1 molecule, 1 fondaparinux molecule, 2,600 water molecules, and 7 Na$^+$ atoms.

reaction coordinate R for each system was defined as the distance between the centroid of the fondaparinux molecule and the centroid of the binding group. A harmonic restraint constant of 100 kcal mol^{-1} Å^2 was applied over this distance. The simulation time at each interval was 60 ps, which consisted of 10 ps of equilibration followed by 50 ps of trajectory generation [27] for a total simulation time of 2.5 ns. The equilibration time was found to be sufficient for properties to equilibrate at each interval. The same NVT conditions, temperature, and time step previously described were used for the free-energy calculations, and each simulation was repeated 5, 4, 7, 8, and 9 times for R1-1, R3-1, R4-1, R4-2, and R6-1, respectively.

3. Electrostatic Model for Polymer Antidote and Fondaparinux

We have previously described an electrostatic model that provided an indication of the binding affinity between fondaparinux and a polymer structure with R4-1 groups [13]. Since the association rate constant, k_a, is determined by diffusion and can be increased by favourable electrostatic forces [28–31] whereas the dissociation rate constant, k_d, is determined by short-range interactions between the molecules and is independent of long-range electrostatic forces [28, 32], the overall association constant, K_a, and thus the affinity of a complex can be increased by optimizing the electrostatic interactions between the molecules [28]. Our model predicts k_a based on these electrostatic interactions and on the following equations, first derived by Schreiber and coworkers [28, 33, 34]:

$$\ln k_a = \ln k_a^o - \frac{\Delta U}{k_B T}\left(\frac{1}{1+ka}\right), \qquad (2)$$

where k_a and k_a^o are the association rate constants in the presence and absence of long-range electrostatic forces, respectively; ΔU is the electrostatic energy of interaction; k_B is the Boltzmann constant; T is the temperature of the solution; a is the minimal distance of approach between the

molecules; and k is the Debye-Hückel parameter. The Debye-Hückel parameter is defined as [35]

$$k = \sqrt{\frac{2F^2 I}{\epsilon_o \epsilon_r RT}}, \qquad (3)$$

where F is the Faraday constant, I is the ionic strength of the solution, ϵ_o is the vacuum permittivity, ϵ_r is the dielectric constant of the solution, and R is the gas constant. The electrostatic energy of interaction is defined as [28, 32]

$$\Delta U = U_{\text{complex}} - U_{\text{molecule A}} - U_{\text{molecule B}}, \qquad (4)$$

where U, the Debye-Hückel energy of a molecule, can be calculated from

$$U = \frac{1}{2}\sum_{i,j} \frac{q_i q_j}{4\pi\epsilon_o \epsilon_r r_{ij}} \frac{e^{-k(r_{ij}-a)}}{1+ka}. \qquad (5)$$

In this equation, q_i and q_j are the charges of the atoms in the molecules, and r is the distance between the charges.

As we have shown previously [13], we extended the empirically proven model to determine k_a for the interactions of fondaparinux and a polymer structure with R4-1 groups. This model considers the polymer and fondaparinux, due to their structures, as a sphere and a rod, respectively. Based on the Smoluchowski limit for the diffusion-controlled association of two uniformly reactive molecules with these geometries, k_a^o can be calculated as [29]

$$k_a^o = 4\pi N_A (D_A + D_B) R_x. \qquad (6)$$

Here, N_A is the Avogadro constant, D_A and D_B are the diffusion constants of molecules A and B, respectively, and R_x is the interaction radius. R_x is defined as

$$R_x = \frac{l}{\ln(2l/w)}, \qquad (7)$$

where l and w are the major and minor semiaxes of the ellipsoid. The diffusion constants are calculated as

$$D_A = \frac{k_B T}{6\pi\eta r_A'}, \qquad D_B = \frac{k_B T}{6\pi\eta r_B'}, \qquad (8)$$

Use of Molecular Dynamics for the Refinement of an Electrostatic Model for the In Silico Design of a Polymer
Antidote for the Anticoagulant Fondaparinux

97

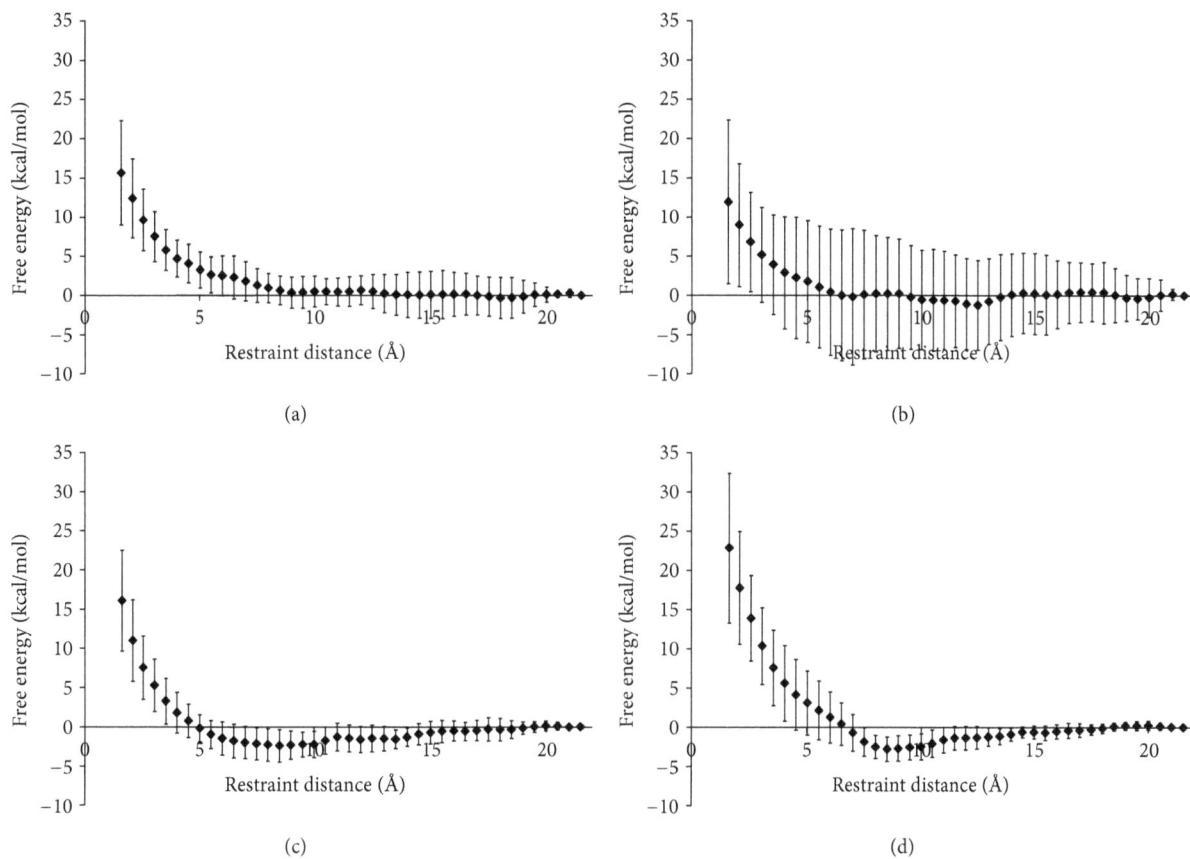

FIGURE 3: Free-energy profile (PMF) for the interaction of fondaparinux with (a) R1-1, ($n = 5$); (b) R3-1, ($n = 4$); (c) R4-1, ($n = 7$); and (d) R6-1, ($n = 9$). Calculated values are the Helmholtz free energy which approximates the Gibbs free energy for systems in the condensed phase [39, 40]. Simulations were performed using a step size of 0.5 Å and a simulation time of 60 ps at each interval. The error bars represent the 95% confidence intervals for n number of replicates.

where r_A' and r_B' are the hydrodynamic radii and η is the viscosity of the solvent.

For modeling purposes, the interaction between the polymer and fondaparinux was assumed to be entirely electrostatic in nature [13]. Also, to simulate the functionality of the polymer, cationic binding groups on the polymer were modeled as randomly distributed on the surface of a sphere with a given radius, r_H, and a minimum distance between the binding groups, r_{min}, was introduced. The purpose of this r_{min} was to account for electrostatic repulsions between the binding groups and to avoid placing these binding groups at the same location, which is physically impossible [13]. In order to represent the solution conditions used in experiments, the following parameters were defined: $I = 150$ mM, $T = 25°$C, $\eta = 0.90 \times 10^{-3}$ kgs^{-1} m^{-1}, and $\epsilon_r = 80$.

To capture the average properties of a large group of individual molecules, each simulation for a given condition consisted of constructing 1,000 unique polymers by randomly attaching the desired number of binding groups over the surface of the polymer. The model equations were then solved for each of the 1,000 polymers, and the average association rate constant was given by the geometric mean of these 1,000 runs. In the case of no polymer-fondaparinux binding, k_a was set to a value of 1.

4. Results and Discussions

4.1. MD Interaction between Fondaparinux and Candidate Binding Groups.
MD simulations were performed to calculate the potential of mean force (PMF)—the free-energy profile along the reaction coordinate [36] that yields the difference in free energy between the two states of interest [26]. These free-energy differences ($\sim\Delta G$) between the unbound and bound state of the cationic groups are directly related to binding constants [26].

4.1.1. Effect of Cationic Charge per Binding Group on Free Energy.
In order to investigate the effect of the binding group's charge on fondaparinux binding, MD simulations were run to follow the interaction of fondaparinux and each of the cationic binding groups (Figure 1) in a solution of Na$^+$ ions. As shown in their respective free-energy profiles, both of the cationic binding groups R1-1 and R3-1 did not have distinctive energy minima (Figures 3(a) and 3(b)). The lack of free-energy wells indicates weak binding of both R1-1 and R3-1 to fondaparinux. The poor binding of R1-1 was to be expected since it has been shown that the electrostatic interactions of a single protonated amine with a polyanionic molecule are weak and have to compete with salt binding

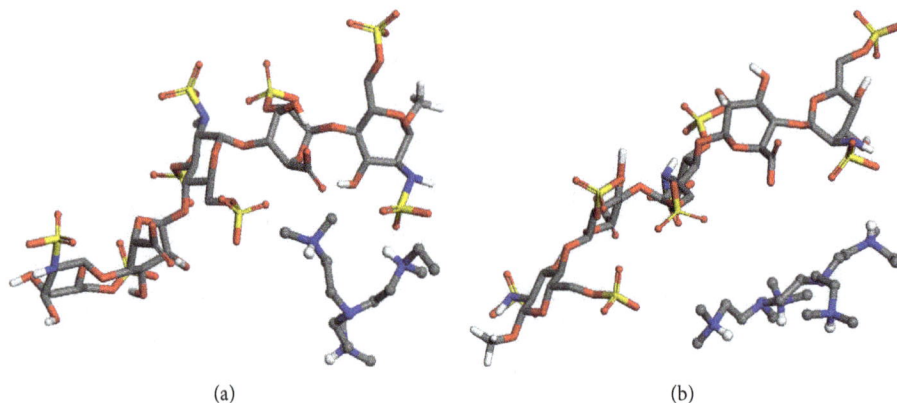

FIGURE 4: Snapshot of the complexes: (a) R4-1 (ball and stick) and fondaparinux (stick) in a model system consisting of 1 R4-1 molecule, 1 fondaparinux molecule, 2,600 water molecules, and 7 Na$^+$ atoms; and (b) R6-1 (ball and stick) and fondaparinux (stick) in a model system consisting of 1 R6-1 molecule, 1 fondaparinux molecule, 2,600 water molecules, and 6 Na$^+$ atoms. In both cases the centroid-centroid distance is 8.5 Å. Water molecules and Na$^+$ atoms are deleted for clarity.

under physiological conditions [37]. R3-1 has an increased charge compared to R1-1 and would therefore be assumed to display improved binding; however, geometry and chemical structure also play a role in complex formation. Therefore, the MD results suggest that R3-1 shows some binding to fondaparinux although it is not sufficient to overcome unfavourable orientations, thus the large degree of variability seen in the PMF for R3-1 compared to R1-1.

Considering the aforementioned weak electrostatic interactions between single protonated amines and polyanionic molecules [37], it is not surprising that the PMF of both R4-1 and R6-1 displayed deeper free-energy wells than the lesser charged R1-1 and R3-1 (Figures 3(c) and 3(d)). In addition, the energy wells of both R4-1 and R6-1 were wide, spanning for 10 Å and 7 Å, respectively, and plateauing at a centroid-centroid distance of approximately 17.5 Å in both cases. This indicates that interactions between R4-1 and R6-1 and fondaparinux were much stronger and therefore felt over a larger distance than those seen in R3-1. Moreover, the variabilities in the PMFs of R4-1 and R6-1 were reduced compared to that of R3-1 which indicates that the electrostatic interactions between fondaparinux and the higher charged binding groups were strong enough to overcome unfavourable orientations.

Since binding affinity can be improved by increasing the cationic charges within the binding group, R6-1 could be expected to show a more favourable energy of interaction with fondaparinux than R4-1. However, both of the binding groups had their local minima at a centroid-centroid distance of 8.5 Å, and both PMF profiles yielded comparable calculated ΔG values of -2.394 kcal mol^{-1} for R4-1 and -2.768 kcal mol^{-1} for R6-1. These results suggest that there is not a significant difference between the binding of individual R4-1 and R6-1 binding groups to fondaparinux. However, because of its additional cationic charge it could be hypothesized that when many R6-1 binding groups are working in concert on the surface of the polymer, the small improvement in binding affinity they show compared to R4-1 will be amplified and will provide stronger electrostatic

interactions with fondaparinux. This is investigated further below with the electrostatic model. The complexes formed by each of these binding groups and fondaparinux can be observed in Figure 4.

4.1.2. Effect of Binding Group Structure on Free Energy. An alternate method of improving binding affinity between binding groups and fondaparinux is to change the spacing of the cationic charges within the binding groups [38]. Shortening the linkage between the N atoms in R4-1 from –CH$_2$–CH$_2$– to –CH$_2$– resulted in the protonation of only one of the amines at pH 7.4 (as calculated with ChemAxon pKa Calculator Plugin (Marvin 5.5.5, 2011) [15]), and therefore, this molecule was deemed unsuitable for further study. Conversely, the theoretical charge of R4-1 was maintained at +3 when the linkage between the N atoms was lengthened from –CH$_2$–CH$_2$– (3.84 Å) to –CH$_2$–CH$_2$–CH$_2$– (4.97 Å) to form the R4-2 binding group (Figure 1).

The free-energy profile for R4-2 (Figure 5) displayed a shallow and poorly defined well. Comparing the free-energy profiles of R4-1 (Figure 3(c)) and R4-2 suggests that increasing the spacing of cationic charges for this binding group will not generate an improvement in the binding affinity to fondaparinux. In fact such a structural change is shown to inhibit fondaparinux binding. Therefore, the binding groups R4-1 and R6-1 are considered the most promising binding groups for the effective neutralization of fondaparinux and will be investigated using the electrostatic model.

4.2. Refinement of Electrostatic Model to Optimize Polymer Antidote Structure. The electrostatic model first developed in our previous work [13] is herein refined based on the microscopic information gathered from the MD simulations described above. In the simplified model, the minimal distance of approach, a, was set to 6 Å since this value had been shown to give the best fit for experimental data collected for a variety of protein systems [28]. However, the free-energy profiles of the interaction of binding groups R4-1 and

Use of Molecular Dynamics for the Refinement of an Electrostatic Model for the In Silico Design of a Polymer
Antidote for the Anticoagulant Fondaparinux

99

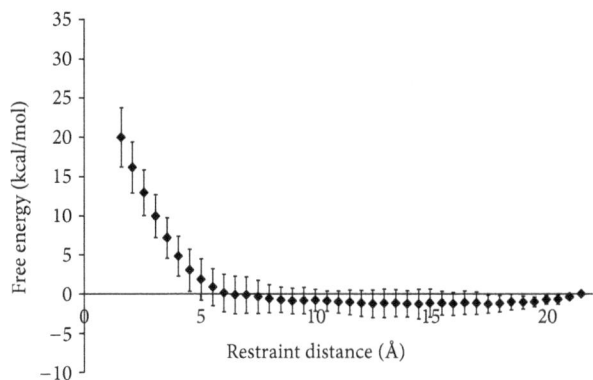

FIGURE 5: Free-energy profile (PMF) for the interaction of fonda-parinux with R4-2 ($n = 8$). Calculated values are the Helmholtz free energy which approximates the Gibbs free energy for systems in the condensed phase [39, 40]. Simulations were performed using a step size of 0.5 Å and a simulation time of 60 ps at each interval. The error bars represent the 95% confidence intervals for n number of replicates.

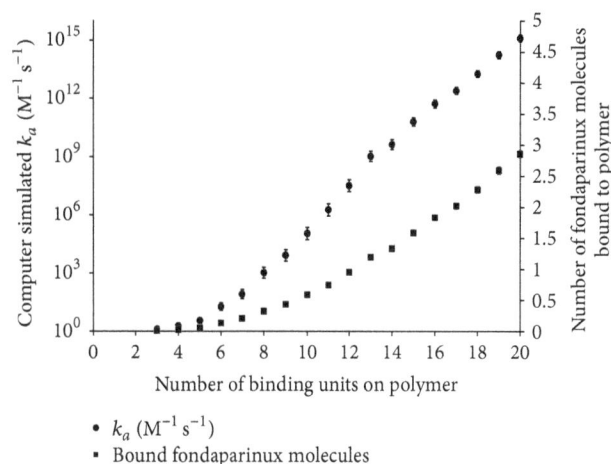

- k_a ($M^{-1}\,s^{-1}$)
- Bound fondaparinux molecules

FIGURE 6: Computer simulated k_a and number of fondaparinux molecules bound per polymer for polymers with 3 to 20 attached R4-1 binding groups. The HBSPCMs had an r_H of 4.0 nm and an r_{min} of 9.16 Å. k_a and number of fondaparinux molecules bound to a polymer are the geometric and arithmetic means, respectively, of 1,000 calculated values. The error bars represent the 95% confidence intervals.

R6-1 to fondaparinux (Figures 3(c) and 3(d)) showed that the centroid-centroid distance between a binding group and fondaparinux was 8.5 Å at binding. Since the refined model focused on these binding groups, the value for a was changed to 8.5 Å.

The binding criterion used in the simplified model was based on the direct contact of fondaparinux with each of the three closely spaced binding groups that formed a binding site [13]. However, the MD simulation results suggest that during the binding of fondaparinux to either R4-1 or R6-1 the ionic sites of fondaparinux remain at a distance from the protonated N atoms of the binding groups. In addition, it was found that the molecules in a bound complex show significant and constant relative motion to one another. Therefore, the model was refined to incorporate a more realistic binding criterion. A binding site was redefined to be three or more binding groups that are within an area that allows them to come within a prescribed distance (8.5 Å) of a fondaparinux molecule, centered around one of the binding groups. Upon binding, a single charge was then assigned to this binding site (the sum of the charges of fondaparinux and the associated binding groups) at the location of the first binding group that formed the binding site.

In order to account for bridging effects, excluding associated structural modifications that might occur, the model was altered to consider system neutralization rather than fondaparinux neutralization by allowing fondaparinux molecules to bind without the requirement for almost complete neutralization of a bound fondaparinux molecule. The fondaparinux charges that are not involved in binding to the generated polymer are therefore available for binding to another polymer in solution.

The minimum distance between binding groups, r_{min}, for the refined model was determined as described in our previous publication [13]. The new value of r_{min} was found to be 9.16 Å, which is larger than the radius of R4-1 (6.61 Å) and R6-1 (7.10 Å). Since the purpose of introducing an r_{min} was

to account for electrostatic repulsions and steric interactions between binding groups, this new r_{min} is a more realistic constraint than the previously reported value [13], which was smaller than the radius of R4-1.

4.2.1. Effect of Number of Binding Groups on k_a and Number of Molecules of Fondaparinux Bound per Polymer.
The metric for binding affinity used in this work is k_a since, as previously described, an increase in k_a would increase K_a. With the refined model, k_a and the number of fondaparinux bound per polymer increased with the number of attached R4-1 binding groups in similar fashions to what was observed using the simplified model (Figure 6). However, the range for k_a was reduced by 6 orders of magnitude with the refined model compared to the results obtained with the simplified form. These results indicate that although the simplified form of the model predicts overall trends in k_a and thus represents an effective but rough tool for antidote discovery, the refined model with the aid of MD simulation results yields more quantitatively accurate information.

4.2.2. Effect of HBSPCM Size on k_a and Number of Molecules of Fondaparinux Bound per Polymer.
The effect of polymer size and number of R4-1 binding groups on the number of fondaparinux molecules bound per polymer and k_a was investigated with the improved binding model (Figures 7 and 8). As with the simplified model, it was observed that as the surface area of the polymer core increases, the probability of finding binding groups sufficiently close to each other to form a binding site decreases. The improvements to the model have a greater impact for smaller polymers and as shown in Figure 7, the refined model does not predict a linear increase in the number of fondaparinux molecules bound per polymer with a radius of 2 nm as was the case for the simplified model.

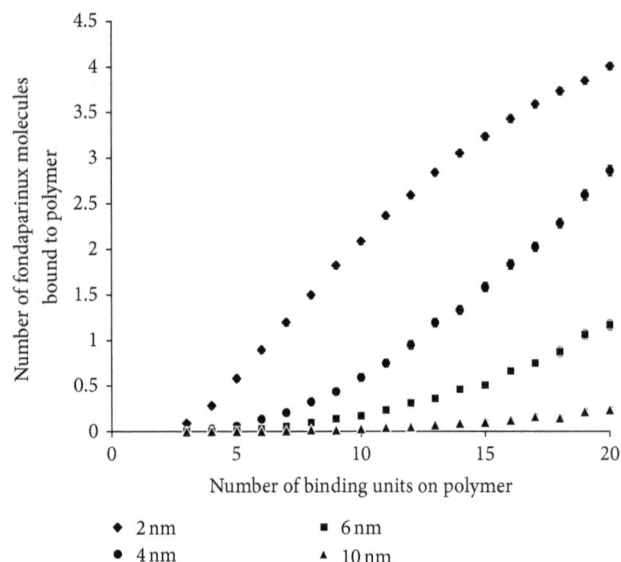

FIGURE 7: Computer simulated number of fondaparinux molecules bound per polymer for polymers with radii, r_H, of 2.0 nm, 4.0 nm, 6.0 nm, and 10.0 nm. All different sized polymers were tested with a number of R4-1 binding groups ranging from 3 to 20 using an r_{min} of 9.16 Å. The number of fondaparinux molecules bound to a polymer is the arithmetic mean of 1,000 calculated values. The error bars represent the 95% confidence interval.

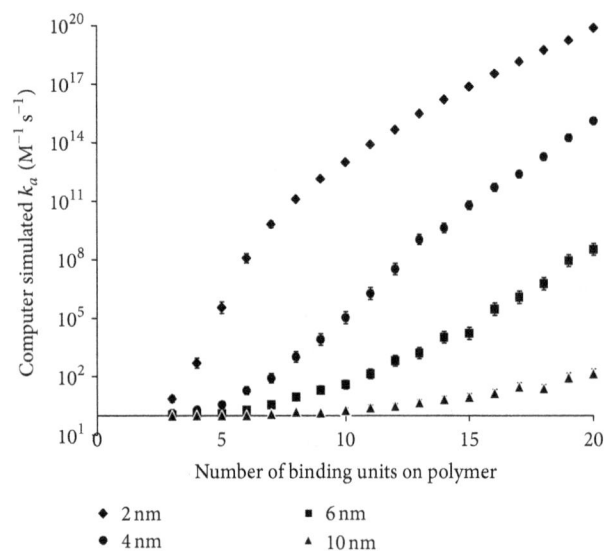

FIGURE 8: Computer simulated k_a for polymers with radii, r_H, of 2.0 nm, 4.0 nm, 6.0 nm, and 10.0 nm. All different sized polymers were tested with a number of R4-1 binding groups ranging from 3 to 20 using an r_{min} of 9.16 Å. k_a is the geometric mean of 1,000 calculated values. The error bars represent the 95% confidence interval.

Instead, the number of fondaparinux molecules bound per polymer starts to plateau at a high number of binding groups. Similar trends were observed for k_a. These results are to be expected as the number of fondaparinux bound to a polymer of a given surface area will eventually reach a maximum due to space limitations.

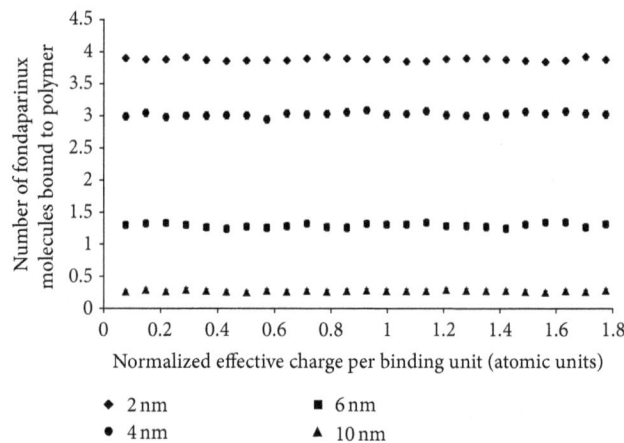

FIGURE 9: Computer simulated number of fondaparinux molecules bound per polymer for polymers with radii, r_H, of 2.0 nm, 4.0 nm, 6.0 nm, and 10.0 nm. All different sized polymers were tested using an r_{min} of 9.16 Å with 20 R6-1 binding groups with different effective charges. The number of fondaparinux molecules bound to a polymer is the arithmetic mean of 1,000 calculated values. The error bars represent the 95% confidence interval. The effective charges shown were normalized against the undisclosed effective charge of the R4-1 binding groups.

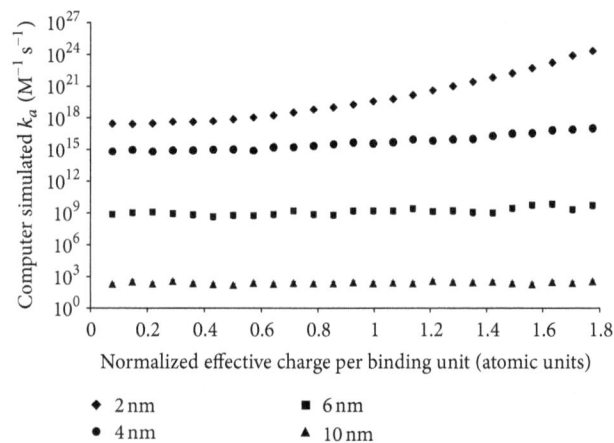

FIGURE 10: Computer simulated k_a for polymers with radii, r_H, of 2.0 nm, 4.0 nm, 6.0 nm, and 10.0 nm. All different sized polymers were tested using an r_{min} of 9.16 Å with 20 R6-1 binding groups with different effective charges. k_a is the geometric mean of 1,000 calculated values. The error bars represent the 95% confidence interval. The effective charges shown were normalized against the undisclosed effective charge of the R4-1 binding groups.

4.2.3. Impact of Binding Group Effective Charge on k_a and Number of Molecules of Fondaparinux Bound per Polymer. Although the effective charge for R4-1 was known *a priori*, it was not known for the R6-1 binding groups. Therefore, the impact of the effective charge of attached R6-1 binding groups on k_a and the number of molecules of fondaparinux bound per polymer was also investigated *in silico*. For the purposes of this investigation, polymers with 20 R6-1 binding groups on their surface were used because charge effects are more easily observed at a high number of binding groups. As seen

Use of Molecular Dynamics for the Refinement of an Electrostatic Model for the In Silico Design of a Polymer
Antidote for the Anticoagulant Fondaparinux

101

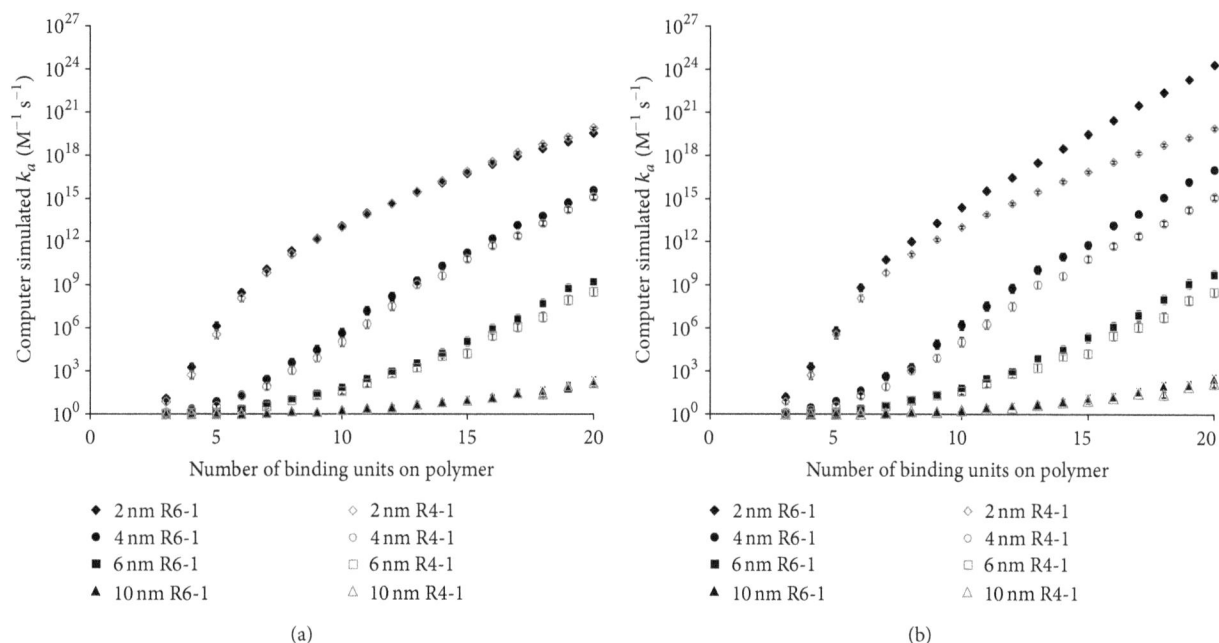

FIGURE 11: Comparison of computer simulated k_a values for polymers consisting of R4-1 binding groups and R6-1 binding groups with effective charges of (a) a value equivalent to the undisclosed effective charge of the R4-1 binding groups and (b) a value +1 greater than that of R4-1. Polymers of radii, r_H, of 2.0 nm, 4.0 nm, 6.0 nm, and 10.0 nm were tested with the number of binding groups ranging from 3 to 10 and using an r_{min} of 9.16 Å. k_a is the geometric mean of 1,000 calculated values. The error bars represent the 95% confidence interval.

in Figure 9, the number of fondaparinux molecules bound per polymer was not dependent on the effective charge of the binding groups. This was to be expected as the charge of the binding group does not affect the model's placement of binding groups on the surface of the polymer nor the binding criterion and, thus, will have no impact on an individual fondaparinux molecule finding an appropriate binding site.

Conversely, it was expected that k_a would increase as the effective charge of the binding groups was increased (Figure 10) since k_a is dependent on electrostatic forces and can be increased by increasing the magnitude of these forces [28, 31, 32]. The results show that the increase in k_a with effective charge was most pronounced at an r_H of 2.0 nm and that for polymers with larger radii the effective charge did not have a large impact on k_a. This can be explained by the fact that electrostatic energy is inversely proportional to the distance between two charges. As the binding groups become spaced farther apart with increasing polymer radius, the impact of any increase in effective charge becomes dampened.

4.2.4. Effect of Binding Groups on k_a and Number of Molecules of Fondaparinux Bound per Polymer. The improved model can also be used to determine the effect of replacing R4-1 binding groups with R6-1. As mentioned above, the effective charge of R6-1 is unknown; therefore, model predictions for k_a were made over a range of R6-1 effective charges (Figures 11(a) and 11(b)). When R6-1 has the same effective charge as R4-1 (Figure 11(a)), the model predicts no difference in the calculated values of k_a for R4-1 and R6-1. This indicates that the slightly larger radius of R6-1 does not impact the formation of binding sites to a degree that an effect in the

affinity of the polymer to fondaparinux is observed. However, it is likely that the effective charge of the R6-1 binding groups would be higher than that of R4-1 since the slightly larger R6-1 binding groups would most likely experience less charge shielding associated with the polymer's core. The effect of having R6-1 binding groups with higher effective charge is an increase in k_a values, especially for polymers with an r_H of 2.0 nm and 4.0 nm (Figure 11(b)). This result suggests that the binding of the polymer to fondaparinux can be improved by using R6-1 binding groups on small polymer cores. This is consistent with the results of the simplified model, which recommended the synthesis of polymers with low hydrodynamic size in order to increase the charge density of the polymer and, therefore, enhance k_a [13].

5. Conclusions

MD simulations have enabled the refinement of a model that has previously been shown to qualitatively but not quantitatively characterize the binding of fondaparinux to a polymer-based antidote molecule. In particular, the free-energy differences calculated from MD simulations have been used to determine those potential binding groups that will not bind to fondaparinux with high affinity and those that will provide strong binding. The MD simulation results suggest that both R4-1 and R6-1 binding groups show high affinity for fondaparinux.

The refined electrostatic model was also extended to polymers with R6-1 binding groups. It was found that if, as would be expected, the R6-1 binding groups have a higher effective charge than the R4-1 binding groups, improved binding affinity between the polymer and fondaparinux can be achieved. The results of this work therefore indicate that increasing the charge density of the polymer with favourable binding groups will improve complex formation between the polymer and fondaparinux. Therefore, it is recommended to synthesize small polymers containing as many R6-1 binding groups as possible.

The strengths of the model, both in its simplified and refined forms, are twofold: (a) minimal experimental data is required—data that is difficult to obtain for these polymers—and (b) the polymer structural adjustments can be determined *in silico* thus avoiding the costly and time-intensive synthesis of new polymers. In fact, because the synthesis and testing of even one of these polymers is such an arduous task, the model represents the only viable method for comprehensive antidote candidate screening; even without unfeasible, rigorous experimental validation. The research presented in this work is thus a major contribution to the process of finding an antidote for fondaparinux and, therefore, will greatly impact the therapeutic field.

Acknowledgments

Adriana Cajiao acknowledges the WestGrid Bugaboo facility for their technical assistance and their computer resources to run all of the molecular dynamics simulations. Jayachandran N. Kizhakkedathu acknowledges the funding from Canadian Institutes of Health Research (CIHR) and Natural Science and Engineering Council of Canada (NSERC). Jayachandran N. Kizhakkedathu is a recipient of CIHR/Canadian Blood Services New Investigator in Transfusion Science and a Career Investigator Scholar award from Michael Smith Foundation for Health Research (MSFHR). The authors declare no competing financial conflict of interests.

References

[1] S. Schulman and N. R. Bijsterveld, "Anticoagulants and their reversal," *Transfusion Medicine Reviews*, vol. 21, no. 1, pp. 37–48, 2007.

[2] J. H. Levy and K. A. Tanaka, "The anticoagulated patient: strategies for effective blood loss management," *Surgery*, vol. 142, no. 4, pp. S71–77, 2007.

[3] I. Capila and R. J. Lindhardt, "Heparin-protein interactions," *Angewandte Chemie International Edition*, vol. 41, no. 3, pp. 390–412, 2002.

[4] J. I. Weitz, "Emerging anticoagulants for the treatment of venous thromboembolism," *Thrombosis and Haemostasis*, vol. 96, no. 3, pp. 274–284, 2006.

[5] J. I. Weitz and L.-A. Linkins, "Beyond heparin and warfarin: the new generation of anticoagulants," *Expert Opinion on Investigational Drugs*, vol. 16, no. 3, pp. 271–282, 2007.

[6] L.-A. Linkins and J. I. Weitz, "New anticoagulants," *Seminars in Thrombosis and Hemostasis*, vol. 29, no. 6, pp. 619–631, 2003.

[7] S. Alban, "From heparins to factor Xa inhibitors and beyond," *European Journal of Clinical Investigation*, vol. 35, no. 1, pp. 12–20, 2005.

[8] B. K. Adler, "Unfractionated heparin and other antithrombin mediated anticoagulants," *Clinical Laboratory Science*, vol. 17, no. 2, pp. 113–117, 2004.

[9] E. P. Bianchini, J. Fazavana, V. Picard, and D. Borgel, "Development of a recombinant antithrombin variant as a potent antidote to fondaparinux and other heparin derivatives," *Blood*, vol. 117, no. 6, pp. 2054–2060, 2011.

[10] J. Fazavana, S. F. Bianchini, E. P. C. Smadja, V. Picard, M. Taverna, and D. Borgel, "A chemically-modified inactive antithrombin as a potent antagonist of fondaparinux and heparins anticoagulant activity," *Journal of Thrombosis and Haemostasis*, 2013.

[11] M. Remko, "Molecular structure, lipophilicity, solubility, absorption, and polar surface area of novel anticoagulant agents," *Journal of Molecular Structure*, vol. 916, no. 1–3, pp. 76–85, 2009.

[12] G. Scapin, "Structural biology and drug discovery," *Current Pharmaceutical Design*, vol. 12, no. 17, pp. 2087–2097, 2006.

[13] A. Cajiao, B. Gopaluni, E. Kwok, and J. N. Kizhakkedathu, "*In silico* design of polymeric antidote for anticoagulant fondaparinux," *Journal of Medical and Biological Engineering*, vol. 31, no. 2, pp. 129–134, 2011.

[14] DrugBank, "Fondaparinux sodium," 2010, http://www.drugbank.ca/drugs/DB00569.

[15] ChemAxon. Marvin 5. 5. 5, 2011, http://www.chemaxon.com.

[16] Accelrys Software, Materials Studio, Online Help, Release 5. 5, 5. 5 edition, San Diego, Calif, USA, 2010.

[17] H. Sun, "COMPASS: an ab initio force-field optimized for condensed-phase applications: overview with details on alkane and benzene compounds," *Journal of Physical Chemistry B*, vol. 102, no. 38, pp. 7338–7364, 1998.

[18] M. Remko and C.-W. von der Lieth, "Conformational structure of some trimeric and pentameric structural units of heparin," *Journal of Physical Chemistry A*, vol. 111, no. 51, pp. 13484–13491, 2007.

[19] M. Remko, M. Swart, and F. M. Bickelhaupt, "Conformational behavior of basic monomeric building units of glycosaminoglycans: isolated systems and solvent effect," *Journal of Physical Chemistry B*, vol. 111, no. 9, pp. 2313–2321, 2007.

[20] R. Tyler-Cross, M. Sobel, L. E. McAdory, and R. B. Harris, "Structure-function relations of antithrombin III-heparin interactions as assessed by biophysical and biological assays and molecular modeling of peptide-pentasaccharide-docked complexes," *Archives of Biochemistry and Biophysics*, vol. 334, no. 2, pp. 206–213, 1996.

[21] M. P. Allen and D. J. Tildesley, *Computer Simulation of Liquids*, Oxford University Press, 1991.

[22] S.-T. Lin, P. K. Maiti, and W. A. Goddard III, "Dynamics and thermodynamics of water in PAMAM dendrimers at subnanosecond time scales," *Journal of Physical Chemistry B*, vol. 109, no. 18, pp. 8663–8672, 2005.

[23] P. K. Maiti, T. Çağin, S.-T. Lin, and W. A. Goddard III, "Effect of solvent and pH on the structure of PAMAM dendrimers," *Macromolecules*, vol. 38, no. 3, pp. 979–991, 2005.

[24] Y. Cui, "Parallel stacking of caffeine with riboflavin in aqueous solutions: the potential mechanism for hydrotropic solubilization of riboflavin," *International Journal of Pharmaceutics*, vol. 397, no. 1-2, pp. 36–43, 2010.

[25] T. Baştuğ and S. Kuyucak, "Free energy simulations of single and double ion occupancy in gramicidin A," *Journal of Chemical Physics*, vol. 126, no. 10, Article ID 105103, 2007.

[26] D. Trzesniak, A.-P. E. Kunz, and W. F. van Gunsteren, "A comparison of methods to compute the potential of mean force," *ChemPhysChem*, vol. 8, no. 1, pp. 162–169, 2007.

[27] T. W. Allen, T. Baştuğ, S. Kuyucak, and S.-H. Chung, "Gramicidin A channel as a test ground for molecular dynamics force fields," *Biophysical Journal*, vol. 84, no. 4, pp. 2159–2168, 2003.

[28] T. Selzer, S. Albeck, and G. Schreiber, "Rational design of faster associating and tighter binding protein complexes," *Nature Structural Biology*, vol. 7, no. 7, pp. 537–541, 2000.

[29] O. G. Berg and P. H. von Hippel, "Diffusion-controlled macromolecular interactions," *Annual Review of Biophysics and Biophysical Chemistry*, vol. 14, pp. 131–160, 1985.

[30] G. Schreiber and A. R. Fersht, "Rapid, electrostatically assisted association of proteins," *Nature Structural Biology*, vol. 3, no. 5, pp. 427–431, 1996.

[31] G. Schreiber, G. Haran, and H.-X. Zhou, "Fundamental aspects of protein: protein association kinetics," *Chemical Reviews*, vol. 109, no. 3, pp. 839–860, 2009.

[32] T. Selzer and G. Schreiber, "New insights into the mechanism of protein-protein association," *Proteins*, vol. 45, no. 3, pp. 190–198, 2001.

[33] H.-X. Zhou, "Disparate ionic-strength dependencies of on and off rates in protein-protein association," *Biopolymers*, vol. 59, pp. 427–433, 2001.

[34] M. Vijayakumar, K.-Y. Wong, G. Schreiber, A. R. Fersht, A. Szabo, and H.-X. Zhou, "Electrostatic enhancement of diffusion-controlled protein-protein association: comparison of theory and experiment on barnase and barstar," *Journal of Molecular Biology*, vol. 278, no. 5, pp. 1015–1024, 1998.

[35] J. T. H.G. Overbeek and B. H. Bijsterbosch, "Electrokinetic separation methods," in *The Electrical Double Layer and the Theory of Electrophoresis*, pp. 1–33, Elsevier; North-Holland Biomedical Press, 1979.

[36] S. Park and K. Schulten, "Calculating potentials of mean force from steered molecular dynamics simulations," *Journal of Chemical Physics*, vol. 120, no. 13, pp. 5946–5961, 2004.

[37] M. A. Kostiainen, J. G. Hardy, and D. K. Smith, "High-affinity multivalent DNA binding by using low-molecular-weight dendrons," *Angewandte Chemie*, vol. 44, no. 17, pp. 2556–2559, 2005.

[38] C. K. Nisha, S. V. Manorama, M. Ganguli, S. Maiti, and J. N. Kizhakkedathu, "Complexes of poly(ethylene glycol)-based cationic random copolymer and calf thymus DNA: a complete biophysical characterization," *Langmuir*, vol. 20, no. 6, pp. 2386–2396, 2004.

[39] J. Kästener, "Umbrella sampling," *Wiley Interdisciplinary Reviews*, vol. 1, no. 6, pp. 932–942, 2011.

[40] J. Kästner, "Umbrella integration in two or more reaction coordinates," *Journal of Chemical Physics*, vol. 131, no. 3, Article ID 034109, 2009.

11

Chondrocyte Behavior on Micropatterns Fabricated Using Layer-by-Layer Lift-Off: Morphological Analysis

Jameel Shaik,[1,2,3] **Javeed Shaikh Mohammed,**[1,4]
Michael J. McShane,[1,2,5] **and David K. Mills**[1,2,6]

[1] *Institute for Micromanufacturing, Louisiana Tech University, Ruston, LA 71272, USA*
[2] *Biomedical Engineering Program, Louisiana Tech University, Ruston, LA 71272, USA*
[3] *School of Bio Sciences & Technology, VIT University, Vellore 632014, India*
[4] *Biomedical Technology Department, King Saud University, Riyadh 11433, Saudi Arabia*
[5] *Biomedical Engineering Program, Texas A&M University, College Station, TX 77843, USA*
[6] *School of Biological Sciences, Louisiana Tech University, Ruston, LA 71272, USA*

Correspondence should be addressed to David K. Mills; dkmills@latech.edu

Academic Editor: Rad Zdero

Cell patterning has emerged as an elegant tool in developing cellular arrays, bioreactors, biosensors, and lab-on-chip devices and for use in engineering neotissue for repair or regeneration. In this study, micropatterned surfaces were created using the layer-by-layer lift-off (LbL-LO) method for analyzing canine chondrocytes response to patterned substrates. Five materials were chosen based on our previous studies. These included: poly(dimethyldiallylammonium chloride) (PDDA), poly(ethyleneimine) (PEI), poly(styrene sulfonate) (PSS), collagen, and chondroitin sulfate (CS). The substrates were patterned with these five different materials, in five and ten bilayers, resulting in the following multilayer nanofilm architectures: $(PSS/PDDA)_5$, $(PSS/PDDA)_{10}$; $(CS/PEI)_4/CS$, $(CS/PEI)_9/CS$; $(PSS/PEI)_5$, $(PSS/PEI)_{10}$; $(PSS/Collagen)_5$, $(PSS/Collagen)_{10}$; $(PSS/PEI)_4/PSS$, $(PSS/PEI)_9/PSS$. Cell characterization studies were used to assess the viability, longevity, and cellular response to the configured patterned multilayer architectures. The cumulative cell characterization data suggests that cell viability, longevity, and functionality were enhanced on micropatterned PEI, PSS, collagen, and CS multilayer nanofilms suggesting their possible use in biomedical applications.

1. Introduction

Replicating the highly structured *in vivo* microenvironment is crucial in understanding cellular behavior [1]. Traditional cell culture surfaces cannot provide sufficient control over the cellular microenvironment [2] for use in studying many anchorage-dependent cellular processes such as cellular differentiation, proliferation, and phenotypic expression. Cell supportive substrates, with the requisite spatiotemporal surface properties, are also a critical feature in designing appropriate biomaterial surfaces for use in cell arrays, bioreactors, biosensors [3], and cocultures [4–6] and for use in engineering new tissues for repair or replacement.

Micropatterned surfaces have been explored as a means not only to answer fundamental questions in cell biology but also to develop cell culture substrates with surface features tailored for specific bio- and tissue engineering applications [2, 3, 7]. This was demonstrated by the growth of hepatocytes on micropatterned surfaces [4, 5]. The authors observed decreased DNA production and increased cellular apoptosis associated with a decrease in the adhesiveness of the surfaces [7]. Cell shape was also found to be the regulatory factor in both cell apoptosis and growth [7, 8]. This was achieved by an increasing restriction of the size of micropatterned islands coated with different densities of ECM and growing bovine and human endothelial cells on these islands [2].

Patterning cells using cell-adhesive [9–13] or cell-repulsive [14–20] surfaces or combinations [21, 22] of adhesive and nonadhesive surfaces have been developed, and a wide variety of eukaryotic cells have been grown and studied on these micropatterned surfaces [2, 5, 8, 20, 23–25].

A broad range of materials have been used in creating these micropatterned cell culture surfaces [3, 8, 26, 27].

Micropatterned substrates have also lent credence to the important understanding that the degree of cellular contraction is crucial in determining a cells fate during differentiation, especially in the case of stem cells [1]. This has been demonstrated in several studies that showed that variation in micropattern size directed stem cell differentiation into different cell lineages. For example, human mesenchymal stem cells (hMSCs) cultured in differentiating medium exhibited differences in the contraction levels and also exhibited different lineages—those hMSCs grown on 1,000 μm^2 micropatterns had low contraction levels and differentiated into adipocytes, while hMSCs plated on 10,000 μm^2 micropatterns were highly contracted and differentiated into osteoblasts [28]. Similarly, hMSCs treated with transforming growth factor β (TGF-β) exhibited differential behavior dependent upon the size of micropatterns—hMSCs plated on small micropatterns differentiated into chondrocytes, while hMSCs plated on large micropatterns differentiated into myocytes [29].

The use of layer by layer (LbL) nanoassembly for creating micropatterned surfaces brings in all the advantages offered by LbL—simplicity and excellent control over surface properties such as thickness, roughness, and porosity [3]. LbL surfaces can potentially be used in obtaining the precise cellular microenvironment as the surfaces can be tuned to release the factors necessary for the growth and regulation of cells [22]. Polyelectrolytes and proteins deposited through the LbL technique can be used to create either cell-resistant or cell-adhesive micropatterns. Our previous studies focused on the growth and behavior of bovine articular chondrocytes [30], human chondrosarcoma cells, and canine chondrocytes [31] on LbL-assembled nanothin films of varying configurations. We chose chondrocytes as our model cell type as they have a very plastic phenotype. Cell characterization studies were used to assess chondrocyte viability, longevity, and functionality in response to the configured architectures. Cell adhesion, shape, and functionality are linked to the nature of the underlying culture substrate [32, 33].

Our goal in this study was to expand our previous work by examining interspecies differences in chondrocyte behavior on micropatterned substrates created using the LbL-LO method. Our expectation was that difference in nanofilm architectures atop micropatterned substrates would evoke variations in chondrocyte behavior. Different micropatterned surfaces were created using the LbL-LO technique [6, 34–36]. Based on our previous studies, five polyelectrolytes/proteins were used to construct the nanofilms [31]. These were poly(dimethyldiallylammonium chloride) (PDDA), poly(ethyleneimine) (PEI), poly(styrene sulfonate) (PSS), collagen, and chondroitin sulfate (CS). The substrates were patterned, in five and ten bilayers, resulting in the following multilayer nanofilm architectures: (PSS/PDDA)$_5$, (PSS/PDDA)$_{10}$; (CS/PEI)$_4$/CS, (CS/PEI)$_9$/CS; (PSS/PEI)$_5$, (PSS/PEI)$_{10}$; (PSS/Collagen)$_5$, (PSS/Collagen)$_{10}$; (PSS/PEI)$_4$/PSS, (PSS/PEI)$_9$/PSS.

2. Materials and Methods

2.1. Substrates. Microscope cover slips (Thickness number 2, 18 × 18 mm^2, Electron Microscopy Sciences, Hatfield, PA, USA) were used as the substrates for deposition of the micropatterns. These substrates were chosen for ease in optical characterization.

2.2. Chemicals. Nano-Strip from CYANTEK Corporation (Fremont, CA); positive photoresist S1813 and positive resist developer MF-319 from the Shipley Corporation (Marlboro, Massachusetts) were used. All the chemicals were purchased from Sigma-Aldrich unless otherwise specified. All commercial chemicals were used following manufacturer's directions.

2.3. Preparation of Polyelectrolyte, Polypeptide, and Protein Solutions. PDDA (Mw ~ 150 kDa), PSS (Mw ~ 1 MDa) solutions were prepared at concentrations of 2 mg mL^{-1} with 0.5 M KCl, and a PEI (Mw ~ 750 kDa) solution of 2 mg mL^{-1} was prepared in deionized (DI) H$_2$O for use in LbL nanoassembly. Chondroitin sulfate (Mw ~ 500 Da) and type I collagen (Mw ~ 100 kDa) (Cohesion, Palo Alto, CA, USA) were prepared at a concentration of 120 μg mL^{-1}. All solutions were prepared using DI water with a resistivity of 18.2 MΩ cm (Millipore systems, Burlington, MA, USA).

2.4. Substrate Pretreatment. The substrates were first incubated in Nano-Strip at 70°C for 1 h followed by rinsing in DI water anddried in a N$_2$ stream to remove any organic materials and to create a uniform negative charge on the substrates. A precursor layer of PDDA was then deposited onto the substrates to render a cytophobic background on the substrates. This was based on our previous results with smooth muscle and neuronal cells [35, 37]. PDDA application is not exclusive as any cytophobic material other than PDDA can also be used.

2.5. Photolithography. To help withstand the centrifugal forces during spin coating, the PDDA-coated glass substrates were attached to silicon wafer pieces using photoresist S1813 and heated at 165°C for 5 min to hard bake the photoresist. Next, positive photoresist S1813 was spun (1000 rpm-100 r s^{-1}-10 s, 3000 rpm-500 r s^{-1}-50 s) on the PDDA-coated substrates, soft baked at 115°C for 1 min, and photo-patterned using UV radiation (400 nm, 7 mW cm^{-2}) applied for 18 s. The mask used for pattern transfer contained 80 μm wide stripe patterns separated by 240 μm and 100 μm wide stripe patterns separated by 300 μm. Finally, the patterns were developed for 15 s using MF-319, and the substrates quickly rinsed in DI water and dried using N$_2$.

2.6. Layer-by-Layer (LbL) Self-Assembly. Micropatterned substrates were then modified using LbL nanoassembly. The substrates were dipped in polyelectrolyte and protein solutions for 10 min and 30 min, respectively. PSS or PEI was

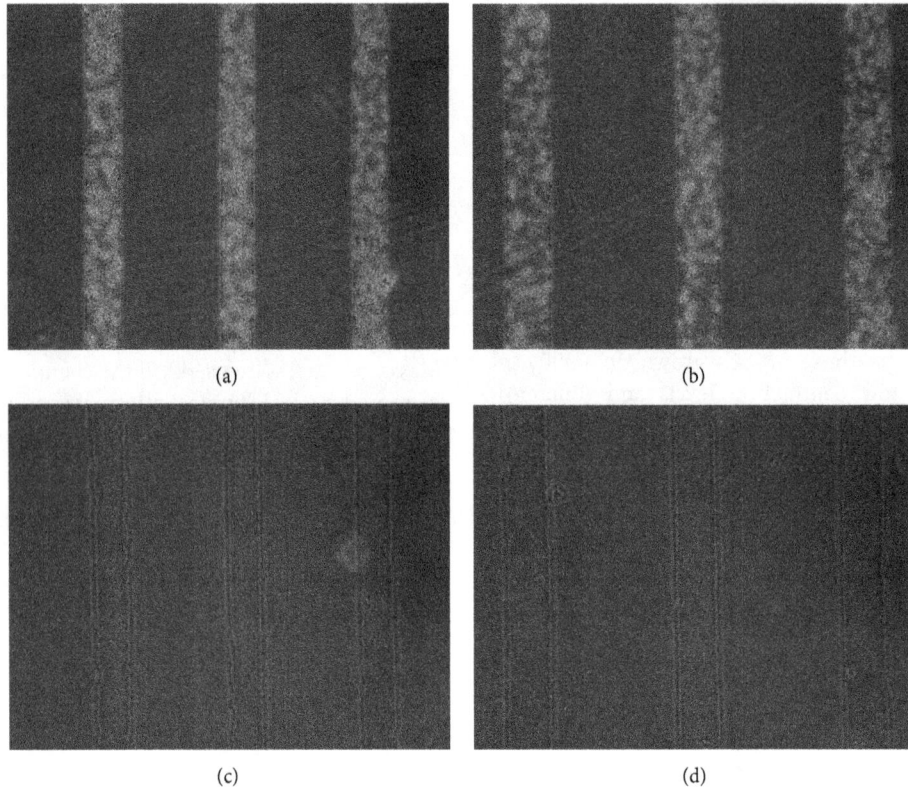

FIGURE 1: Micropatterned substrates with PDDA as the outermost layer: (a) (PSS/PDDA)$_5$—80 μm, (b) (PSS/PDDA)$_5$—100 μm, (c) (PSS/PDDA)$_{10}$—80 μm, and (d) (PSS/PDDA)$_{10}$—100 μm.

used as the polyanions in all multilayer nanofilm configurations. After every deposition step, substrates were rinsed in DI water and then dried using N$_2$. Substrates were patterned with PDDA, PSS, PEI, collagen, and CS as either five- or ten-bilayer nanofilms. Thus, the following configurations were fabricated: (PSS/PDDA)$_5$, (PSS/PDDA)$_{10}$; (CS/PEI)$_4$/CS, (CS/PEI)$_9$/CS; (PSS/PEI)$_5$, (PSS/PEI)$_{10}$; (PSS/Collagen)$_5$, (PSS/Collagen)$_{10}$; (PSS/PEI)$_4$/PSS, (PSS/PEI)$_9$/PSS.

2.7. Lift-Off.

Lift-off was performed by sonicating the substrates in acetone for 5 to 10 minutes. The photoresist and the nanofilms deposited on the photoresist were removed during the lift-off process. Also, the cover slip glass was detached from the silicon wafer. Surprisingly, the use of acetone was shown not to affect the biological functions of the molecules used in the LbL-LO process [38].

2.8. Cell Culture.

Canine chondrocytes (CnC) were obtained from Cell Applications, Inc. (San Diego, CA, USA). Chondrocytes were isolated from normal canine articular cartilage and obtained at second passage. Their phenotype is preserved through ten population doublings. Chondrocytes were grown as monolayers and maintained in Chondrocyte Growth Medium (Cell Applications, Inc., San Diego, CA, USA) until the necessary cell numbers were obtained. Canine chondrocytes from passage three were used for the cell characterization studies on the micropatterned surfaces.

2.9. Cell Characterization.

Phase-contrast microscopy was used to demonstrate the successful creation of the micropatterns. Phase-contrast microscopy was also used for the characterization of CnC on the micropatterned surfaces.

3. Results

3.1. Phase-Contrast Microscopy of Micropatterned Substrates.

Figures 1, 2, 3, 4, and 5 contain phase-contrast images of the micropatterned substrates of five different materials, in five and ten-bilayer nanofilm configurations. In all the multilayer nanofilm architectures, the terminating nanofilm layer was one of the five different materials studied here. The images show 80 μm wide stripe patterns separated by 240 μm or 100 μm wide stripe patterns separated by 300 μm.

All the micropatterns, with the exception of collagen, had high edge resolution. There could be several factors contributing to the low resolution of edges in collagen micropatterns. Some of the factors affecting the edge resolution of the collagen micropatterns could be deposition time of collagen, pH of collagen solution, sonication time during lift-off, and the height of photoresist used to define the stripe micropatterns in LbL-LO. From Figure 4, it also appears that the 100 μm collagen micropatterns have better edge resolution compared to the 80 μm micropatterns. One of the reasons for this difference in edge resolution of collagen micropatterns could be due to the differences in the spacing between the stripe micropatterns, which is 240 μm for the

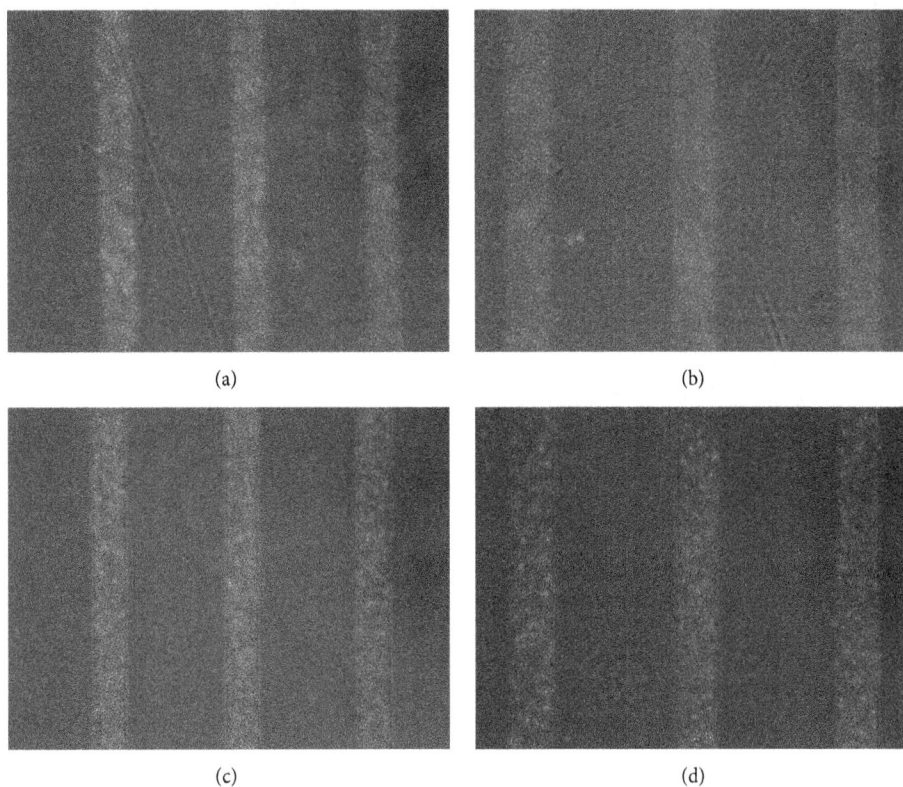

FIGURE 2: Micropatterned substrates with CS as the outermost layer: (a) $(CS/PEI)_4/CS$—$80\,\mu m$, (b) $(CS/PEI)_4/CS$—$100\,\mu m$, (c) $(CS/PEI)_9/CS$—$80\,\mu m$, and (d) $(CS/PEI)_9/CS$—$100\,\mu m$.

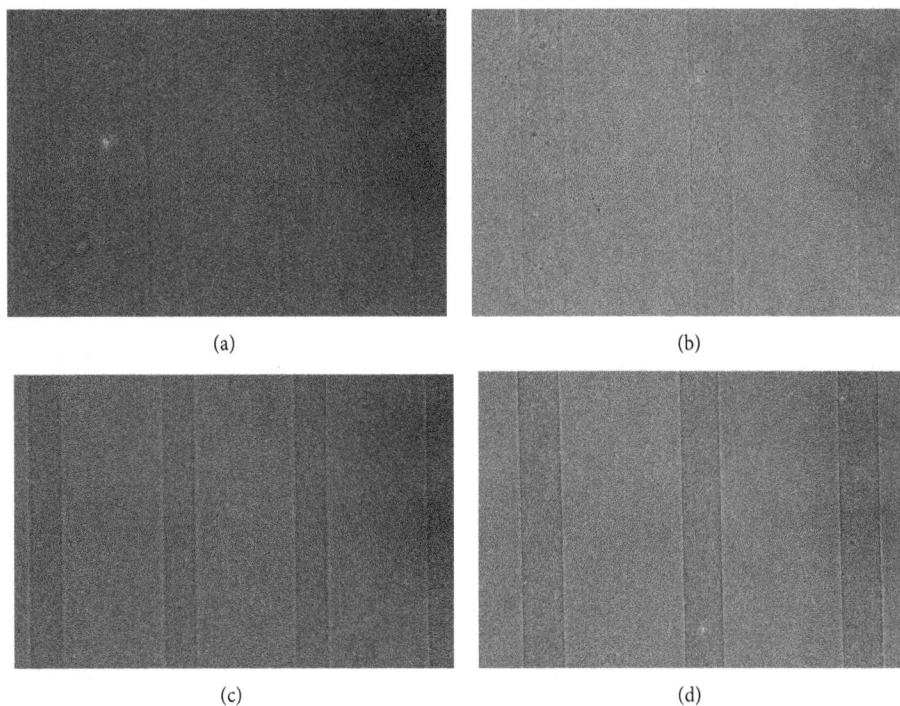

FIGURE 3: Micropatterned substrates with PEI as the outermost layer: (a) $(PSS/PEI)_5$—$80\,\mu m$, (b) $(PSS/PEI)_5$—$100\,\mu m$, (c) $(PSS/PEI)_{10}$—$80\,\mu m$, and (d) $(PSS/PEI)_{10}$—$100\,\mu m$.

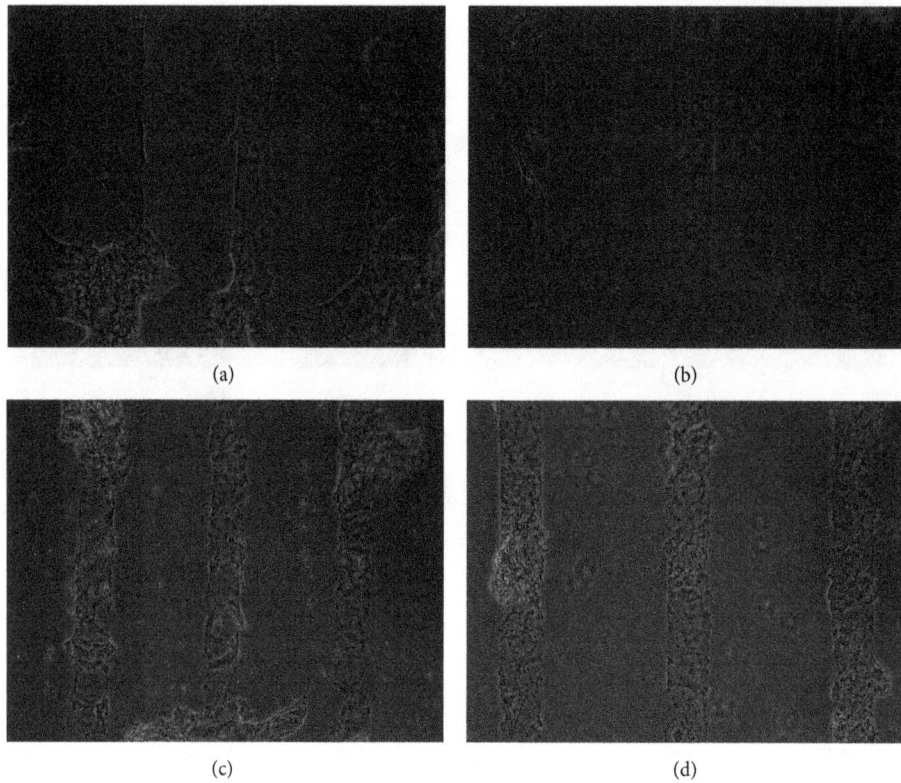

FIGURE 4: Micropatterned substrates with collagen as the outermost layer: (a) (PSS/Collagen)$_5$—80 μm, (b) (PSS/Collagen)$_5$—100 μm, (c) (PSS/Collagen)$_{10}$—80 μm, and (d) (PSS/Collagen)$_{10}$—100 μm.

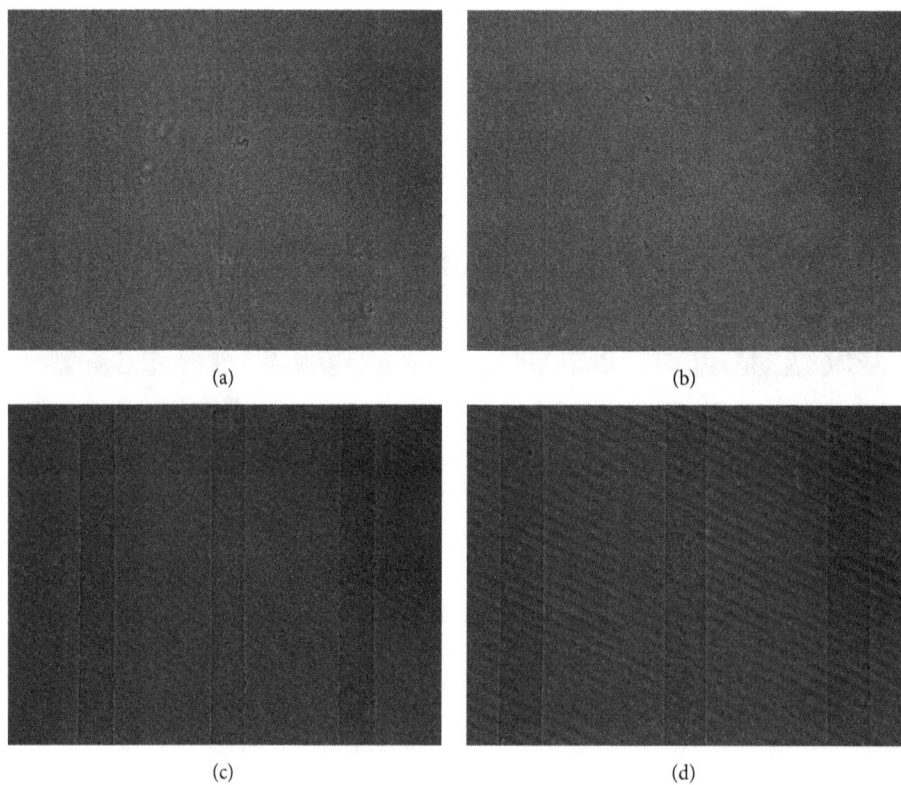

FIGURE 5: Micropatterned substrates with PSS as the outermost layer: (a) (PSS/PEI)$_4$/PSS—80 μm, (b) (PSS/PEI)$_4$/PSS—100 μm, (c) (PSS/PEI)$_9$/PSS—80 μm, and (d) (PSS/PEI)$_9$/PSS—100 μm.

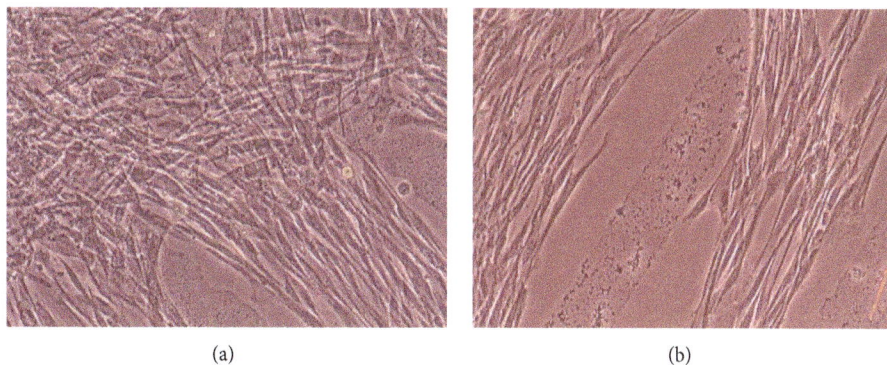

(a)

(b)

Figure 6: CnC on PDDA micropatterns at 3 days after seeding: (a) $(PSS/PDDA)_5$—100 μm, (b) $(PSS/PDDA)_{10}$—100 μm.

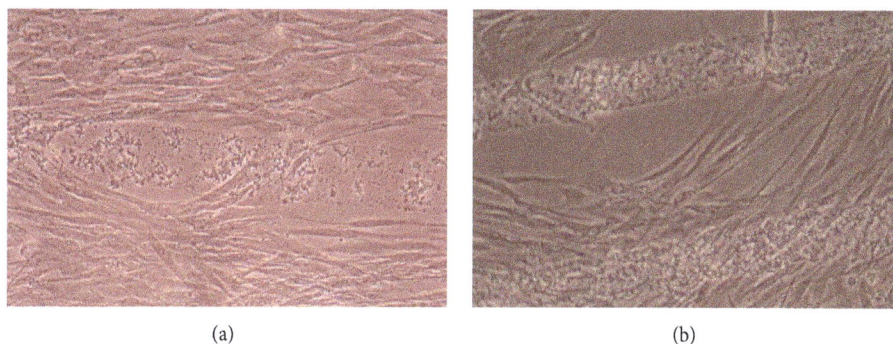

(a)

(b)

Figure 7: CnC on PDDA micropatterns at 10 days after seeding: (a) $(PSS/PDDA)_5$—100 μm, (b) $(PSS/PDDA)_{10}$—80 μm.

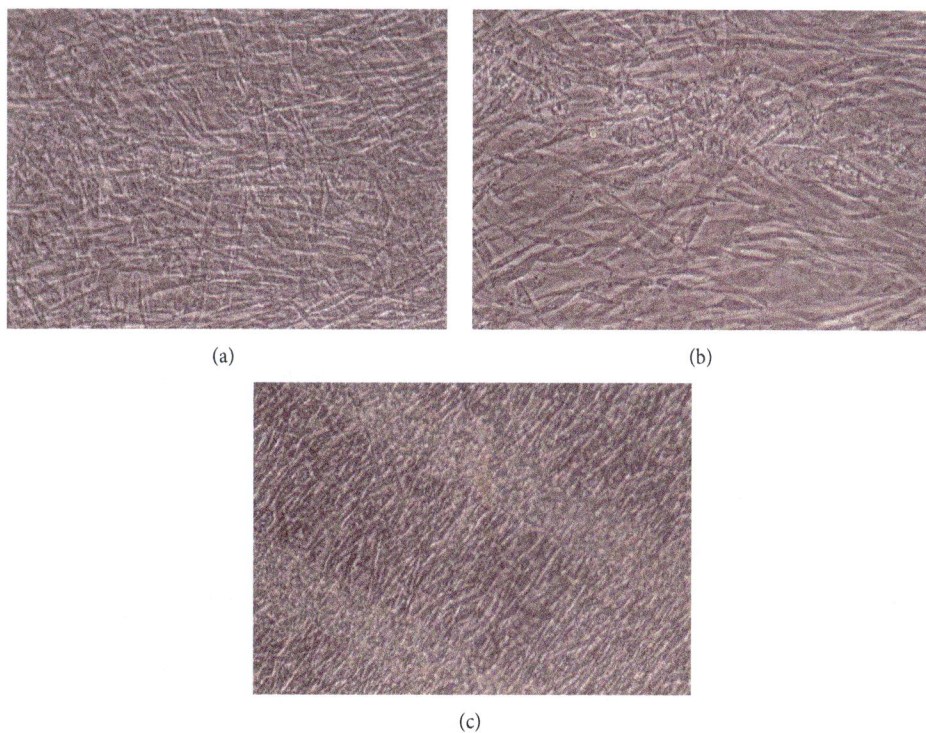

(a)

(b)

(c)

Figure 8: CnC on PDDA micropatterns at 11 days after seeding: (a) $(PSS/PDDA)_5$—80 μm, (b) $(PSS/PDDA)_5$—100 μm, and (c) $(PSS/PDDA)_{10}$—80 μm.

(a)

(b)

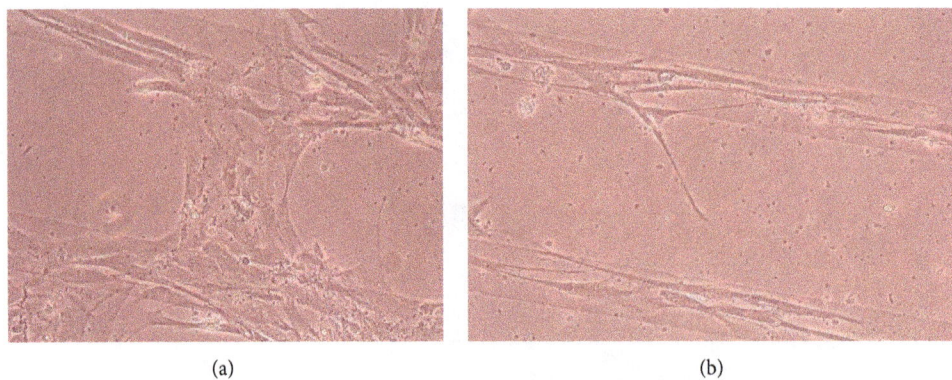

Figure 9: CnC on PEI micropatterns at 9 days after seeding: (a) (PSS/PEI)$_{10}$—100 μm, (b) (PSS/PEI)$_{10}$—100 μm.

(a)

(b)

(c)

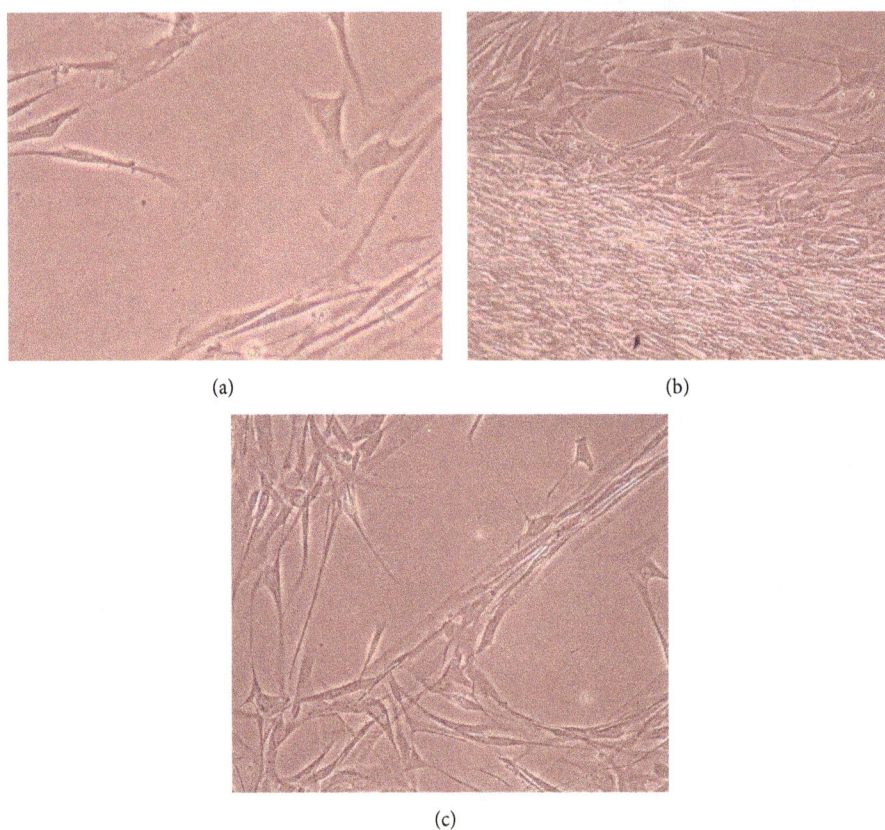

Figure 10: CnC on PEI micropatterns at 11 days after seeding: (a) (PSS/PEI)$_5$—100 μm, (b) (PSS/PEI)$_{10}$—80 μm, and (c) (PSS/PEI)$_{10}$—100 μm.

(a)

(b)

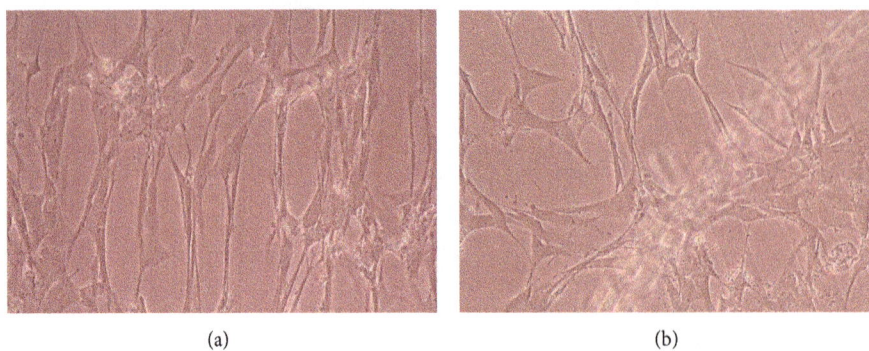

Figure 11: CnC on PSS micropatterns at 9 days after seeding: (a) (PSS/PEI)$_4$/PSS—80 μm, (b) (PSS/PEI)$_4$/PSS—80 μm.

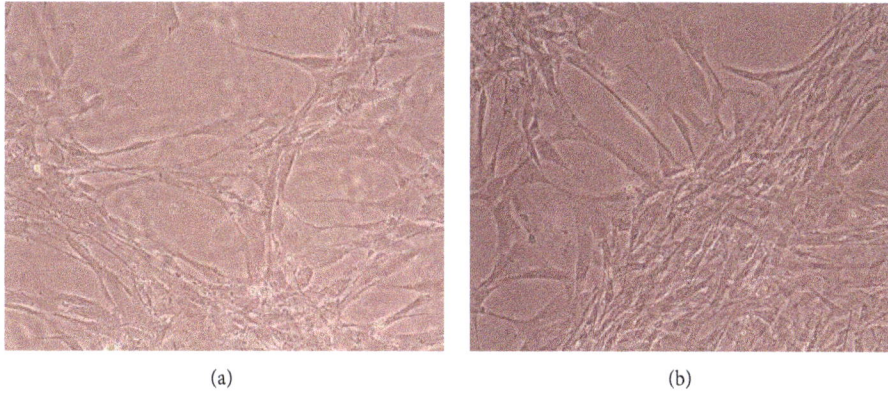

(a) (b)

FIGURE 12: CnC on PSS micropatterns at 10 days after seeding: (a) (PSS/PEI)$_4$/PSS—80 μm, (b) (PSS/PEI)$_4$/PSS—100 μm.

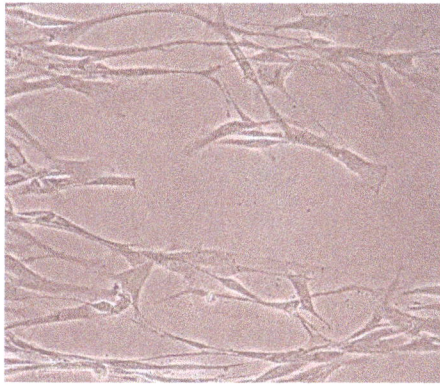

FIGURE 13: CnC on PSS micropatterns at 11 days after seeding: (PSS/PEI)$_4$/PSS—100 μm.

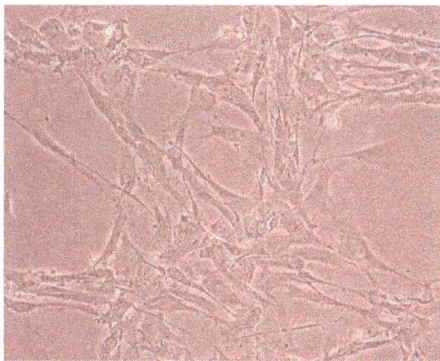

FIGURE 15: CnC on collagen micropatterns at 11 days after seeding: (PSS/Collagen)$_5$.

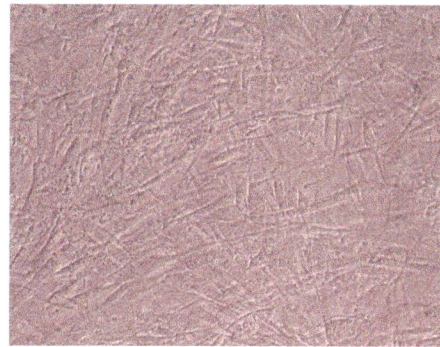

FIGURE 14: CnC on collagen micropatterns at 10 days after seeding: (PSS/Collagen)$_5$—100 μm.

80 μm stripe patterns and 300 μm for the 100 μm stripe patterns. From Figures 1–5, it is also clear that the 10-bilayer micropatterns appear to have better surface coverage of materials compared to their 5-bilayer counterparts.

3.2. Phase-Contrast Microscopy of CnC on Micropatterned Substrates. Figures 6, 7, 8, 9, 10, 11, 12, 13, 14, 15, 16, 17, and 18 contain representative phase-contrast images of the

chondrocytes on micropatterned substrates of five different materials, in five- and ten-bilayer nanofilm configurations. In all the multilayer nanofilm architectures, the terminating nanofilm layer was one of the five different materials studied here, and the background material is a single layer of PDDA.

Figure 6 shows CnC on PDDA-terminating substrates 3 days after cell seeding. The cells appear to be avoiding the PDDA stripe patterns and are showing defined growth on the surfaces between the stripe patterns coated with a single nanofilm layer of PDDA. PDDA was shown to be a cytophobic material that inhibited smooth muscle cells and neuronal attachment [35, 37]. From our AFM measurements [39], it was observed that the thicknesses of 5 and 10 bilayers of PDDA are roughly 59.61 ± 4.07 and 251.58 ± 6.28 nm (mean ± standard deviation), respectively. Both these thicknesses are considerably greater than the thickness of a PDDA monolayer. Chondrocytes seem to prefer a thinner PDDA layer for their attachment as compared to the thicker-layered micropatterned PDDA.

From Figure 7 it can be seen that some CnC are growing on the PDDA micropatterns. However, the growth of CnC between the micropatterns is still greater than that on the micropatterns, and it is clear that the cells have reached confluence on the PDDA micropatterns (as seen in Figure 8). From Figures 9 and 10, it can be observed that CnC are preferentially growing on the PEI micropatterns.

(a)   (b)

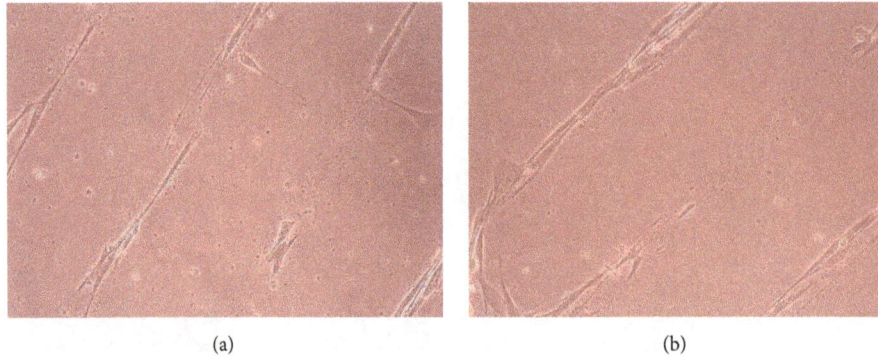

FIGURE 16: CnC on CS micropatterns at 9 days after seeding: (a) $(CS/PEI)_4/CS$—80 μm, (b) $(CS/PEI)_9/CS$—80 μm.

(a)   (b)

(c)   (d)

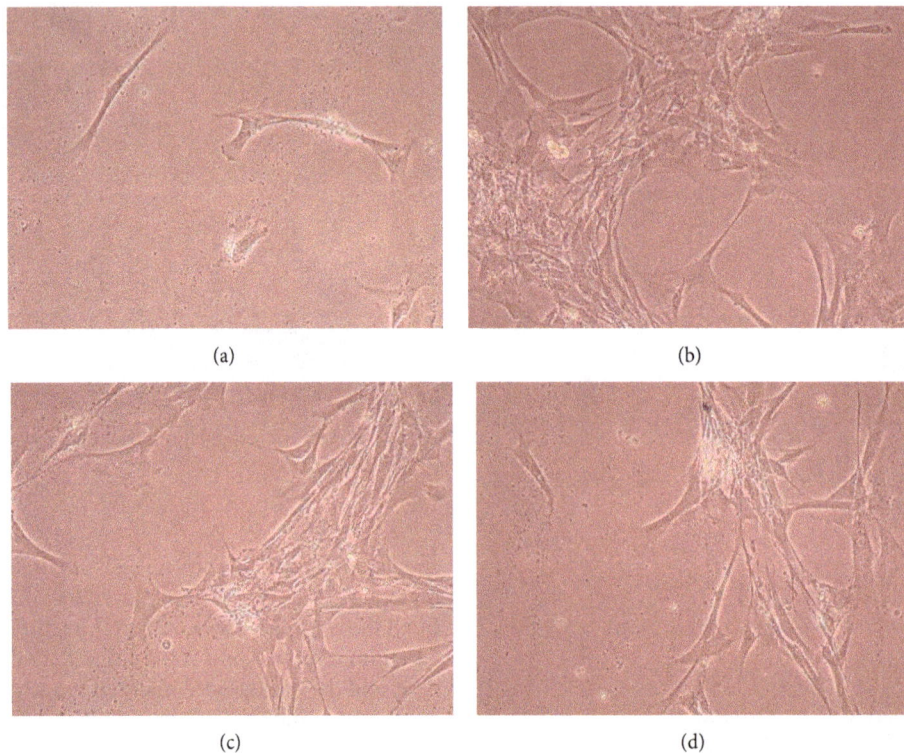

FIGURE 17: CnC on CS micropatterns at 10 days after seeding: (a) $(CS/PEI)_4/CS$—80 μm, (b) $(CS/PEI)_4/CS$—100 μm, (c) $(CS/PEI)_9/CS$—100 μm, and (d) $(CS/PEI)_9/CS$—80 μm.

From Figure 10(b), it is clearly shown that CnC growing on the micropatterns were aligned along the length of the stripe patterns, whereas CnC growing on unpatterned substrate are attached and growing on the monolayer of PDDA.

Figures 11–13 demonstrate that CnC growth is mainly restricted to the PSS micropatterns. The CnC on the micropatterns were aligned along the length of the stripe patterns. From Figures 14 and 15, CnC have also grown to confluence. The confluent CnC make the collagen micropatterns very hard to be discerned in the images. From Figures 16–18, it can be observed that CnC are preferentially growing on the CS micropatterns. Also, it can be observed that CnC growing on the micropatterns are aligned along the length of the stripe patterns.

4. Discussion

Several similar studies have been conducted in other eukaryotic cells [11, 14, 24, 35, 40–45], and a few studies have also focused on different types of chondrocytes [46]. Also, diverse methods have been adapted to create the micropatterns required for the studies. To our knowledge, ours is the first study using canine chondrocytes as a model for directed growth on micropatterned substrates. The majority of the above-mentioned studies have reported constrained and preferential cell growth on micropatterned substrates. Specifically, protein micropatterns (bone morphogenetic protein 2 (BMP-2) printed on polystyrene (PS)) fabricated by microcontact printing significantly influenced the adhesion,

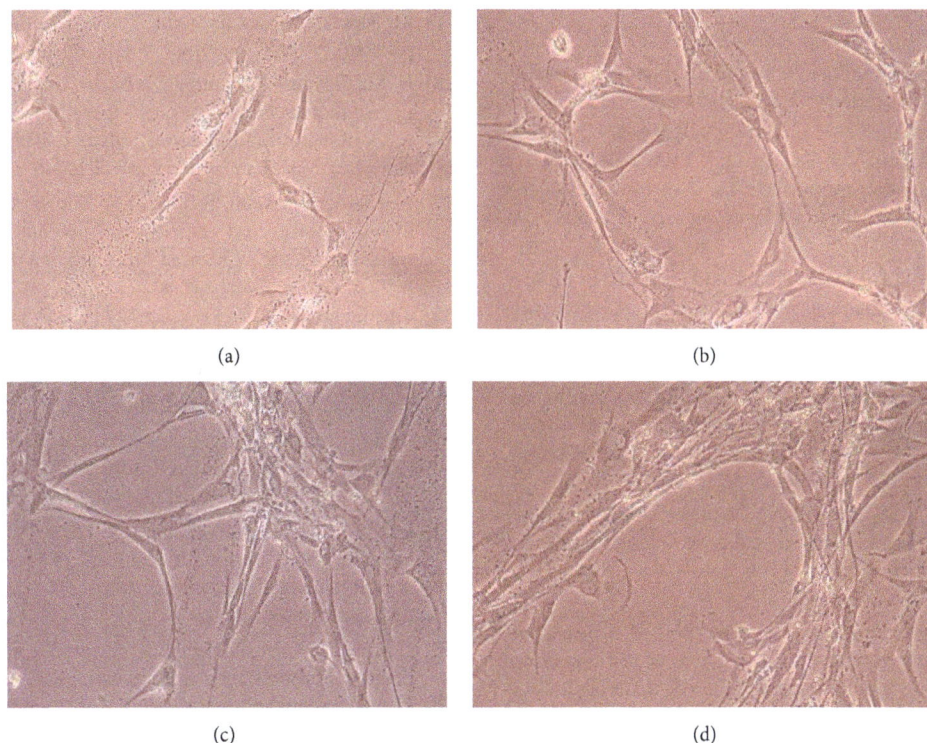

FIGURE 18: CnC on CS micropatterns at 11 days after seeding: (a) $(CS/PEI)_4/CS—80\,\mu m$, (b) $(CS/PEI)_4/CS—100\,\mu m$, (c) $(CS/PEI)_9/CS—80\,\mu m$, and (d) $(CS/PEI)_9/CS—100\,\mu m$.

spread, alignment, and functions of human chondrocytes. As in our studies and several similar studies, human chondrocytes showed preferential adhesion on the BMP-2 micropatterns. Both the shapes and sizes of the micropatterns were instrumental in influencing cell adhesion, cell morphology, the degree of spreading of the cells, and more significantly type II and VI collagen expression, thus emphasizing the importance of protein micropatterns in influencing the growth and functionality of human chondrocytes [47].

While our studies were not focused on expression of different collagen and proteoglycan types, future studies will be directed towards analysis of gene expression and protein synthesis including the level of phenotypic protein marker expression as well as the potential for long-term chondrocyte functionality on micropatterned substrates. These studies would be beneficial in understanding the influence of micropatterns generated from proteins and other biomaterials on the growth and behavior of canine chondrocytes. Incorporation of growth factors and other bioactive factors that may modulate the behavior of canine chondroctyes should also be addressed. Micropatterns generated using the LbL-LO technique used in the current study can be helpful in creating *in vitro* drug-delivery models for studying the effects of different drugs on chondrocytes of varying types (growth versus articular cartilage). Our previous work, current study, and the suggested future studies (short-term and long-term) would also be extremely useful in cartilage tissue engineering and also for creating disease-study models, studying chondrocyte involvement in degenerative changes in articular cartilage, for example.

5. Conclusions

Defined and restrained growth of canine chondrocytes was achieved on 5 and 10 bilayer micropatterns fabricated using the LbL-LO technique. From the morphological observations, the 5- and 10-bilayer nanofilms do not produce any apparent differences in the growth pattern of CnC.

CnC appeared to remain confluent for a longer period of time on the thinner monolayer PDDA surface between the micropatterns compared to CnC grown on the thicker PDDA micropatterns. This suggests that 5 and 10 bilayers of micropatterned PDDA might act as cell-resistant surfaces; further studies are needed to understand this observation. CnC exhibited preferential attachment on micropatterns of PEI, PSS, collagen, and CS multilayer nanofilms. CnC growth was stable for an extended period of time on micropatterned PEI, PSS, collagen, and CS suggesting their possible use in biomedical applications.

Conflict of Interests

The authors declare that there is no conflict of interest.

Acknowledgments

This work was supported by DARPA and NAVY SPAWAR SC N66001-05-1-8903 Grant awarded to Dr. D. K. Mills and by the National Science Foundation (Grant no. 0092001).

References

[1] M. Théry, "Micropatterning as a tool to decipher cell morphogenesis and functions," *Journal of Cell Science*, vol. 123, no. 24, pp. 4201–4213, 2010.

[2] C. S. Chen, M. Mrksich, S. Huang, G. M. Whitesides, and D. E. Ingber, "Micropatterned surfaces for control of cell shape, position, and function," *Biotechnology Progress*, vol. 14, no. 3, pp. 356–363, 1998.

[3] M. C. Berg, S. Y. Yang, P. T. Hammond, and M. F. Rubner, "Controlling mammalian cell interactions on patterned polyelectrolyte multilayer surfaces," *Langmuir*, vol. 20, no. 4, pp. 1362–1368, 2004.

[4] S. N. Bhatia, M. L. Yarmush, and M. Toner, "Controlling cell interactions by micropatterning in co-cultures: Hepatocytes and 3T3 fibroblasts," *Journalof Biomedical Materials Research*, vol. 34, pp. 189–199, 1997.

[5] Y. S. Zinchenko, L. W. Schrum, M. Clemens, and R. N. Coger, "Hepatocyte and Kupffer cells co-cultured on micropatterned surfaces to optimize hepatocyte function," *Tissue Engineering*, vol. 12, no. 4, pp. 751–761, 2006.

[6] J. Shaikh Mohammed, M. A. DeCoster, and M. J. McShane, "Fabrication of interdigitated micropatterns of self-assembled polymer nanofilms containing cell-adhesive materials," *Langmuir*, vol. 22, no. 6, pp. 2738–2746, 2006.

[7] R. Singhvi, A. Kumar, G. P. Lopez et al., "Engineering cell shape and function," *Science*, vol. 264, no. 5159, pp. 696–698, 1994.

[8] C. S. Chen, M. Mrksich, S. Huang, G. M. Whitesides, and D. E. Ingber, "Geometric control of cell life and death," *Science*, vol. 276, no. 5317, pp. 1425–1428, 1997.

[9] M. Morra and C. Cassinelli, "Cell adhesion micropatterning by plasma treatment of alginate coated surfaces," *Plasmas and Polymers*, vol. 7, no. 2, pp. 89–101, 2002.

[10] M. Li, D. K. Mills, T. Cui, and M. J. McShane, "Cellular response to gelatin- and fibronectin-coated multilayer polyelectrolyte nanofilms," *IEEE Transactions on Nanobioscience*, vol. 4, no. 2, pp. 170–179, 2005.

[11] J. Hyun, H. Ma, P. Banerjee, J. Cole, K. Gonsalves, and A. Chilkoti, "Micropatterns of a cell-adhesive peptide on an amphiphilic comb polymer film," *Langmuir*, vol. 18, no. 8, pp. 2975–2979, 2002.

[12] D. Falconnet, G. Csucs, H. Michelle Grandin, and M. Textor, "Surface engineering approaches to micropattern surfaces for cell-based assays," *Biomaterials*, vol. 27, no. 16, pp. 3044–3063, 2006.

[13] D. L. Elbert and J. A. Hubbell, "Conjugate addition reactions combined with free-radical cross-linking for the design of materials for tissue engineering," *Biomacromolecules*, vol. 2, pp. 430–441, 2001.

[14] A. Tourovskaia, T. Barber, B. T. Wickes et al., "Micropatterns of chemisorbed cell adhesion-repellent films using oxygen plasma etching and elastomeric masks," *Langmuir*, vol. 19, no. 11, pp. 4754–4764, 2003.

[15] S. Y. Yang, J. D. Mendelsohn, and M. F. Rubner, "Patternable, cell-resistant surfaces prepared with H-bonded polyelectrolyte multilayers," *Polymer Preprints (American Chemical Society, Division of Polymer Chemistry)*, vol. 43, pp. 723–724, 2002.

[16] S. Y. Yang, J. D. Mendelsohn, and M. F. Rubner, "New class of ultrathin, highly cell-adhesion-resistant polyelectrolyte multilayers with micropatterning capabilities," *Biomacromolecules*, vol. 4, no. 4, pp. 987–994, 2003.

[17] M. Morra and C. Cassinelli, "Surface studies on a model cell-resistant system," *Langmuir*, vol. 15, no. 13, pp. 4658–4663, 1999.

[18] H. Ma, J. Hyun, Z. Zhang, T. P. Beebe, and A. Chilkoti, "Fabrication of biofunctionalized quasi-three-dimensional microstructures of a nonfouling comb polymer using soft lithography," *Advanced Functional Materials*, vol. 15, no. 4, pp. 529–540, 2005.

[19] A. Khademhosseini, S. Jon, K. Y. Suh et al., "Direct patterning of protein- and cell-resistant polymeric monolayers and microstructures," *Advanced Materials*, vol. 15, no. 23, pp. 1995–2000, 2003.

[20] S. G. Olenych, M. D. Moussallem, D. S. Salloum, J. B. Schlenoff, and T. C. S. Keller, "Fibronectin and cell attachment to cell and protein resistant polyelectrolyte surfaces," *Biomacromolecules*, vol. 6, no. 6, pp. 3252–3258, 2005.

[21] A. Tourovskaia, X. Figueroa-Masot, and A. Folch, "Differentiation-on-a-chip: a microfluidic platform for long-term cell culture studies," *Lab on a Chip*, vol. 5, no. 1, pp. 14–19, 2005.

[22] J. D. Mendelsohn, S. Y. Yang, J. A. Hiller, A. I. Hochbaum, and M. F. Rubner, "Rational design of cytophilic and cytophobic polyelectrolyte multilayer thin films," *Biomacromolecules*, vol. 4, no. 1, pp. 96–106, 2003.

[23] J. Tan and W. M. Saltzman, "Topographical control of human neutrophil motility on micropatterned materials with various surface chemistry," *Biomaterials*, vol. 23, no. 15, pp. 3215–3225, 2002.

[24] L. Lu, L. Kam, M. Hasenbein et al., "Retinal pigment epithelial cell function on substrates with chemically micropatterned surfaces," *Biomaterials*, vol. 20, no. 23-24, pp. 2351–2361, 1999.

[25] J. T. Groves, L. K. Mahal, and C. R. Bertozzi, "Control of cell adhesion and growth with micropatterned supported lipid membranes," *Langmuir*, vol. 17, no. 17, pp. 5129–5133, 2001.

[26] J. Lahann, M. Balcells, T. Rodon et al., "Reactive polymer coatings: a platform for patterning proteins and mammalian cells onto a broad range of materials," *Langmuir*, vol. 18, no. 9, pp. 3632–3638, 2002.

[27] E. Ostuni, R. Kane, C. S. Chen, D. E. Ingber, and G. M. Whitesides, "Patterning mammalian cells using elastomeric membranes," *Langmuir*, vol. 16, no. 20, pp. 7811–7819, 2000.

[28] R. McBeath, D. M. Pirone, C. M. Nelson, K. Bhadriraju, and C. S. Chen, "Cell shape, cytoskeletal tension, and RhoA regulate stem cell lineage commitment," *Developmental Cell*, vol. 6, no. 4, pp. 483–495, 2004.

[29] L. Gao, R. McBeath, and C. S. Chen, "Stem cell shape regulates a chondrogenic versus myogenic fate through rac1 and N-cadherin," *Stem Cells*, vol. 28, no. 3, pp. 564–572, 2010.

[30] J. Shaik, J. S. Mohammed, M. J. McShane, and D. K. Mills, "Growth and behaviour of bovine articular chondrocytes on nanoengineered surfaces: part I," *International Journal of Nanotechnology*, vol. 8, no. 8-9, pp. 679–699, 2011.

[31] J. Shaik, J. S. Mohammed, M. J. McShane et al., "*In vitro* evaluation of chondrosarcoma cells and canine chondrocytes on layer-by-layer (LbL) self-assembled multilayer nanofilms," *Biofabrication*, vol. 5, Article ID 015004, 2013.

[32] J. C. Daniel and D. K. Mills, "Proteoglycan synthesis by cells cultured from regions of the rabbit flexor tendon," *Connective Tissue Research*, vol. 17, no. 3, pp. 215–230, 1988.

[33] J. Glowacki, E. Trepman, and J. Folkman, "Cell shape and phenotypic expression in chondrocytes," *Proceedings of the Society for Experimental Biology and Medicine*, vol. 172, no. 1, pp. 93–98, 1983.

[34] M. Li, K. K. Kondabatni, T. Cui, and M. J. McShane, "Fabrication of 3-D gelatin-patterned glass substrates with layer-by-layer and lift-off (LbL-LO) technology," *IEEE Transactions on Nanotechnology*, vol. 3, no. 1, pp. 115–123, 2004.

[35] J. S. Mohammed, M. A. DeCoster, and M. J. McShane, "Micropatterning of nanoengineered surfaces to study neuronal cell attachment *in vitro*," *Biomacromolecules*, vol. 5, no. 5, pp. 1745–1755, 2004.

[36] J. Shaik and D. K. Mills, "Micropatterned antibody-terminated nanocomposites (MANs) fabricated using layer-by-layer lift-off (LBL-LO) technique," *Journal of Biomedical Materials Research B*, vol. 100, pp. 1411–1415, 2012.

[37] M. Li, T. Cui, D. K. Mills, Y. M. Lvov, and M. J. McShane, "Comparison of selective attachment and growth of smooth muscle cells on gelatin- and fibronectin-coated micropatterns," *Journal of Nanoscience and Nanotechnology*, vol. 5, no. 11, pp. 1809–1815, 2005.

[38] J. Shaikh Mohammed, *Multicomponent patterning of nanocomposite polymer and nanoparticle films using photolithography and layer-by-layer self-assembly [Ph.D. dissertation]*, Louisiana Tech University, Ruston, La, USA, 2006.

[39] J. Shaik, *Growth and behavior of chondrocytes on nano engineered surfaces and construction of micropatterned co-culture platforms using layer-by-layer platforms using layer-by-layer assembly lift-off method [Ph.D. dissertation]*, Louisiana Tech University, Ruston, La, USA, 2007.

[40] Y. S. Zinchenko and R. N. Coger, "Engineering micropatterned surfaces for the coculture of hepatocytes and Kupffer cells," *Journal of Biomedical Materials Research A*, vol. 75, no. 1, pp. 242–248, 2005.

[41] J. Shaikh-Mohammed, M. A. DeCoster, and M. J. McShane, "Cell adhesion testing using novel testbeds containing micropatterns of complex nanoengineered multilayer films," in *Proceedings of the 26th Annual International Conference of the IEEE Engineering in Medicine and Biology Society (EMBC '04)*, vol. 4, pp. 2671–2674, September 2004.

[42] H. Ma, D. Li, X. Sheng, B. Zhao, and A. Chilkoti, "Protein-resistant polymer coatings on silicon oxide by surface-initiated atom transfer radical polymerization," *Langmuir*, vol. 22, no. 8, pp. 3751–3756, 2006.

[43] L. Lu, K. Nyalakonda, L. Kam, R. Bizios, A. Göpferich, and A. G. Mikos, "Retinal pigment epithelial cell adhesion on novel micropatterned surfaces fabricated from synthetic biodegradable polymers," *Biomaterials*, vol. 22, no. 3, pp. 291–297, 2001.

[44] J. Y. Lee, C. Jones, M. A. Zern, and A. Revzin, "Analysis of local tissue-specific gene expression in cellular micropatterns," *Analytical Chemistry*, vol. 78, no. 24, pp. 8305–8312, 2006.

[45] A. Khademhosseini, K. Y. Suh, J. M. Yang et al., "Layer-by-layer deposition of hyaluronic acid and poly-L-lysine for patterned cell co-cultures," *Biomaterials*, vol. 25, no. 17, pp. 3583–3592, 2004.

[46] N. A. Bullett, D. P. Bullett, F. E. Truica-Marasescu, S. Lerouge, F. Mwale, and M. R. Wertheimer, "Polymer surface micropatterning by plasma and VUV-photochemical modification for controlled cell culture," *Applied Surface Science*, vol. 235, no. 4, pp. 395–405, 2004.

[47] C. J. Pan, Y. X. Dong, Y. D. Nie, and Y. L. Wang, "Effects of protein micropatterns of biomaterials surfaces on human chondrocytes morphology and protein expression," *Progress in Biochemistry and Biophysics*, vol. 37, no. 12, pp. 1296–1302, 2010.

12

The Effects of Laser Marking and Symbol Etching on the Fatigue Life of Medical Devices

P. J. Ogrodnik, C. I. Moorcroft, and P. Wardle

Staffordshire University, Stafford ST18 0AD, UK

Correspondence should be addressed to P. J. Ogrodnik; p.j.ogrodnik@staffs.ac.uk

Academic Editor: Michel Labrosse

This paper examines the question; "does permanent laser marking affect the mechanical performance of a metallic medical component?" The literature review revealed the surprising fact that very little has been presented or studied even though intuition suggests that its effect could be detrimental to a component's fatigue life. A brief investigation of laser marking suggests that defects greater than 25 μm are possible. A theoretical investigation further suggests that this is unlikely to cause issues with relation to fast fracture but is highly likely to cause fatigue life issues. An experimental investigation confirmed that laser marking reduced the fatigue life of a component. This combination of lines of evidence suggests, strongly, that positioning of laser marking is highly critical and should not be left to chance. It is further suggested that medical device designers, especially those related to orthopaedic implants, should consider the position of laser marking in the design process. They should ensure that it is in an area of low stress amplitude. They should also ensure that they investigate worst-case scenarios when considering the stress environment; this, however, may not be straightforward.

1. Introduction

It is compulsory to have medical devices permanently marked. Commonly, this information is the part number for the component, the manufacturer's trademark, and the CE symbol (in Europe) and a lot number; to ensure legibility, the size of the lettering is further governed. This means that the positioning of the marking can be arbitrary and may be positioned purely because of available free space away from holes, changes in profile, and so forth. For metallic components, this marking is commonly produced using laser techniques. Most commonly, the laser is used to etch the surface of the component producing a defect whose surface finish is different from the rest of the body, hence making a discernible area. Controlling the position of the laser using CNC techniques can produce letters and graphics. The power and duration of the laser dictate the depth of the etching. The technical details [1] for a specific laser marking device (designed to minimize fatigue issues in safety critical components) state the following:

"Laser etching markers work by focusing energy directly on the surface to be marked. The heat generated by the beam actually alters the surface of the part or vaporizes surface material... in other metals (than steel), the surface is etched when material is removed by high temperature vaporization... it modifies the metal alloying and etches the surface in a way that degrades the part's strength and can lead to fatigue or stress corrosion crack failure."

This is not a bold statement; it is highlighted in several commercial publications. Rosecrans [2], for example, investigated the effect of laser marking on typical materials used in components for the space shuttle. Rosecrans found that the effects on fatigue life were not predictable; in some materials, the fatigue life was markedly reduced, and in others, there was no effect. In 2007, Grivas et al. [3] presented a case report following the failure of a total hip arthroplasty; in this paper, they highlighted the laser marking as a potential source for the fatigue failure of the component. However, this was a forensic study after failure. Prior to this paper [4] had presented an examination of the effect of laser marking on stress corrosion resistance of stainless steels. They stated that the use of a nano-second-based system has a marked

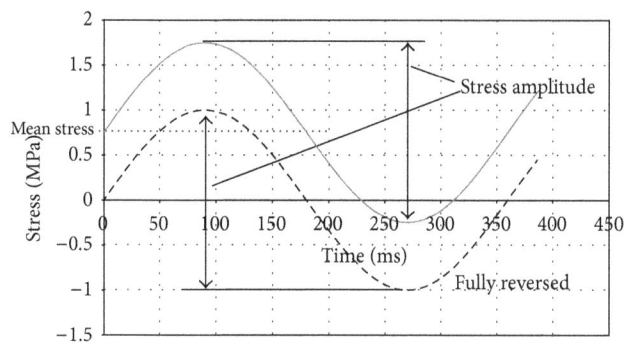

FIGURE 1: Simplified cyclic stress illustration ($\Delta\sigma = 2\,\mathrm{MPa} : f = 1.4\,\mathrm{Hz}$).

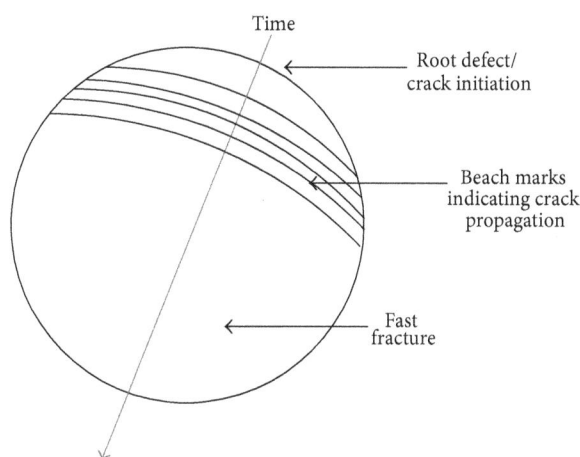

FIGURE 2: Representation of a typical fatigue failure.

deleterious effect on the material's performance. They further highlighted that little research had been conducted in this area which is rather surprising considering the implications. Although one may infer the effects of laser marking from classic fatigue theory, it is surprising that there is little research verifying or disproving the inferences.

Modern implants are designed to last for many years. In the case of an orthopaedic implant in a leg, this could mean 3 million loading cycles per year [5], and hence, it must be designed to be able to withstand fatigue failure. For example, a plate used to treat a plateau fracture of the tibia may be required to withstand cyclic loading for up to 2 years (before the fracture has fully healed), and hence it may be required to withstand 6 million cycles. It is, therefore, in a category that requires fatigue failures to be analysed and minimised. To this extent, the surface finish of these components is such that they are, to all intents and purposes, defect-free. Indeed, it is common for these components to have a virtual mirror finish (a finish commonplace in motorsports specifically to minimize fatigue failures). Unfortunately, as described earlier, they also require permanent marking; and as discussed earlier, this induces a surface defect that can initiate fatigue failure. Recently, it has been suggested that all devices should have a unique bar-code indelibly marked too; the number of defects has suddenly increased by the number of lines on the bar-code. Therefore, this paper examines, through a controlled case study, whether marking has any effect on fatigue life and whether the position and effects of marking on an implantable medical device are worthy of further investigation.

2. Fatigue and Component Failure

Fatigue failure is a well-known phenomenon, but it is worth describing the basics at this point to facilitate later discussions. It occurs in components subject to cyclic loading. The loading may cycle between compression and tension (so-called full reversal), or it can cycle between low and high values of tensile loads (Figure 1). The variation between the maxima and minima is called the stress amplitude, and it is this value that dictates the component's fatigue life. The component's existence is one of the constantly changing stresses, and this can lead to crack propagation. Normally, an initial defect will act as the source of the crack (hence, the mirror finish is described earlier). As time goes on, the crack will grow in length, normally in discernible steps (Figure 2). At some stage, the crack is so large that the component can no longer withstand the loads imparted in it, and it will suddenly fail (often by fast fracture). Thus, a fatigue failure has three, distinct, and observable regions: the point of crack initiation, a region of crack growth (highlighted by "beach marks"), and finally a region of fast fracture.

For components that are easily visible, crack propagation can be investigated and observed. In highly critical areas of engineering, this is conducted using nondestructive testing techniques (NDTs). However, implants in the human body do not avail themselves to gamma irradiation for crack detection, nor are they easily visible for inspection. Hence, implants must be designed to ensure that fatigue failure is minimized; and hence, they are routinely polished to a mirror finish. But they are then subjected to a defect in the form of laser marking. Recently, it has been suggested that class II (and higher) devices should be indelibly marked with a unique identifying bar code and increasing the number of defects by the number of lines in the bar-code.

3. Does a Laser Mark Constitute a Root Defect?

Before this question can be answered, the concept of a root defect must be evaluated. In fact, there are two concepts that are hidden within: the first is crack depth; the second is stress concentration.

In relation to crack depth, this relates to the remaining life of the component and has two main issues. The first issue is critical crack length. This relates to the size of crack that leads to fast fracture of the component (the ultimate failure mode) and is determined by the fracture toughness of the material. For most materials, the critical crack length can be determined using [6]:

$$K_c = \sigma\sqrt{\pi a} \qquad (1)$$

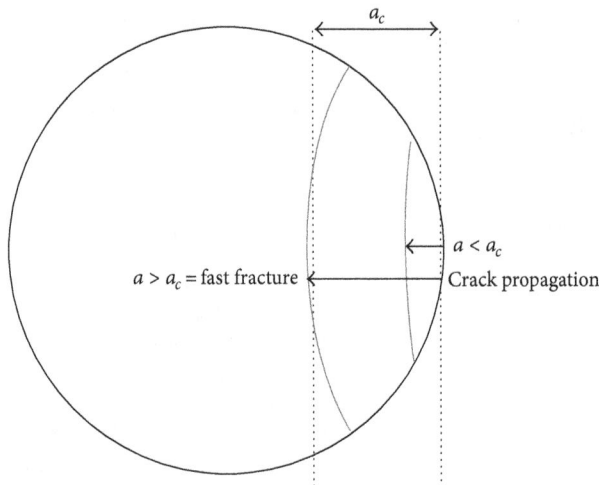

FIGURE 3: Critical crack length for fast fracture.

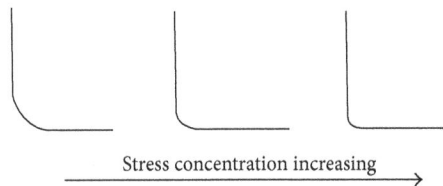

Stress concentration increasing

FIGURE 4: Illustration of relationship between sharp corners and stress concentration.

or

$$a_c = \frac{1}{\pi}\left[\frac{K_c}{\sigma}\right]^2, \qquad (2)$$

where K_c is a material constant (fracture toughness) and σ is the applied stress. If the depth of the root defect is greater than a_c, then the component will fail due to fast fracture. The common process is that the root defect is much smaller than a_c but, through fatigue and crack propagation, the crack finally achieves a value near to a_c and the component fails [6]. This is illustrated in Figure 3.

Stress concentrations are associated with the shape of the defect. In general, sharper defects cause greater concentrations of stress. Peterson's stress concentration factors [7] illustrate that concentration factors of 3 are highly plausible (this means that the actual stress at the root is 3 times the average stress) and the stress concentration factor is never less than 1. Hence, this means that the stress amplitude discussed previously can be much greater than initial estimates, thereby reducing fatigue life considerably. It is, therefore, a commonplace to insert a radius in a corner to alleviate stress concentrations. (as illustrated in Figure 4).

The obvious question is: what does this have to do with laser marking? As previously presented, the process of marking can result in material removal. This is, in effect, the same as producing a small crack. The depth of the crack will depend on the power of the laser, the duration of exposure to the beam, and the number of passes. However, unlike physical

machining, one has no control of the radius at the corners. This is illustrated in Figure 5.

Whilst the illustration in Figure 5 is theoretically plausible, this does not represent factual existence. Figure 6 illustrates the marking of a typical titanium (Ti 6AL 4V) medical component. Figure 6(a) illustrates the marking, and Figure 6(b) illustrates a magnification of the lettering.

Figure 6 clearly demonstrates that the laser marked lettering can be considered to be a defect. In the case of number "1", the transverse nature of the letter makes this highly dangerous. Qi et al. [8] examined the effect of frequency on the depth of laser marking of stainless steels. They demonstrated that 25 μm is achievable with ease. This exceeds the defect size of 0.1–0.8 μm expected from electropolishing or even the 1.6 μm expected from simple grinding. A selection of components from a variety of manufacturers had the laser marking assessed using a Mitutoyo surface roughness analyzer. As expected, there was a great deal of variability due to the variability of substrate and laser etching methodology. However, these initial investigations illustrated that laser marking can exceed 25 μm with ease.

It is, therefore, argued that laser marking does constitute a defect.

4. Case Study to Determine Effect

For the purposes of this case study, a monolateral fixation is assumed (as illustrated in Figure 7). Under normal, static, body weight, the applied force F would be in the region of 900 N (clearly dependent on body mass and gait—when walking the peak load would increase to 1.1 kN). This applied force creates an overall value compressive stress, but the stress in the system is dominated by the bending stress generated by the applied moment.

4.1. Fast Fracture. The component is made from Ti 6Al 4V, then its fracture toughness is, approximately, $K_c = 75\,\mathrm{MPa\,m}^{1/2}$ [9]. Further, assuming the maximum stress is 600 MPa, then the critical crack length is given (from (2)) as

$$a_c = \frac{1}{\pi}\left[\frac{75}{600}\right]^2 = 0.0049\,\mathrm{m}. \qquad (3)$$

Hence, the deformity of 25 μm is not a cause for concern.

4.2. Fatigue Life. From Basquin's law [6]

$$\Delta\sigma = C_1 N_f{}^B \qquad (4)$$

and the data for Ti 6Al 4V from Casavola et al. [10] where $B = -0.09$ and $C_1 = 1068$, the uncracked fatigue life for the component can be estimated. Assuming that the stress amplitude is 340 MPa, hence, the number of cycles to failure is

$$340 = 1068 N_f^{-0.09},$$
$$N_f = 333.6 \times 10^3 \text{ cycles}, \qquad (5)$$

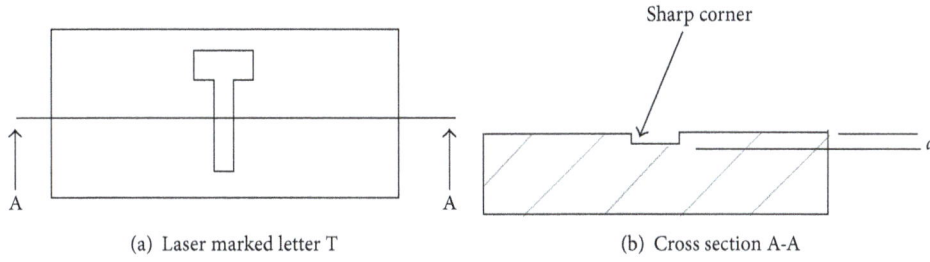

(a) Laser marked letter T   (b) Cross section A-A

FIGURE 5: Representation of laser marking creating a defect.

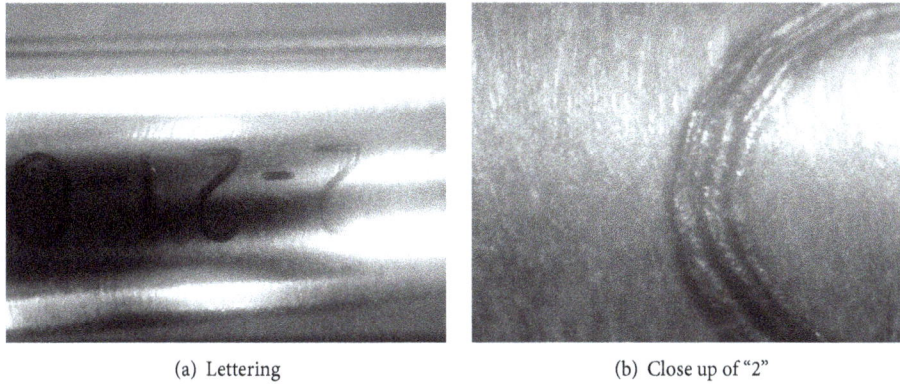

(a) Lettering   (b) Close up of "2"

FIGURE 6: Laser marked lettering of a component.

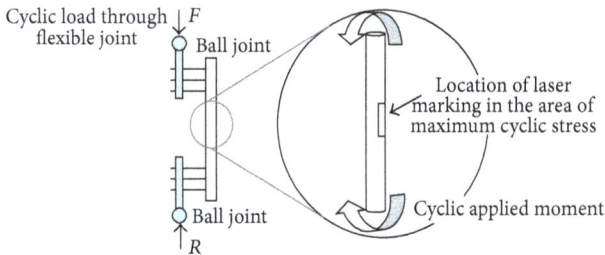

FIGURE 7: Schematic of experiment.

a figure that would be slightly concerning for a long-term implant but not for something that is going to be used in short term. Let us now consider the effect of the defect. Using the Paris' law fatigue crack growth model [6]:

$$\frac{da}{dN} = C(\Delta K)^n, \tag{6}$$

where the constants C and n are material dependent, Ghidini et al. [11] present fatigue data for Ti 6AL 4V for space applications and their data suggests that $C = 6 \times 10^{-13}$ and $n = 4$. This means that (6) can be rearranged and integrated to yield the number of cycles for a crack to grow from one size to another:

$$N = \frac{1}{6.13 \times 10^{-4}(\Delta\sigma)^4\pi^2} \left\{ \frac{1}{a_1} - \frac{1}{a_2} \right\}. \tag{7}$$

TABLE 1: Comparison between fatigue life and initial marking depth.

Initial depth	$a_1 = 25\,\mu m$	$a_1 = 50\,\mu m$	$a_1 = 75\,\mu m$	$a_1 = 37.6\,\mu m$
Life (cycles)	503×10^3	250×10^3	165×10^3	333.6×10^3
Reduction in life	N/A	25%	50%	0%

The starting crack size $a_1 = 0.0025$ mm. For this to grow to the critical crack size $a_2 = 4.9$ mm, the number of cycles can be estimated as

$$N = \frac{1}{6 \times 10^{-13}(340)^4\pi^2} \left\{ \frac{1}{0.000025} - \frac{1}{0.0049} \right\}, \tag{8}$$

$$N = 503 \times 10^3 \text{ cycles.}$$

Table 1 presents fatigue lives when the initial depths are 25, 50, and 75 μm, respectively. It demonstrates that setting the maximum marking depth to be 25 mm would create no further reduction in fatigue life compared to that of an unmarked component.

Interestingly use of this calculation enables the designer to determine the maximum marking depth. The other values all reduce fatigue life by up to 50% compared with the uncracked component. This estimate has not allowed for the effect of stress concentrations nor for the effect of increasing stress as crack length increases. Clearly, increasing the starting depth of the deformity reduces the fatigue life further.

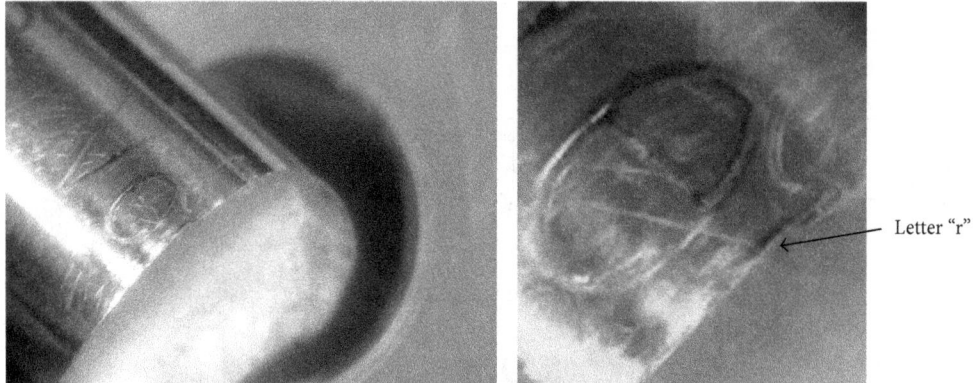

FIGURE 8: Magnified view of the root.

TABLE 2: Experimental fatigue lives.

Theory	Unmarked specimen	Laser marked specimen	Reduction
333,617	315,789	99,745	70%

Once again, this suggests that laser marking does constitute a deformity that has the ability to reduce fatigue life.

5. Experimental Investigation

A controlled experiment was conducted on a simple component manufactured from Ti6AL4V. The component was a single piece monolateral external fixator used in the treatment of distal tibial fractures. Both components were of identical manufacture, but one was laser marked and one not. Because the design of the component is mechanically simple, it was easy to ascertain the location of maximum stress, and as a consequence, the laser marking was placed where its effect was likely to be the greatest. Figure 7 illustrates a schematic of the experiment.

The component was cyclically loaded from 0 to 900 N using a sine function of frequency 3 Hz, causing a cyclic stress amplitude of 340 MN/m^2. Using Basquin's law and the data presented earlier, the estimated life is 333,192 cycles. The actual experimental results are illustrated in Table 2.

Although one could argue that the results may be misleading, an investigation of the root source of the crack demonstrates that it was the laser marking that gave rise to the fatigue crack (Figure 8). This further suggests that laser marking can reduce fatigue life.

6. The Impact on Medical Device Design

Most engineers would not be surprised by the results illustrated above. It is well known that the addition of a defect on a component reduces its fatigue life. So, why is this fact not well documented in the literature associated with medical implants? Why do we find references associated with space applications and not with medical applications? Why is it that the position of laser marking is often left to chance rather than to design? The results of this paper suggest that the placement of laser marking is highly critical. Select the wrong position and trouble, in the form of premature failure, will follow. The main issue, however, is not selecting a position but actually determining where marking cannot be placed. For example, minimally invasive plates are becoming more widespread in orthopaedic trauma. Being minimally invasive means that they are by definition slender. Which, in turn, means they are subject to high stresses. However, design on the CAD system does not fully represent how the loading actually occurs in real life. The placement of the plate and the position of fixings are, almost, random, but there will be a worst-case scenario that does not assume perfect placement. It is, therefore, incumbent on the designer to ensure that they imagine worst case scenarios and determine the location of maximum stresses. Only in this way can the location of laser marking be determined. Almost certainly, the location of the laser marking cannot be left to the discretion of the marker, nor to the discretion of the draftsman; the location where marking cannot be made must be analysed and specified in the design process. It is almost certain that it is incumbent on the design engineer to specify the position where markings cannot be made.

The work also demonstrated that there is a lack of papers/specific research in this important area. Stress corrosion has not been considered, nor has there been a full analysis to determine if pulse rate can be used to mitigate effects or whether different marking techniques could be adopted or developed. It would seem that the effect of laser marking has been treated in a "matter of fact" way when its effects can be disastrous (as in [3]). It is not possible to, simply, extract data from the aerospace industries (where loading patterns can be prescribed) and assume that this must be the case for medical devices (where the loading regime is not prescribed). Certainly, the first important study is to determine suitable marking depths for critical materials (to ensure legibility) and for clinical evaluations to consider this depth in the analysis. Certainly, future clinical evaluations should cover marking location in the risk analysis. For long term implants, more important are the consequences of stress corrosion.

It is suggested, due to the lack of research papers in this field, that further research should be conducted on the effects of laser marking on the life of implantable and nonimplantable medical devices.

7. Conclusions

The limited study presented in this paper suggests that laser marking can be deleterious to the fatigue life of a medical device. The use of laser marking is widespread and is, probably, the only commercially viable permanent marking system available. Hence, it is incumbent on medical device designers to document and justify the position of laser marking to ensure that it does not, by pure chance or mishap, become located in a region where high cyclic stresses are experienced. This may not be as simple as it first seems as many implantable medical components may not, actually, be used in the exact configuration they were designed at. This is not misused by the surgeon or clinician; this is a fact which is wholly due to the variability associated with humans and their range of potential ailments. This is particularly true in orthopaedic trauma where, for example, the range of fracture patterns for one particular device can be highly variable.

Notations

A: Paris' law constant
a: Crack length (m)
a_c: Critical crack length (m)
B: Basquin's law constant
C_1: Basquin's law constant
K: Stress intensity Factor ($\mathrm{MN\,m^{-3/2}}$)
K_c: Fracture toughness ($\mathrm{MN\,m^{-3/2}}$)
M: Paris' law constant
N: Number of cycles
N_f: Number of cycles to failure
σ: Stress ($\mathrm{MN\,m^{-2}}$).

References

[1] C. B. Dane, L. Hackel, J. Honig et al., *Lasershot Marking System:High-Volume Labeling For Safety-Critical Parts*, Lawrence Livermore National Laboratory: U.S. Department of Energy, 2001.

[2] L Rosecrans, "Effects of laser marking on fatigue strengths of selected space shuttle main engine materials," in *The Changing Frontiers of Laser Materials Processing*, pp. 223–230, Arlington, Va, USA, November 1986.

[3] T. B. Grivas, O. D. Savvidou, S. A. Psarakis et al., "Neck fracture of a cementless forged titanium alloy femoral stem following total hip arthroplasty: a case report and review of the literature," *Journal of Medical Case Reports*, vol. 1, article 174, 2007.

[4] S. Valette, P. Steyer, L. Richard, B. Forest, C. Donnet, and E. Audouard, "Influence of femtosecond laser marking on the corrosion resistance of stainless steels," *Applied Surface Science*, vol. 252, no. 13, pp. 4696–4701, 2006.

[5] C. I. Moorcroft, *Control and monitoring of fracture movement in fractures of the human tibia [Ph.D. thesis]*, Staffordshire University, 1998.

[6] M. F. Ashby and D. R. H. Jones, *Engineering Materials 1: An Introduction to Properties, Applications and Design: v. 1*, Butterworth-Heinemann, 2005.

[7] W. D. Pilkey and D. F. Pilkey, *Peterson's Stress Concentration Factors*, John Wiley & Sons, Chichester, UK, 2008.

[8] J. Qi, K. L. Wang, and Y. M. Zhu, "A study on the laser marking process of stainless steel," *Journal of Materials Processing Technology*, vol. 139, no. 1–3, pp. 273–276, 2003.

[9] MATWEB, Material Properties: Ti-6AL-4V, 2012, http://www.matweb.com/.

[10] C. Casavola, C. Pappalettere, and G. Pluvinage, "Fatigue resistance of titanium laser and hybrid welded joints," *Materials and Design*, vol. 32, no. 5, pp. 3127–3135, 2011.

[11] T. Ghidini, A. de Rooij, A. Graham, M. Nikullainen, and G. Bussu, "The role of failure analysis and fracture mechanics in the spacecraft systems safety," in *European Space Agency, Workshop on Fracture Control of Spacecraft, Launchers and their Payloads and Experiments*, Noordwijk, The Netherlands, February 2009.

13

A Hybrid Image Filtering Method for Computer-Aided Detection of Microcalcification Clusters in Mammograms

Xiaoyong Zhang,[1] Noriyasu Homma,[1] Shotaro Goto,[2] Yosuke Kawasumi,[3] Tadashi Ishibashi,[3] Makoto Abe,[2] Norihiro Sugita,[2] and Makoto Yoshizawa[1]

[1] *Research Division on Advanced Information Technology, Cyberscience Center, Tohoku University, 6-6-05 Aoba, Aramaki, Aoba-ku, Sendai 980-8579, Japan*
[2] *Graduate School of Engineering, Tohoku University, 6-6-05 Aoba, Aramaki, Aoba-ku, Sendai 980-8579, Japan*
[3] *Tohoku University Graduate School of Medicine, Tohoku University, 2-1 Seiryo-mashi, Aoba-ku, Sendai 980-8575, Japan*

Correspondence should be addressed to Xiaoyong Zhang; xiaoyong@ieee.org

Academic Editor: Valentina Camomilla

The presence of microcalcification clusters (MCs) in mammogram is a major indicator of breast cancer. Detection of an MC is one of the key issues for breast cancer control. In this paper, we present a highly accurate method based on a morphological image processing and wavelet transform technique to detect the MCs in mammograms. The microcalcifications are firstly enhanced by using multistructure elements morphological processing. Then, the candidates of microcalcifications are refined by a multilevel wavelet reconstruction approach. Finally, MCs are detected based on their distributions feature. Experiments are performed on 138 clinical mammograms. The proposed method is capable of detecting 92.9% of true microcalcification clusters with an average of 0.08 false microcalcification clusters detected per image.

1. Introduction

Breast cancer is one of the major causes of mortality in middle-aged women, especially in developed countries [1]. At present, there are no effective ways to prevent breast cancer since its cause remains unknown [2]. Therefore, early detection becomes the key to improving the breast cancer prognosis and reducing the mortality rates. Mammography has been widely recognized as being one of the most effective imaging modalities for early detection of breast cancer. However, it is a hard work for radiologists to provide both accurate and uniform evaluation for the enormous number of mammograms generated in widespread screening. A computer-aided detection or diagnosis (CAD) system, which uses computer technologies to detect the typical signs

of breast cancer, has been developed to provide a "second opinion" for radiologists and to improve the accuracy and stability of diagnosis.

In general, there are three signs of breast cancer in a mammogram: microcalcification clusters (MCs), architectural distortions, and masses [2]. In this paper, we particularly focus on the detection of MCs since they appear in 30–50% of mammographic diagnosed cases and show a high correlation with breast cancer [3]. According to the Breast Image Reporting and Data System (BI-RADS) lexicon [4], MCs are tiny calcium deposits that appear as small bright spots in mammograms. As an example, Figure 1 shows an MC in a mediolateral-oblique (MLO) mammogram. It is often hard for radiologists to find individual MCs in mammograms because they are very small (typically, 0.05–1 mm [3]) in the

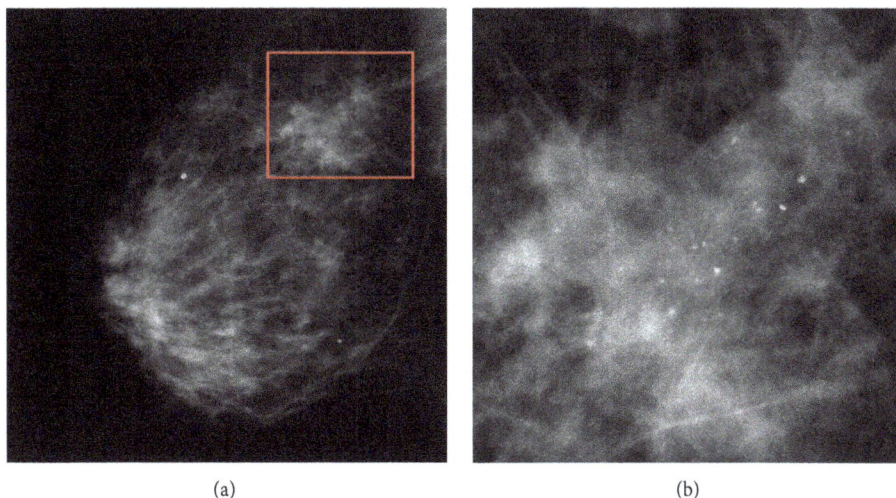

(a)   (b)

FIGURE 1: An example of an MC. (a) A mediolateral-oblique (MLO) mammogram. (b) Expanded view showing the MC.

size and the contrast between the MCs and the surrounding breast tissue is not high enough.

Over the past two decades, there has been extensive research focused on developing the CAD tools for automatic detection of MCs in mammograms. Several review papers have also been published on this topic [2, 4–8]. As described in [8], MC detection can be classified into four categories: (1) image enhancement methods [9–13]; (2) multiscale decomposition methods [14–17]; (3) stochastic modeling methods [18, 19]; and (4) machine learning methods [20–28]. In [27], a performance evaluation was summarized using free-response receiver operating characteristic (FROC) in which a support vector machine (SVM) approach [27] is found to be superior to conventional methods and is capable of achieving a true positive rate of approximately 90% when the false-positive (FP) rate is on average of 1.1 FP clusters per image. However, such FP is not low enough for clinical application.

In this paper, we present a new method based on a hybrid method which combines a morphological technique [29] and wavelet decomposition processing for detecting the MCs in mammograms. Compared with previous methods, the proposed method can achieve not only a high sensitivity but also a lower FP. In the proposed method, we first use a multistructuring-elements (SEs-) based top-hat transform to enhance the intensity of microcalcification. Subsequently, we employ a wavelet decomposition to refine the enhanced results for removing the false enhanced microcalcifications. Then, based on the feature of malignant MCs, threshold processing is used to segment the MCs from mammograms.

The rest of paper is organized as follows: Section 2 presents a morphological processing technique to enhance the MCs in mammogram. In Section 3, a wavelet decomposition method is employed to refine the MCs candidates. Section 4 presents the detection of the MCs and gives the experimental results by using the proposed method. A conclusion is given in Section 5.

2. Enhancement of MCs Using Top-Hat Transform

As mentioned in Section 1, the MCs appear as small bright spots in mammograms. In this sense, MCs can be directly segmented by using a threshold process. However, since most mammograms have a low dynamical rang and the intensity contrast between MCs and surrounding tissue is quite low, selection of a threshold for the whole image is not an easy task. As a solution, the difference of Gaussian (DoG-) based method which approximates the individual microcalcification as a two-dimensional (2D) Gaussian kernel has been reported in [13], but this method only is suitable for detecting the microcalcifications with approximate circle shape.

In this section, we propose a new method which is based on a morphological filtering technique [29] to enhance the individual microcalcifications for MC detection. The basic idea of the method is to use a set of top-hat transforms based on multi-structuring elements (SEs) of which sizes and shapes are fitted to the individual microcalcifications to enhance them. The top-hat transform of a gray-scale image f is defined as f minus its opening by structuring element b [30]:

$$T = f - (f \circ b), \qquad (1)$$

where $\circ$ denotes the opening operation; the difference operation yields an image in which only the components fitting to the SE remain. Figure 2 illustrates the concept of the top-hat transform in one dimension.

In the top-hat transform, selection of an appropriate SE fitting to the target objects is the key. Since individual microcalcifications in mammograms frequently vary both in size and shape, it is imposable to use a single SE to remove all of them. To solve this problem, we use a multi-SEs-based method which uses eight different flat SEs, denoted as b_i ($i = 1, 2, \ldots, 8$) to remove the individual microcalcifications in

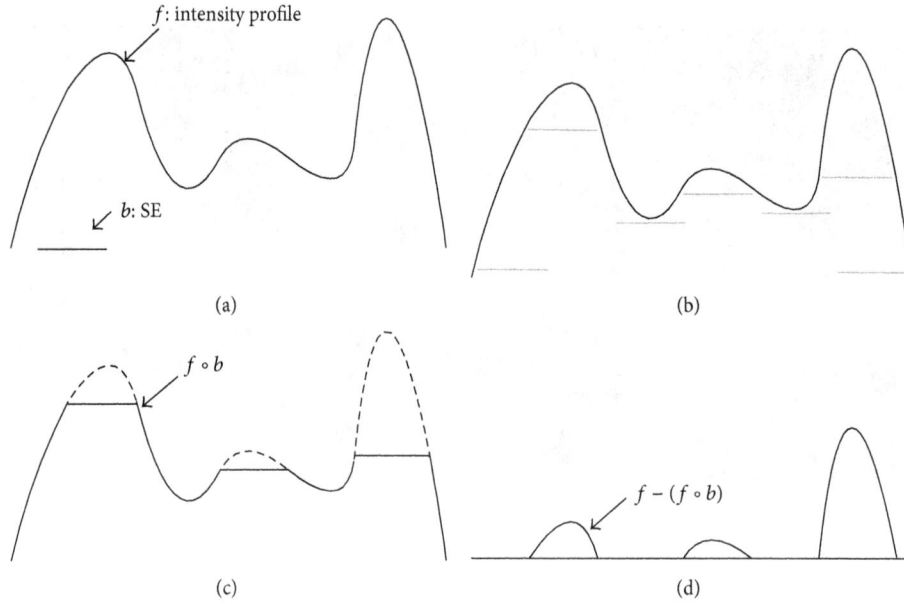

FIGURE 2: Top-hat transform in one dimension. (a) Original 1D signal and SE. (b) SE pushed up underneath the signal. (c) Opening. (d) Signal subtraction with opening result.

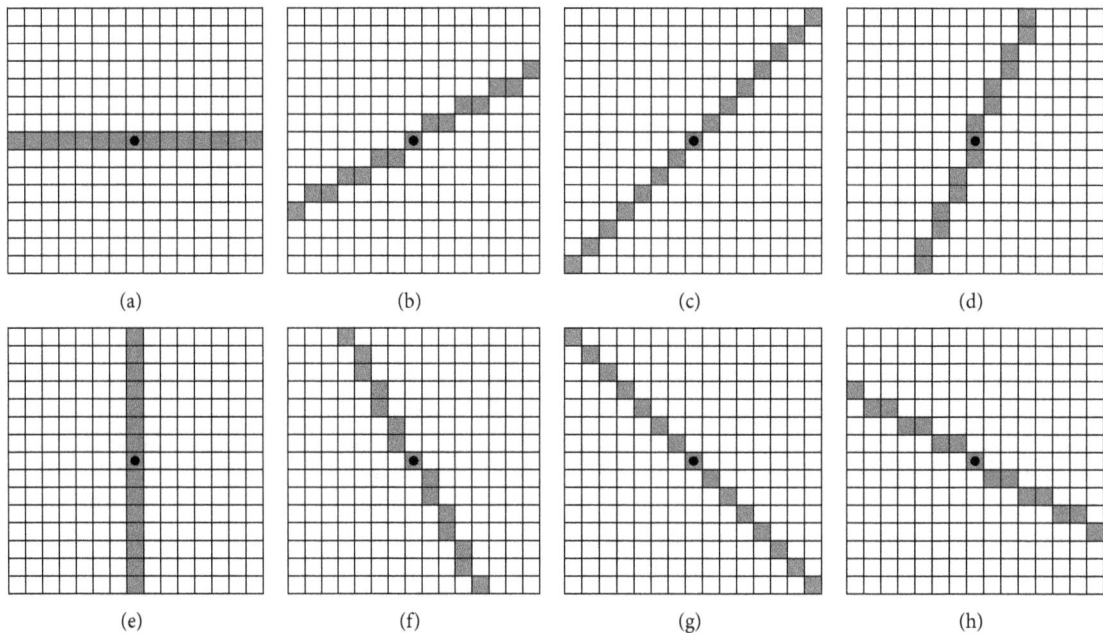

FIGURE 3: Multi-SEs converted to rectangular arrays. The dots denote the center of the SEs. These SEs are designed for fitting to the individual microcalcifications with different shapes.

the opening processing. In this paper, the SEs are a set of lines revolving around its center in a 15 × 15 array. For a mammogram image with resolution 0.05 mm/pixel, the 15 × 15 array indicates that the objects whose size are larger than $0.75 \times 0.75\,\text{mm}^2$ will be removed in the opening operation. Figure 3 shows the flat SEs used in the top-hat transform where the dots denote the centers of the SEs.

For each SE b_i, $i = 1, 2, \ldots, 8$, the opening operation yields an image, denoted by $(f \circ b_i)$, in which the individual

microcalcifications fitting to the SE are removed. Then, in order to enhance these individual microcalcifications removed by the opening operation, we use a subtraction between the original image and the maximum of the opening results to obtain an image, given by

$$E = f - \arg\max\left(f \circ b_i\right). \tag{2}$$

Figure 4 shows an original mammogram and an enhanced image obtained from (2). Comparing Figure 4(b) with

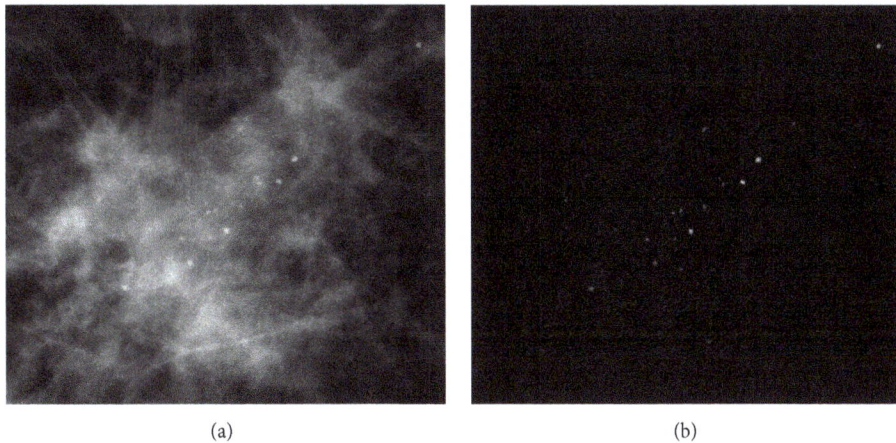

FIGURE 4: Top-hat transform enhancement of microcalcification. (a) An original mammogram image. (b) The results of the top-hat transform.

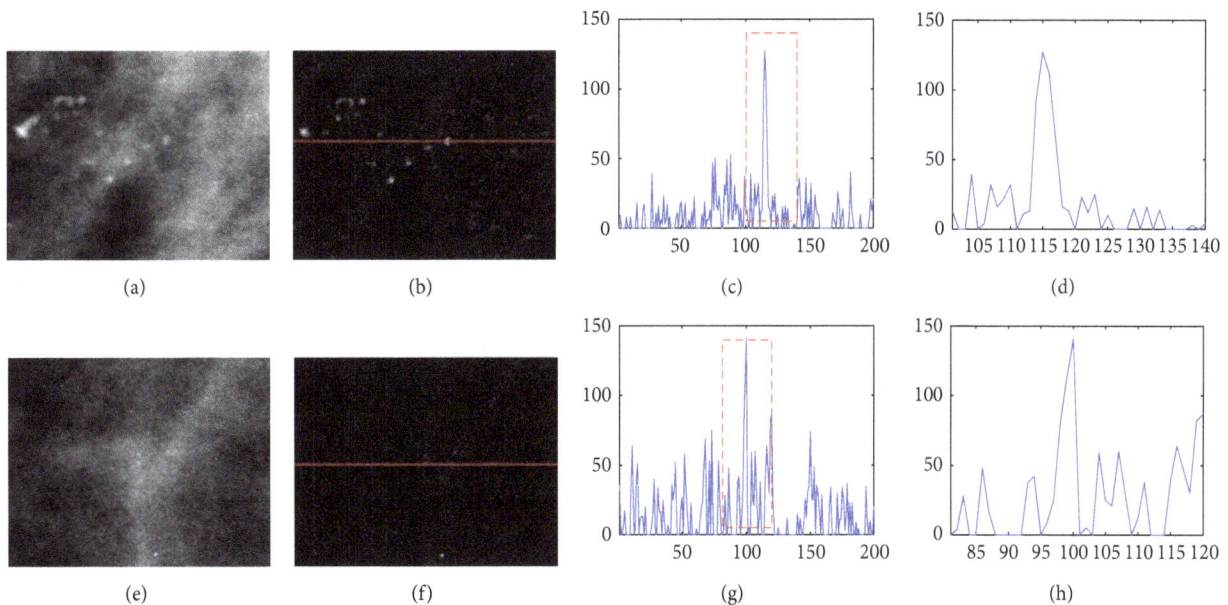

FIGURE 5: Comparison of intensity profile between microcalcification and mammary gland. (a) Microcalcifications in original mammogram image. (b) The result of the top-hat transform of (a). (c) Intensity profile of a row in (b). (d) Expanded view of the dashed rectangle in (c). (e) Mammary glands in original mammogram image. (f) The result of the top-hat transform of (e). (g) Intensity profile of the row in (f). (h) Expanded view of the dashed rectangle in (g).

Figure 4(a), we can see that individual microcalcifications appearing on a complex background were enhanced successfully. These results indicate that the microcalcifications can be easily segmented by using a threshold process. However, the top-hat transform also enhances some undesired objects in the mammograms, such as mammary glands, vessels, and so on. Therefore, a refinement processing is required to remove these undesirable objects from the enhanced mammogram images.

3. Denoising Using Wavelet Decomposition

As mentioned above, although the top-hat transform is capable of enhancing microcalcifications varying in size and shape, a side effect is that the undesirable objects are also enhanced. Based on an investigation, we found that these unwanted objects are mainly caused by the soft tissues, such as mammary glands and vessels [4]. The typical feature of them is that they have a relatively higher intensity compared with their surrounding area as well as an inner nonhomogeneous intensity. Figure 5 shows a comparison between the microcalcification and the mammary gland after the top-hat transform. Comparing Figures 5(c) and 5(g), we find that the intensity of microcalcification in the enhanced image approximately equals that of the mammary glands. However, since the individual microcalcification has almost homogeneous intensity, the size of the enhanced microcalcification is generally larger than that of the mammary gland.

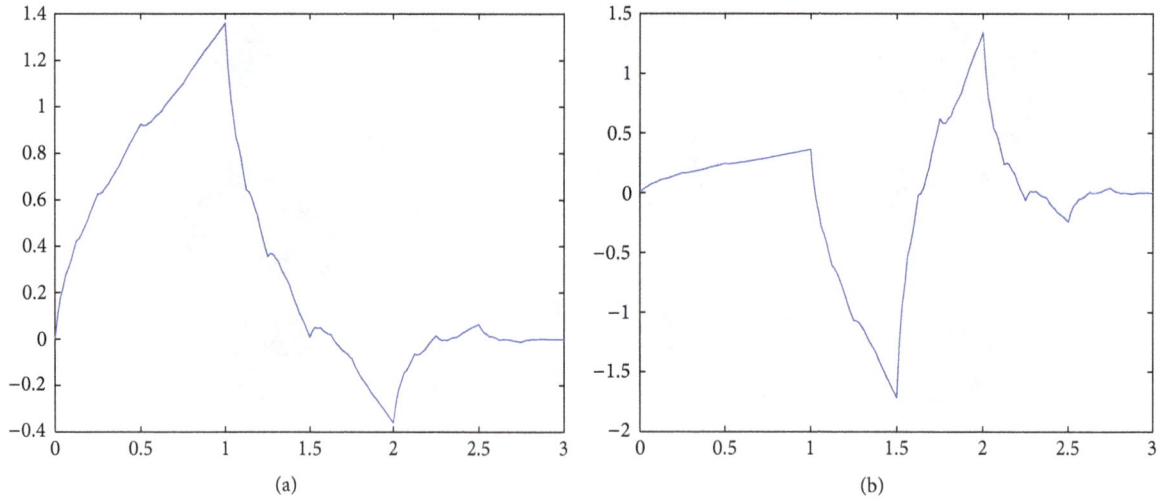

FIGURE 6: The second-order symmetrical wavelets. (a) Scaling function. (b) Wavelet function.

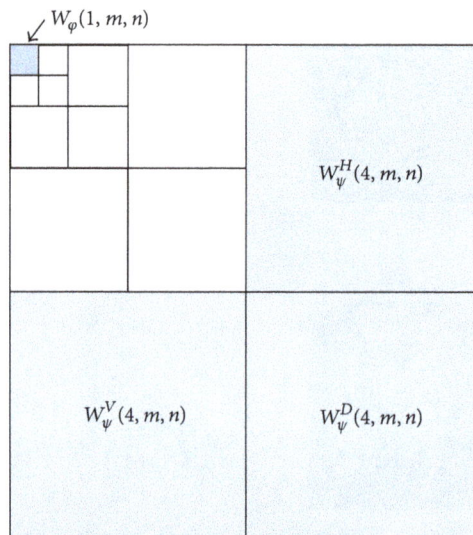

FIGURE 7: Thresholding the approximation and detail coefficients in the wavelet decomposition.

Figures 5(d) and 5(e) show the expanded view of the dashed rectangles in Figures 5(c) and 5(g), respectively. We can find that the size of the microcalcification is almost twice as large as the size of the mammary gland. Therefore, the enhanced mammary glands can be treated as noise that can be removed according to their size. In this paper, we employ a wavelet denoising method to remove the mammary glands because the wavelet decomposition can easily separate them according to their particular size.

The wavelet-based procedure for denoising the image consists of the following four steps.

Step 1. A four-scale wavelet transform is used to decompose the image obtained from the top-hat transform. We select the second-order symmetrical wavelets, short by "symlet-2," as the decomposition filters since they have the least asymmetry and highest number of vanishing moments [31].

Figures 6(a) and 6(b) show the scaling function and wavelet function used in the decomposition filtering. Figure 7 shows the decomposition results in which $W_\psi^H(4, m, n)$, $W_\psi^V(4, m, n)$, and $W_\psi^D(4, m, n)$ denote the detail coefficients at level 4 size of half the original image.

Step 2. Thresholding the approximation and detail coefficients. Since the size of the microcalcification is almost twice as large as the size of the mammary gland, the noise caused by the mammary gland can be decomposed in level 4. Therefore, we set the detail coefficients at level 4 as well as the approximation coefficients at level 1 to zeros. Figure 7 shows the result of the four-scale wavelet transform in which the shaded coefficients marked are set to zero.

Step 3. Compute the inverse wavelet transform using the modified detail coefficient. Figures 8(a) and 8(b) show the

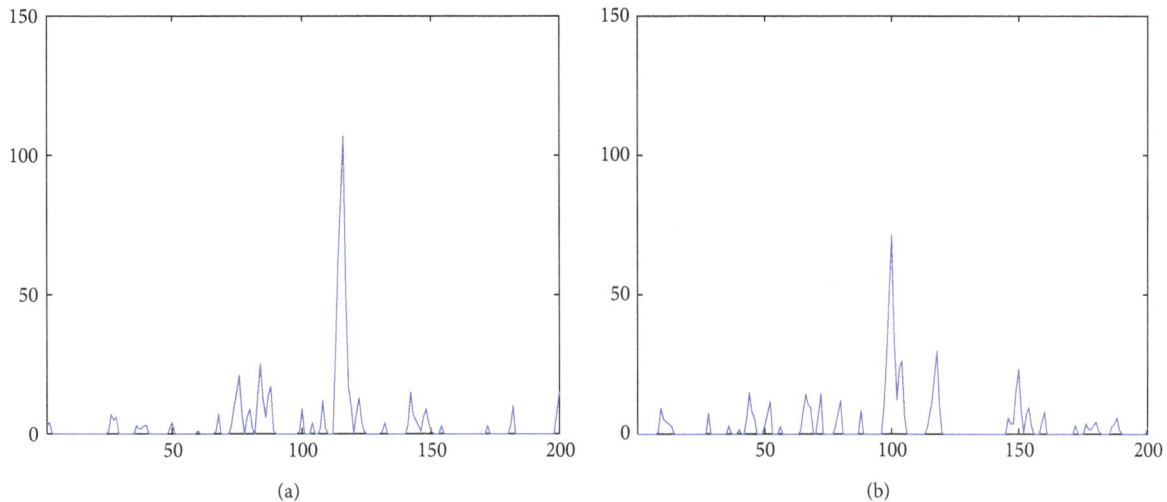

FIGURE 8: Results of the inverse wavelet transform. (a) Intensity profiles of microcalcification. (b) The intensity profiles of mammary gland candidates.

TABLE 1: Experimental results of MCs detection.

Methods	TP (True Positive)	FPs (False positives)/image
Difference-of-Gaussians- (DoG-) based method [13]	60%	1.2
Image-difference-technique- (IDT-) based method [9]	70%	1.5
Neural network-based method [20]	70%	1.1
Wavelet-based method [14]	75%	1.5
Support-vector-machine- (SVM-) based method [8]	90%	1.3
Weighted-local-differences-based method [32]	70%	1.9
Texture-coding-based method [33]	95%	4.3
Proposed method	92.9%	0.08

results of the inverse wavelet transform of the microcalcification and mammary gland in Figures 5(d) and 5(h). We see that the intensity of the mammary gland is clearly reduced so that the microcalcifications and mammary glands can be easily separated using a threshold.

Step 4. Segment the individual microcalcifications. In this step, we use a thresholding processing to obtain a binary image in which each microcalcification is segmented as connected components.

4. Detection of MCs and Experiments

In this section, we introduce a procedure that segments the MCs from the mammogram image obtained from the above sections and give some experimental results by using the proposed method. As mentioned in Section 1, the MCs are tiny calcium deposits clustering together in the mammogram images. Based on an investigation on the BI-RADS [4], we found that a typical MC generally has the following two features:

(1) an MC usually consists of four or more individual microcalcifications;

(2) an MC generally appears within a limited area of size $10 \times 10 \, \text{mm}^2$ in mammogram images.

According to these two features, we use the following two steps to detect the MCs. First, the binary image is subdivided into blocks that overlap their neighbors both horizontally and vertically. The blocks are of size 200×200 pixels; for each pair of overlapping blocks, the overlapping boundary region is of size 100×100 pixels. Second, for each block, if the total number of the connected components is larger than three, this block will be labeled as an MC.

We developed and tested the proposed method using a database collected by the Tohoku University School of Medicine. This data set consists of 138 clinical mammograms in which 14 mammograms contain MCs. These mammograms are of size 4740×3540 pixels, with a spatial resolution of 0.05 mm/pixel and 16-bit grayscale.

The performance of the proposed method is summarized by the true positive rate and false positive cluster per image. Table 1 summarizes the experimental results of the proposed method comparing with several previous methods [8, 9, 13, 14, 32, 33]. The proposed method is capable of detecting 92.9% of true microcalcification clusters with an average of 0.08 false microcalcification clusters detected per image. To our best knowledge, this performance is better than most state-of-the-art methods in MC detection [2, 34].

In the experiments, we found that the FPs are mainly caused by the linear-structure tissues interlacing with each

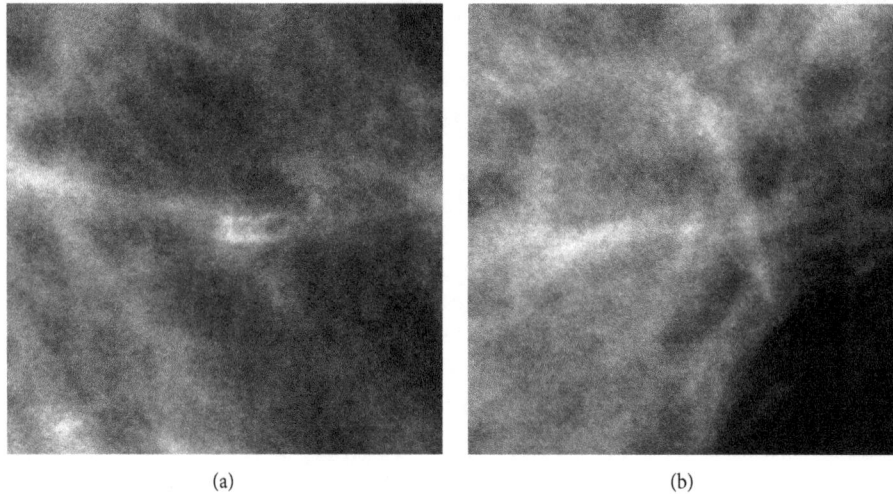

(a)   (b)

FIGURE 9: Objects causing FPs of MCs detection. (a), (b) Linear-structure tissues interlacing with each other.

other or benign calcifications with nonhomogenous intensity. Figure 9 shows three examples of these objects in mammograms. Figures 9(a) and 9(b) show the linear-structure tissues. Since the characteristics of linear-structure tissues are quite different with that of the MCs, we consider that FPs can still be reduced by using a statistic classifier, such as a support vector machine (SVM).

5. Conclusion

In this paper, we presented a high-accuracy method for the detection of MCs in mammograms. The proposed method combined a multi-SEs top-hat transform and a wavelet-based denoising approach to enhance the individual microcalcifications in mammograms. Experimental results demonstrated that the proposed method is capable of detecting the MCs in mammograms with high accuracy.

In further work, we will conduct more experiments on a wider database and improve its performance both in accuracy and computational cost.

References

[1] R. E. Bird, T. W. Wallace, and B. C. Yankaskas, "Analysis of cancers missed at screening mammography," *Radiology*, vol. 184, no. 3, pp. 613–617, 1992.

[2] J. Tang, R. M. Rangayyan, J. Xu, I. E. El Naqa, and Y. Yang, "Computer-aided detection and diagnosis of breast cancer with mammography: recent advances," *IEEE Transactions on Information Technology in Biomedicine*, vol. 13, no. 2, pp. 236–251, 2009.

[3] D. B. Kopans, *Breast Imaging*, J. B. Lippincott Company, 1989.

[4] H. D. Cheng, X. Cai, X. Chen, L. Hu, and X. Lou, "Computer-aided detection and classification of microcalcifications in mammograms: a survey," *Pattern Recognition*, vol. 36, no. 12, pp. 2967–2991, 2003.

[5] M. L. Giger, "Computer-aided diagnosis of breast lesions in medical images," *Computing in Science and Engineering*, vol. 2, no. 5, pp. 39–45, 2000.

[6] N. Karssemeijer and J. H. Hendriks, "Computer-assisted reading of mammograms," *European Radiology*, vol. 7, no. 5, pp. 743–748, 1997.

[7] L. Zhang, R. Sankar, and W. Qian, "Advances in microcalcification clusters detection inmammography," *Computers in Biology and Medicine*, vol. 32, no. 6, pp. 515–528, 2002.

[8] I. El Naqa and Y. Yang, "Techniques in the detection of microcalcification clusters in digital mammograms," *Medical Imaging Systems Technology: Methods in Diagnosis Optimization*, vol. 4, pp. 15–36, 2005.

[9] R. M. Nishikawa, M. L. Giger, K. Doi, C. J. Vyborny, and R. A. Schmidt, "Computer-aided detection of clustered microcalcifications on digital mammograms," *Medical and Biological Engineering and Computing*, vol. 33, no. 2, pp. 174–178, 1995.

[10] K. J. McLoughlin, P. J. Bones, and N. Karssemeijer, "Noise equalization for detection of microcalcification clusters in direct digital mammogram images," *IEEE Transactions on Medical Imaging*, vol. 23, no. 3, pp. 313–320, 2004.

[11] W. Qian, F. Mao, X. Sun, Y. Zhang, D. Song, and R. A. Clarke, "An improved method of region grouping for microcalcification detection in digital mammograms," *Computerized Medical Imaging and Graphics*, vol. 26, no. 6, pp. 361–368, 2002.

[12] M. G. Linguraru, K. Marias, R. English, and M. Brady, "A biologically inspired algorithm for microcalcification cluster detection," *Medical Image Analysis*, vol. 10, no. 6, pp. 850–862, 2006.

[13] J. Dengler, S. Behrens, and J. F. Desaga, "Segmentation of microcalcifications in mammograms," *IEEE Transactions on Medical Imaging*, vol. 12, no. 4, pp. 634–642, 1993.

[14] G. Lemaur, K. Drouiche, and J. DeConinck, "Highly regular wavelets for the detection of clustered microcalcifications in mammograms," *IEEE Transactions on Medical Imaging*, vol. 22, no. 3, pp. 393–401, 2003.

[15] M. G. Mini, V. P. Devassia, and T. Thomas, "Multiplexed wavelet transform technique for detection of microcalcification in digitized mammograms," *Journal of Digital Imaging*, vol. 17, no. 4, pp. 285–291, 2004.

[16] R. Nakayama, Y. Uchiyama, K. Yamamoto, R. Watanabe, and K. Namba, "Computer-aided diagnosis scheme using a filter bank for detection of microcalcification clusters in mammograms,"

IEEE Transactions on Biomedical Engineering, vol. 53, no. 2, pp. 273–283, 2006.

[17] E. Regentova, L. Zhang, J. Zheng, and G. Veni, "Microcalcification detection based on wavelet domain hidden Markov tree model: study for inclusion to computer aided diagnostic prompting system," *Medical Physics*, vol. 34, no. 6, pp. 2206–2219, 2007.

[18] M. N. Gürcan, Y. Yardimci, A. E. Çetin, and R. Ansari, "Detection of microcalcifications in mammograms using higher order statistics," *IEEE Signal Processing Letters*, vol. 4, no. 8, pp. 211–216, 1997.

[19] B. Caputo, E. L. Torre, S. Bouattour, and G. E. Gigante, "A new kernel method for microcalcification detection: spin Glass-Markov random fields," *Studies in Health Technology and Informatics*, vol. 90, pp. 30–34, 2002.

[20] S. Yu and L. Guan, "A CAD system for the automatic detection of clustered microcalcifications in digitized mammogram films," *IEEE Transactions on Medical Imaging*, vol. 19, no. 2, pp. 115–126, 2000.

[21] J. Jiang, B. Yao, and A. M. Wason, "A genetic algorithm design for microcalcification detection and classification in digital mammograms," *Computerized Medical Imaging and Graphics*, vol. 31, no. 1, pp. 49–61, 2007.

[22] Y. Peng, B. Yao, and J. Jiang, "Knowledge-discovery incorporated evolutionary search for microcalcification detection in breast cancer diagnosis," *Artificial Intelligence in Medicine*, vol. 37, no. 1, pp. 43–53, 2006.

[23] L. Bocchi, G. Coppini, J. Nori, and G. Valli, "Detection of single and clustered microcalcifications in mammograms using fractals models and neural networks," *Medical Engineering and Physics*, vol. 26, no. 4, pp. 303–312, 2004.

[24] M. N. Gurcan, H. P. Chan, B. Sahiner, L. Hadjiiski, N. Petrick, and M. A. Helvie, "Optimal neural network architecture selection: improvement in computerized detection of microcalcifications," *Academic Radiology*, vol. 9, no. 4, pp. 420–429, 2002.

[25] A. Papadopoulos, D. I. Fotiadis, and A. Likas, "An automatic microcalcification detection system based on a hybrid neural network classifier," *Artificial Intelligence in Medicine*, vol. 25, no. 2, pp. 149–167, 2002.

[26] P. Sajda, C. Spence, and J. Pearson, "Learning contextual relationships in mammograms using a hierarchical pyramid neural network," *IEEE Transactions on Medical Imaging*, vol. 21, no. 3, pp. 239–250, 2002.

[27] I. El-Naqa, Y. Yang, M. N. Wernick, N. P. Galatsanos, and R. M. Nishikawa, "A support vector machine approach for detection of microcalcifications," *IEEE Transactions on Medical Imaging*, vol. 21, no. 12, pp. 1552–1563, 2002.

[28] S. Singh, V. Kumar, H. K. Verma, and D. Singh, "SVM based system for classification of microcalcifications in digital mammograms," in *Proceedings of the 28th Annual International Conference of the IEEE Engineering in Medicine and Biology Society (EMBS '06)*, pp. 4747–4750, New York, NY, USA, September 2006.

[29] L. Wei, Y. Yang, R. M. Nishikawa, M. N. Wernick, and A. Edwards, "Relevance vector machine for automatic detection a of clustered microcalcifications," *IEEE Transactions on Medical Imaging*, vol. 24, no. 10, pp. 1278–1285, 2005.

[30] R. C. Gonzalez and E. W. Richard, *Digital Image Processing*, Prentice Hall, New York, NY, USA, 3rd edition, 2009.

[31] I. Daubechies, *Ten Lecture on Wavelet*, SIAM, Philadelphia, Pa, USA, 1992.

[32] Y. Kabbadj, F. Regragui, and M. M. Himmi, "Detection of microcalcification in digitized mammograms using weighted local differences and local contrast," *Applied Mathematical Sciences*, vol. 6, no. 131, pp. 6533–6544, 2012.

[33] F. Eddaoudi and F. Regragui, "Microcalcifications detection in mammographic images using texture coding," *Applied Mathematical Sciences*, vol. 5, no. 8, pp. 381–393, 2011.

[34] H. Jing, Y. Yang, and R. M. Nishikawa, "Detection of clustered microcalcifications using spatial point process modeling," *Physics in Medicine and Biology*, vol. 56, no. 1, pp. 1–17, 2011.

14

Detection of Myoglobin with an Open-Cavity-Based Label-Free Photonic Crystal Biosensor

Bailin Zhang, Juan Manuel Tamez-Vela, Steven Solis, Gilbert Bustamante, Ralph Peterson, Shafiqur Rahman, Andres Morales, Liang Tang, and Jing Yong Ye

University of Texas at San Antonio, One UTSA Circle, San Antonio, TX 78249, USA

Correspondence should be addressed to Jing Yong Ye; jingyong.ye@utsa.edu

Academic Editor: Hala Zreiqat

The label-free detection of one of the cardiac biomarkers, myoglobin, using a photonic-crystal-based biosensor in a total-internal-reflection configuration (PC-TIR) is presented in this paper. The PC-TIR sensor possesses a unique open optical microcavity that allows for several key advantages in biomolecular assays. In contrast to a conventional closed microcavity, the open configuration allows easy functionalization of the sensing surface for rapid biomolecular binding assays. Moreover, the properties of PC structures make it easy to be designed and engineered for operating at any optical wavelength. Through fine design of the photonic crystal structure, biochemical modification of the sensor surface, and integration with a microfluidic system, we have demonstrated that the detection sensitivity of the sensor for myoglobin has reached the clinically significant concentration range, enabling potential usage of this biosensor for diagnosis of acute myocardial infarction. The real-time response of the sensor to the myoglobin binding may potentially provide point-of-care monitoring of patients and treatment effects.

1. Introduction

The diagnosis of cardiac disorders becomes more and more important with the incidence of acute myocardial infarction (AMI) commanding one of the highest mortality rates in the US and around the world [1]. Each year, approximately 635,000 people suffer from AMI [2], among whom it is estimated that 50% will die within the first hour of symptoms [3]. For this reason, many studies have been conducted to shorten the time required to diagnose AMI [2–5]. Given the complex pathophysiology of heart disease, interests have intensified in plasma biochemical markers to predict susceptibility and aid in patient management. After an AMI has occurred, cardiac biomarkers, such as myoglobin, troponin I (cTnI), troponin T (cTnT), and creatine kinase (CK-MB), are released into the bloodstream [3, 4, 6]. In 2000, the World Health Organization set a standard allowing physicians to use the troponins and CK-MB levels, in addition to ECG and the patients' history, to diagnose AMI [7]. Although the serum detection of these biomarkers aids in an accurate diagnosis, it is usually time

consuming due to the laborious lab techniques and logistics of sample transportation to a central lab. Both the turnaround time for laboratory diagnosis and the elapsed time for cTnI or CK-MB biomarkers to be released into the body (up to 3 hours for cTnI and 6 hours for CK-MB after an AMI) [3] may lead to a delay in prime-time treatment or hospitalization of a patient with AMI [8]. Myoglobin, however, is one of the earliest biomolecules released into the bloodstream (~1 hour) after the AMI, reaches peak levels after 2 hours, and has been shown to be an early indicator of AMI [3, 6, 9]. Sallach et al. demonstrated that an increase of 20 ng/mL of myoglobin in 90 minutes provided a highly accurate diagnosis of AMI in patients with normal levels of cTnI [6]. Given that these cardiac markers have different characteristics, including clinical sensitivity and specificity, release time after symptom onset, clinical cutoff level (myoglobin 70–200 ng/mL; CK-MB 3.5–10 ng/mL; cTnI 0.06–1.5 ng/mL) [10, 11], and capability to remain elevated for a reasonable length of time, a rapid, accurate, and simultaneous measurement of the cardiac markers is important in reducing detection time, decreasing cost of patient treatment, and saving patient lives. This has led many researchers to study the use of label-free biosensors to detect myoglobin levels.

Since the first discovery of cardiac biomarkers, many detection methods have been developed such as fluorescence immunoassay [12], surface plasmon resonance (SPR) sensor [13], resonant waveguide grating or quartz crystal microbalance [14], electrical signals from nanowire-based biosensors [15], microresonator based such as microcantilever biosensors [16, 17], and two-dimensional photonic-crystal biosensors [18–21]. For the fluorescence-based detection methods, despite its high sensitivity, the process of fluorescence labeling can be complicated and time consuming. The nanowire-based biosensors may offer label-free detection with high sensitivity, but measurement results are to be easily affected by pH values of solutions and charges of molecules [22], because the detection mechanism is based on the measurement of conductance change of a nanowire. For the cantilever-based resonator sensor, the out-of-plane vibration experiences a high viscous damping in liquid environment, thus lowering the Q-factor and mass resolution [23]. Although the SPR-based sensor has been successfully commercialized, it is expensive and has limited use for small molecular binding assays. In this paper, we report label-free bioassays of myoglobin with a novel photonic-crystal-based sensor having a unique open microcavity structure.

2. Material and Experimental Procedures

2.1. Sensor Fabrication. The optical biosensor is designed based on a photonic crystal structure used in a total-internal-reflection configuration [24–28]. Briefly, the sensor is composed of a BK7 glass substrate, five alternating layers of two different dielectric materials (titania and silica), and a cavity layer at the surface. Each of the titania layer and the silica layer is 89.8 nm and 307.2 nm thick, respectively, and is fabricated on the substrate using a vacuum vapor deposition method, which is a well-established fabrication method and relatively inexpensive. The cavity layer of the sensor is formed with 382 nm of silica and 10 nm of silicon. The silicon layer is used to introduce an appropriate amount of absorption that produces a resonant dip in the reflectance spectrum of the sensor. A broadband light is introduced into the sensor substrate at an incidence angle of 64° through a prism. This PC-TIR sensor functions as a high-finesse Fabry-Pérot resonator, which enables it to yield a sharper resonance mode than SPR-based sensors and thus higher detection sensitivity, and yet the sensor surface available for analyte binding is open to free space and allows real-time binding measurements, bypassing the problems of porous structure-based biosensors. Our sensor is unique in the fact that it utilizes an open optical microcavity as opposed to a conventional closed optical microcavity. A traditional closed optical microcavity has a cavity layer sandwiched between two high-reflection surfaces. We create the open cavity by dividing the cavity layer of a traditional closed cavity in half, placing only one half into a TIR configuration. A microcavity is still created as the incident light is confined between the photonic crystal structure and its mirror image due to TIR. The open sensing surface is allows easy immobilization of analyte-recognition molecules on the surface and direct exposure of analyte molecules for real-time bioassays. When molecular bindings occur on the cavity layer surface, the wavelength of the resonant dip in the reflectance spectrum of the sensor shifts in a manner highly sensitive to analyte binding. Through monitoring the shift in the resonant dip, a real-time detection of the molecular binding can be performed.

2.2. Biochemical Modification of Sensor Surface. Similar to the surface treatment protocol described in other papers [29–31], the sensor surface is cleaned and oxidized via immersion in a piranha solution—a mixture of sulfuric acid and hydrogen peroxide (98%-H_2SO_4 : 30%-H_2O_2 = 3 : 1) in a parafilm sealed beaker on a heated plate (80°C) for 1 hour followed by rinsing with a copious amount of deionized water for 2 minutes under sonication and then immersed in deionized water for overnight to completely remove the residue of acid. The sensor surface is further etched with trifluoroacetic acid (THF) for 90 minutes at room temperature. The THF residue is evaporated by putting the sensor in a vacuum chamber pumped overnight. The silanization of the sensor surface is finished by placing it in a freshly prepared 5% (v/v) 3-aminopropyltriethoxysilane (APTES) solution in 95% acetone for 10 minutes, followed by removing excess reaction reagents with acetone on a shaker for twelve times at 5 minute intervals. Finally, the curing of the silane linkage is carried out by drying the substrates on a hot plate at 110°C for 90 minutes.

After the above process, the sensor surface is enriched with amine groups suitable for further conjugation with carboxyl-terminated biomolecules. Before the immobilization of cardiac myoglobin antibodies, the carboxyl methylated (CM) Dextran (150 kDa) is covalently conjugated onto the sensor surface via EDC/NHS chemistry to maximize the binding sites of cardiac myoglobin antibodies. The CM-Dextran (125 mg) is prepared in a HEPES (N-2-hydroxy-ethylpiperazine-N'-2-ethane sulfonic acid) buffer solution (2 mL, pH = 5.5). The carboxyl groups on the CM-Dextran are activated with the aid of 1-ethyl-3(3-dimethyl amino-propyl) carbodiimide (EDC) and N-hydroxysuccinimide (NHS) molecules for 10 minutes, with the molar ratios of −COOH : EDC : NHS = 1 : 10 : 2.5. The CM-Dextran solution with activated carboxyl groups is adjusted to a pH value of 7.3 from 5.5 by adding an appropriate amount of NaOH solution. The activated CM-Dextran then reacts with the amine groups on the sensor surface for 4 hours at room temperature. The CM-Dextran immobilized on the surface is about 2.5 nm, confirmed by the sensor via the method described in Section 2.5. Finally, the immobilized CM-Dextran on the sensor surface is activated with a similar EDC/NHS chemistry and reacts with 200 μL cardiac myoglobin antibodies (Fitzgerald Industrial, 10-M50C) to functionalize the sensor for specific detection of myoglobin (Fitzgerald Industrial, 30C-cp1030u).

2.3. Preparation of a Microfluidic System. To load samples for analysis with the sensor, two microchannels having a

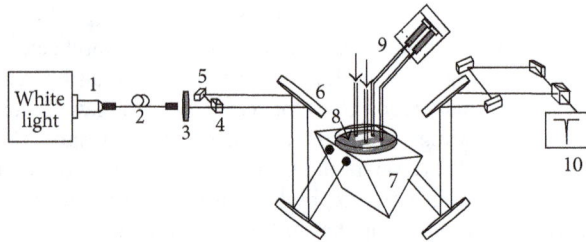

FIGURE 1: Schematic of experimental setup. A broadband white light source is coupled with an objective lens (1) into a single mode fiber (2). S-polarized light from a polarizer (3) goes through a beam splitter (4) and is directed via mirrors (5, 6) to a prism (7), which allows coupling the light into the sensor (8) at a 64° angle. Microfluidic channels attached to the sensor carry the sample solutions with controlled flow rates by a syringe pump (9). The reflected light from the sensor is detected by a high resolution spectrometer (10).

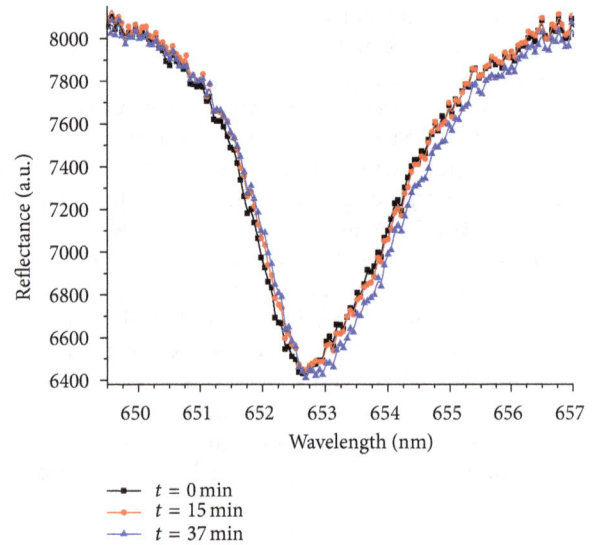

FIGURE 2: Typical reflectance spectra of a PC-TIR sensor when a 70 ng/mL myoglobin solution flows through the microchannel on the sensor surface. As the binding occurs on the surface, the resonant wavelength of the reflectance dip shifts to the longer wavelength with time. Three representative curves at times 0, 15, and 37 min are shown.

width of $500\,\mu m$, height of $380\,\mu m$, and length of $5\,mm$ are formed using a polydimethylsiloxane (PDMS) replica molding process. PDMS base and curing agents (Sylgard184, Dow Corning) are mixed at a ratio of $10:1$. The mixture is degassed in a vacuum chamber for about 10 minutes and then cast on a mold and cured at room temperature. Finally, the two microchannels are sealed on the surface of the functionalized sensor. In order to inject the target sample solution or phosphate buffered saline (PBS) solution, Teflon tubings are used to connect the outlets of the microfluidic channels to a multichannel syringe pump. The inlets of the microfluidic channels are connected to two vials containing the target myoglobin samples and PBS solutions, respectively. The flow of cardiac myoglobin samples or buffer solutions onto the sensing areas is precisely controlled with the syringe pump.

2.4. Experimental Setup Configuration. Our experimental setup is shown in Figure 1. A white light source is coupled into a single-mode optical fiber using an objective lens to obtain a good spatial mode. An aspherical lens is used to collimate the output light from the fiber as it passes through a linear polarizer to select s-polarization. The light is split into two parts and is delivered to the PC-TIR sensor substrate at an incident angle of 64° through a coupling prism. The two beams are carefully aligned to the center of the two microchannels on the sensor surface. The reflected light from the two microchannels on the sensor is collected using a high-resolution spectrometer (HR4000, Ocean Optics) where the resonant wavelength shifts are recorded in real time. One channel is used for flowing cardiac myoglobin samples, while the other is used as a reference channel to compensate for changes in the resonant wavelength due to mechanical drift or temperature fluctuations.

2.5. Experimental Method. The myoglobin at various concentrations from 70 to 1000 ng/mL in PBS is measured using our PC-TIR sensor. Firstly, a PBS solution (pH = 7.4) is injected into both microchannels on the sensor surface. The resonant dip wavelengths in the reflectance spectra corresponding to the two channels are recorded as the detection baselines. Next, $200\,\mu L$ of cardiac myoglobin antibody solution (in PBS) is injected into one of the microchannels, replacing the PBS solution at a speed of $5\,\mu L/min$. With the binding of the myoglobin antibody on the sensor surface, the resonant dip of the sensor shifts accordingly, which is a direct recording of the dynamic binding process. The sensor is then washed by making PBS flow through the both microchannels for 10 minutes at the same flow rate. An additional measurement of the resonant dip is taken after cleaning with PBS wash. The net shift of the resonant wavelength before and after the flow of cardiac myoglobin antibody solution reflects the amount of antibody molecules immobilized on the sensor surface. The binding thickness of the antibodies is confirmed by this method to be ~1 nm. The sensor is prepared for myoglobin assays due to the immobilization of the cardiac myoglobin antibodies on the sensor surface. A $200\,\mu L$ cardiac myoglobin solution in PBS is injected through one microchannel on the newly prepared sensor, while the other microchannel continuously maks PBS as a reference. Different concentrations of cardiac myoglobin samples are measured and the amount of cardiac myoglobin binding to the sensor surface is quantified by analysis of the corresponding resonant dip shift in the reflectance spectra.

3. Results and Discussion

A series of the typical reflectance spectra of the sensor when making $200\,\mu L$ myoglobin solution (in PBS, 70 ng/mL) flow through a microchannel is shown in Figure 2. The full width at half maximum of the resonant dip is only 2.5 nm. The

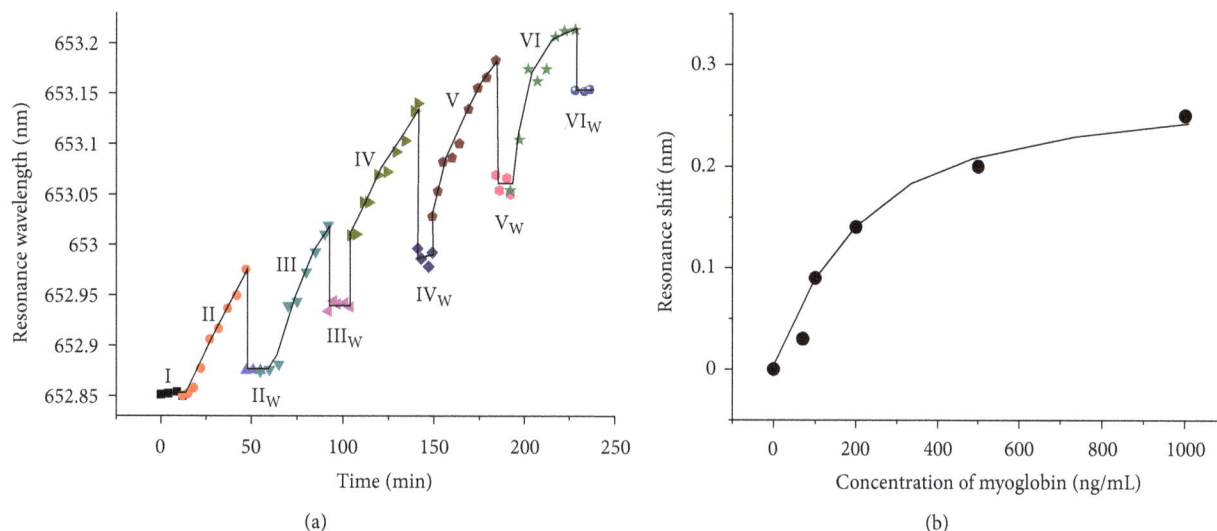

FIGURE 3: (a) Binding kinetics of myoglobin on a functionalized PC-TIR sensor surface. Before detection, a detection baseline is first recorded by making PBS flow through a microchannel on the sensor surface (I). Stages II–VI show the results of binding assays of myoglobin with concentrations of 70, 100, 200, 500, and 1000 ng/mL, respectively. The remaining captured myoglobin on the sensor surface after washing with PBS is then recorded as Stages II_W–VI_W. (b). Resonant wavelength shift of the sensor as a function of myoglobin concentrations.

sharp resonant condition allows precise quantification of the center wavelength of the resonant dip. When the binding of myoglobin molecules to the immobilized antibodies occurs on the sensor surface, the resonant condition of the sensor changes due to an increase of the effective thickness of the cavity layer, thus producing a shift of the resonant wavelength. Although at the low concentration of myoglobin at 70 ng/mL the shift is minimal, it is still clearly measureable (Figure 2). The curves can be fitted with a Lorentzian function, which provides quantitative results of the shift of the center wavelength. At a constant concentration the shift increases with time, indicating that more myoglobin molecules are bound to the antibodies on the sensor surface. When the concentration of myoglobin increases, the shift of the resonant wavelength also increases and can become much bigger than that shown in Figure 2.

The binding kinetics of myoglobin with its antibody is illustrated in Figure 3(a). The time-dependent wavelength shift of the sensor resonant dip is caused by the interaction events between myoglobin and the antibodies immobilized on the sensor surface. The time-dependent response is close to linear for the first concentration of 70 ng/mL. The linear relationship is because the binding sites available on the sensor surface are much larger than the number of myoglobin molecules. Over the time, the available interaction sites decrease until all of the binding sites are occupied, resulting in a saturation of the binding curve for the high concentration (1000 ng/mL). After binding of myoglobin at each concentration, the sensor surface is washed with PBS and a drop of the resonant wavelength is observed, which can be attributed to partial removal of loosely bound molecules. The end point of the resonant wavelength shift after PBS wash is plotted as a function of myoglobin concentrations in Figure 3(b). Saturation of binding has been observed at higher myoglobin concentrations. Based on our transfer matrix simulation of

multilayer interference in our sensor [27], the final resonant wavelength shift of 0.25 nm for 1000 ng/mL of myoglobin corresponds to a binding thickness of myoglobin of 0.30 nm. Our experimental result indicates that a PC-TIR can be functionalized for sensitive detection of myoglobin with a concentration ranging from 70 to 1000 ng/mL. This sensing range of myoglobin meets the present clinical diagnostic requirement for myocardial damage that results in an elevation of myoglobin to more than 110 ng/mL [32, 33].

In order to confirm that the resonant wavelength shift is caused by specific interactions between the antibodies and myoglobin, we conducted a control experiment. For that, myoglobin solutions with the highest (1000 ng/mL) and lowest (70 ng/mL) concentrations used in the binding assay experiment were selected to be injected into a microchannel on a sensor surface without immobilized antibodies. As demonstrated in Figure 4, after PBS wash the average resonant wavelength shift is only ~0.050 (Figure 4(a)) and 0.018 nm (Figure 4(b)), respectively. This result indicates that there is minimal nonspecific binding for a sensor without the immobilization of antibodies, while the resonant wavelength shift for a sensor functionalized with antibodies is caused by specific binding between myoglobin and the antibodies.

To further test the specificity of the sensor to myoglobin, we carried out another control experiment by checking the binding of unrelated protein molecules with the sensor having immobilized myoglobin antibodies. For that, cardiac troponin I (cTnI) was chosen as a random protein molecule. A solution of 200 μL cTnI with a concentration of 200 ng/mL in PBS was flowed across the sensor surface immobilized with myoglobin antibodies. Although we observed a jump of the resonant wavelength when starting injection of cTnI onto the sensor surface, the wavelength shift returned to the baseline level after the sensor was washed with PBS (Figure 5). The initial jump of the resonant wavelength can

(a)

(b)

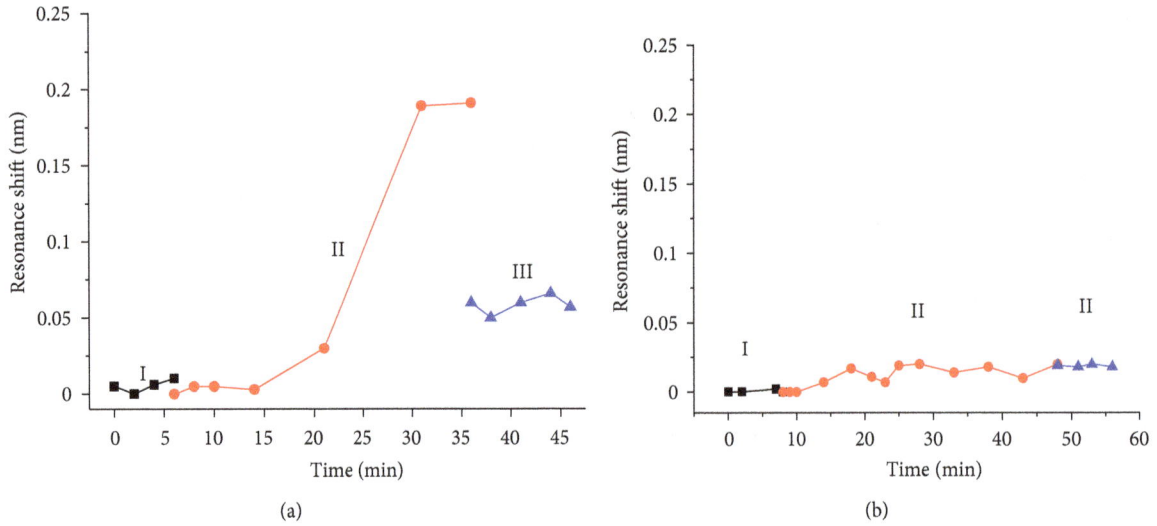

FIGURE 4: Binding kinetics of myoglobin having concentrations of 1000 ng/mL (a) and 70 ng ng/mL (b) in PBS, respectively, on a sensor surface without immobilized antibodies. Stage I: PBS baseline; II: myoglobin solution flowed through the sensing area; III: the remained myoglobin after PBS wash.

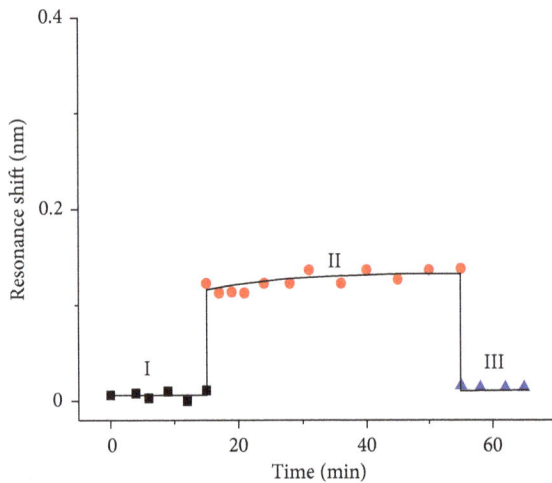

FIGURE 5: Binding kinetics of cTnI on a PC-TIR sensor with immobilized myoglobin antibodies. Before cTnI was injected, a detection baseline was first recorded by making PBS flow through a microchannel on the sensor surface (Stage I). Stage II shows the sensor response when 200 μL of cTnI with a concentration of 200 ng/mL was flowing on the sensor surface. Stage III shows that the resonant wavelength returned to the baseline after a PBS wash.

be attributed to the change of the bulk refractive index caused by replacing PBS with the cTnI solution. Except for the initial jump, the flat sensor response during this period indicated that no real binding occurred when cTnI was flowing on the sensor surface. The fact that the final wavelength shift returned to the baseline after the sensor was rinsed with water further proved that there was no binding of cTnI to the sensor functionalized for myoglobin assays. Therefore, this experimental result confirms the specificity of our sensor for detecting the myoglobin cardiac biomarker.

Besides quantifying myoglobin with different concentrations, the PC-TIR sensor can also be used to obtain the dissociation constant K_d at equilibrium, which is an important parameter in evaluating the biochemical binding activities. Assuming a simple biomolecular binding model where two binding components A and B form a binding complex as AB, [A] + [B] $\rightleftharpoons$ [AB], the kinetic rate constants are described as follows [34]:

$$\frac{dR}{dt} = k_{on} \cdot C \left(R_{max} - R_t \right) - k_{off} \cdot R_t, \tag{1}$$

where R is the response of the sensor, while R_t and R_{max} are the response at time t and the maximum binding response, respectively. C is the concentration of analyte in solution. k_{on} and k_{off} are the binding and dissociation rate constants, respectively. At equilibrium, one obtains

$$K_d R_{eq} = C \left(R_{max} - R_t \right), \tag{2}$$

where K_d is the equilibrium dissociation constant, $K_d = k_{off}/K_{on}$. R_{eq} is the sensor response at equilibrium corresponding to a concentration of C. By measuring the resonance shifts at equilibrium for two different concentrations, one can calculate K_d by solving (2). To ensure binding equilibrium for calculating the equilibrium dissociation constant K_d, we used myoglobin samples at two high concentrations, 8000 ng/mL and 2600 ng/mL, for the binding assays with our PC-TIR sensor. The binding curves are plotted in Figure 6, which shows that equilibrium has been reached at two levels corresponding to the two concentrations used. From the sensor responses at the equilibrium levels and based on (2), the K_d of myoglobin bound with antibodies on the PC-TIR sensor surface was calculated to be 1.2 nM, which is in good agreement with the value of 1.3 nM reported previously in [35].

It should be noted that myoglobin is not specifically indicative of acute myocardial infarction, because it is also

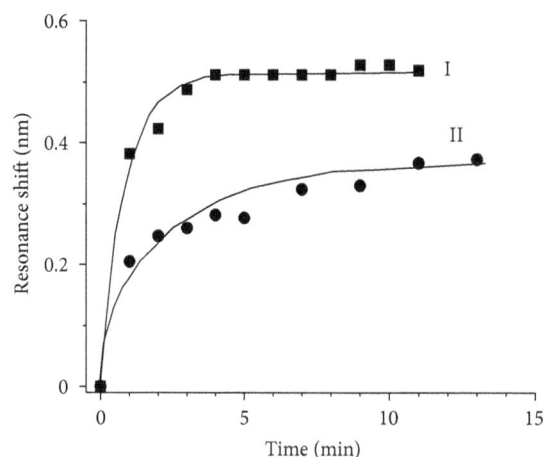

FIGURE 6: Kinetic binding shift curve of myoglobin (I: 8 ug/mL and II: 2.6 ug/mL in PBS solution) on PC-TIR surface immobilized with myoglobin antibody.

present in skeletal muscle; however, it may still provide the earliest indication of myocardial injury and other valuable prognostic information when it is used in combination with other biomarkers [36]. cTnI is the most widely used biomarker, due to its nearly complete cardiac tissue specificity and sensitivity. They remain elevated for 4–10 days after the onset of AMI, indicating their capability to remain elevated for a reasonable length of time to allow a suitable diagnostic window. However, since it takes approximately 4 hrs for cardiac troponins to reach detectable concentrations in blood, they cannot be considered as early markers. On the other hand, myoglobin is one of the earliest biomarkers released into blood circulation after AMI. However, myoglobin is rapidly cleared out from blood compared to other cardiac biomarkers, limiting its diagnostic usefulness in patients after 8–12 hours of presenting symptoms. A combination blood test of cTnI and myoglobin may allow the achievement of high diagnostic sensitivity and specificity [4]. A sensitive biosensor is needed for rapid detection right after the onset of the elevation of myoglobin concentrations. A PC-TIR sensor may potentially address this unmet demand, since it can offer sensitive and rapid detection of myoglobin in the clinically significant concentration range as demonstrated in this study.

4. Conclusion

A photonic crystal structure used in a TIR configuration can form a sensitive biosensor with a unique open microcavity as its sensing surface. Such a sensor is successfully fabricated at low cost and applied for the label-free binding assay of myoglobin for potential early diagnosis of myocardial infarction. The sensor detection limit is as low as 70 ng/mL, which falls in the clinical diagnostic level of AMI patients. The experimental results reported here serve as a stepping stone towards potential applications of the PC-TIR sensor for earliest diagnosis and identification of AMI patients. In the future, more experiments with the PC-TIR sensor will include

the use of other cardiac biomarkers along with actual blood samples.

Acknowledgment

This project was supported in part by a grant from San Antonio Area Foundation.

References

[1] W. H. W. Tang, G. S. Francis, D. A. Morrow et al., "National academy of clinical biochemistry laboratory medicine practice guidelines: clinical utilization of cardiac biomarker testing in heart failure," *Circulation*, vol. 116, no. 5, pp. E99–E109, 2007.

[2] V. L. Roger, A. S. Go, D. M. Lloyd-Jones et al., "Heart disease and stroke statistics—2012 update: a report from the American Heart Association," *Circulation*, vol. 125, no. 1, pp. e2–e220, 2012.

[3] A. L. Straface, J. H. Myers, H. J. Kirchick, and K. E. Blick, "A rapid point-of-care cardiac marker testing strategy facilitates the rapid diagnosis and management of chest pain patients in the emergency department," *American Journal of Clinical Pathology*, vol. 129, no. 5, pp. 788–795, 2008.

[4] J. McCord, R. M. Nowak, P. A. McCullough et al., "Ninety-minute exclusion of acute myocardial infarction by use of quantitative point-of-care testing of myoglobin and troponin I," *Circulation*, vol. 104, no. 13, pp. 1483–1488, 2001.

[5] S. P. J. Macdonald and Y. Nagree, "Rapid risk stratification in suspected acute coronary syndrome using serial multiple cardiac biomarkers: a pilot study," *Emergency Medicine Australasia*, vol. 20, no. 5, pp. 403–409, 2008.

[6] S. M. Sallach, R. Nowak, M. P. Hudson et al., "A change in serum myoglobin to detect acute myocardial infarction in patients with normal troponin I levels," *American Journal of Cardiology*, vol. 94, no. 7, pp. 864–867, 2004.

[7] J. S. Alpert, K. Thygesen, E. Antman, and J. P. Bassand, "Myocardial infarction redefined—a consensus document of the joint European society of cardiology/American college of cardiology committee for the redefinition of myocardial infarction," *Journal of the American College of Cardiology*, vol. 36, no. 3, pp. 959–969, 2000.

[8] G. X. Brogan, S. Friedman, C. McCuskey et al., "Evaluation of a new rapid quantitative immunoassay for serum myoglobin versus CK-MB for ruling out acute myocardial infarction in the emergency department," *Annals of Emergency Medicine*, vol. 24, no. 4, pp. 665–671, 1994.

[9] F. S. Apple, R. H. Christenson, R. Valdes et al., "Simultaneous rapid measurement of whole blood myoglobin, creatine kinase MB, and cardiac troponin I by the triage cardiac panel for detection of myocardial infarction," *Clinical Chemistry*, vol. 45, no. 2, pp. 199–205, 1999.

[10] U. Friess and M. Stark, "Cardiac markers: a clear cause for point-of-care testing," *Analytical and Bioanalytical Chemistry*, vol. 393, no. 5, pp. 1453–1462, 2009.

[11] R. H. Christenson and H. M. E. Azzazy, "Cardiac point of care testing: a focused review of current National Academy of Clinical Biochemistry guidelines and measurement platforms," *Clinical Biochemistry*, vol. 42, no. 3, pp. 150–157, 2009.

[12] F. Darain, P. Yager, K. L. Gan, and S. C. Tjin, "On-chip detection of myoglobin based on fluorescence," *Biosensors and Bioelectronics*, vol. 24, no. 6, pp. 1744–1750, 2009.

[13] R. Kurita, Y. Yokota, Y. Sato, F. Mizutani, and O. Niwa, "On-chip enzyme immunoassay of a cardiac marker using a microfluidic device combined with a portable surface plasmon resonance system," *Analytical Chemistry*, vol. 78, no. 15, pp. 5525–5531, 2006.

[14] Y. Arntz, J. D. Seelig, H. P. Lang et al., "Label-free protein assay based on a nanomechanical cantilever array," *Nanotechnology*, vol. 14, no. 1, pp. 86–90, 2003.

[15] T. W. Lin, P. J. Hsieha, C. L. Linc et al., "Label-free detection of protein-protein interactions using a calmodulin-modified nanowire transistor," *Proceedings of the National Academy of Sciences of the United States of America*, vol. 107, no. 3, pp. 1047–1052, 2010.

[16] K. W. Wee, G. Y. Kang, J. Park et al., "Novel electrical detection of label-free disease marker proteins using piezoresistive self-sensing micro-cantilevers," *Biosensors and Bioelectronics*, vol. 20, no. 10, pp. 1932–1938, 2005.

[17] A. Kooser, K. Manygoats, M. P. Eastman, and T. L. Porter, "Investigation of the antigen antibody reaction between anti-bovine serum albumin (a-BSA) and bovine serum albumin (BSA) using piezoresistive microcantilever based sensors," *Biosensors and Bioelectronics*, vol. 19, no. 5, pp. 503–508, 2003.

[18] S. Chakravarty, Y. Zoua, W. C. Laia et al., "Slow light engineering for high Q high sensitivity photonic crystal microcavity biosensors in silicon," *Biosensors & Bioelectronics*, vol. 38, no. 1, pp. 170–176, 2012.

[19] W. C. Lai, S. Chakravarty, Y. Zou et al., "Silicon photonic crystal microcavity biosensors for label free highly sensitive and specific lung cancer detection," *IEEE Photonics Conference*, vol. 2012, pp. 443–444, 2012.

[20] S. Pal, E. Guillermain, R. Sriram, B. L. Miller, and P. M. Fauchet, "Silicon photonic crystal nanocavity-coupled waveguides for error-corrected optical biosensing," *Biosensors & Bioelectronics*, vol. 26, no. 10, pp. 4024–4031, 2011.

[21] R. Sriram, E. Baker, P. M. Fauchet et al., "Two dimensional photonic crystal biosensors as a platform for label-free sensing of biomolecules," *International Society For Optics and Photonics*, vol. 8570, 2013.

[22] I. L. Lee, X. Luo, and J. Huang, "Detection of cardiac biomarkers using single polyaniline nanowire-base conductometric biosensors," *Biosensors*, vol. 2, pp. 205–220, 2012.

[23] C. Vančura, J. Lichtenberg, A. Hierlemann, and F. Josse, "Characterization of magnetically actuated resonant cantilevers in viscous fluids," *Applied Physics Letters*, vol. 87, no. 16, Article ID 162510, pp. 1–3, 2005.

[24] J. Y. G. Ye, T. B. Norris, and J. Baker, "Patent issued," Patent No. 7, 639, 632, 2009.

[25] B. L. Zhang, S. Dallo, R. Peterson et al., "Detection of anthrax lef with DNA-based photonic crystal sensors," *Journal of Biomedical Optics*, vol. 16, no. 12, 2011.

[26] S. F. Dallo, B. Zhang, J. Denno et al., "Association of acinetobacter baumannii EF-Tu with cell surface, outer membrane vesicles, and fibronectin," *Scientific World Journal*, vol. 2012, Article ID 128705, 2012.

[27] Y. Guo, C. Divin, A. Myc et al., "Sensitive molecular binding assay using a photonic crystal structure in total internal reflection," *Optics Express*, vol. 16, no. 16, pp. 11741–11749, 2008.

[28] Y. Guo, J. Y. Ye, C. Divin et al., "Real-time biomolecular binding detection using a sensitive photonic crystal biosensor," *Analytical Chemistry*, vol. 82, no. 12, pp. 5211–5218, 2010.

[29] S. Flink, F. C. J. M. Van Veggel, and D. N. Reinhoudt, "Functionalization of self-assembled monolayers on glass and oxidized silicon wafers by surface reactions," *Journal of Physical Organic Chemistry*, vol. 14, no. 7, pp. 407–415, 2001.

[30] R. H. Aebersold, D. B. Teplow, L. E. Hood, and S. B. H. Kent, "Electroblotting onto activated glass. High efficiency preparation of proteins from analytical sodium dodecyl sulfate-polyacrylamide gels for direct sequence analysis," *The Journal of Biological Chemistry*, vol. 261, no. 9, pp. 4229–4238, 1986.

[31] S. H. Jung, J. W. Jung, I. B. Suh et al., "Analysis of C-reactive protein on amide-linked N-hydroxysuccinimide-dextran arrays with a spectral surface plasmon resonance biosensor for sero-diagnosis," *Analytical Chemistry*, vol. 79, no. 15, pp. 5703–5710, 2007.

[32] S. M. S. Diagnostics, *ADVIA Centaur Assay Mannual Myoglobin Review E*, 2004.

[33] S. M. S. Diagnostics, *ADVIA Centaur Mannual CKMB, 111803. REView G*, 2003.

[34] M. A. Cooper and D. H. Williams, "Kinetic analysis of antibody-antigen interactions at a supported lipid monolayer," *Analytical Biochemistry*, vol. 276, no. 1, pp. 36–47, 1999.

[35] T. M. A. D. Myszka, Kinetic & Affinity Characterization—Monocolonal Antibodies, http://www.attana.com/wp-content/uploads/2013/01/AE01-03-KineticaAffinity_MAbs.pdf Attana, Editor, 2002.

[36] M. P. Hudson, R. H. Christenson, L. K. Newby, A. L. Kaplan, and E. M. Ohman, "Cardiac markers: point of care testing," *Clinica Chimica Acta*, vol. 284, no. 2, pp. 223–237, 1999.

15

Transplantation of Nonexpanded Adipose Stromal Vascular Fraction and Platelet-Rich Plasma for Articular Cartilage Injury Treatment in Mice Model

Phuc Van Pham,[1] **Khanh Hong-Thien Bui,**[2] **Dat Quoc Ngo,**[3]
Lam Tan Khuat,[1] **and Ngoc Kim Phan**[1]

[1] *Laboratory of Stem Cell Research and Application, University of Science, Vietnam National University, Ho Chi Minh City, Vietnam*
[2] *University of Medical Center, Ho Chi Minh University of Medicine and Pharmacy, Ho Chi Minh City, Vietnam*
[3] *Department of Pathology, University of Medicine and Pharmacy, Ho Chi Minh City, Vietnam*

Correspondence should be addressed to Phuc Van Pham; pvphuc@hcmuns.edu.vn

Academic Editor: Ayako Oyane

Stromal vascular fraction (SVF) combined with platelet-rich plasma (PRP) is commonly used in preclinical and clinical osteoarthritis as well as articular cartilage injury treatment. However, this therapy has not carefully evaluated the safety and the efficacy. This research aims to assess the safety and the efficacy of SVF combined with PRP transplantation. Ten samples of SVFs and PRPs from donors were used in this research. About safety, we evaluate the expression of some genes related to tumor formation such as Oct-4, Nanog, SSEA3, and SSEA4 by RT-PCR, flow cytometry, and tumor formation when injected in NOD/SCID mice. About efficacy, SVF was injected with PRP into murine joint that caused joint failure. The results showed that SVFs are negative with Oct-4, Nanog, SSEA-3, and SSEA-4, as well as they cannot cause tumors in mice. SVFs combined with PRP can improve the joint regeneration in mice. These results proved that SVFs combined with PRP transplantation is a promising therapy for articular cartilage injury treatment.

1. Introduction

Stem cell therapy is considered as a promising therapy for degenerative disease treatment, especially articular cartilage injury as well as osteoarthritis. Osteoarthritis was treated by stem cell transplantation for a few years ago. Stem cells from various sources were used to treat this disease. However, the mesenchymal stem cells (MSCs) are considered as most suitable candidates. MSCs are multipotential cells capable of differentiation into bone, cartilage, fat, and some other cells [1]. MSCs could be isolated from bone marrow [2], adipose tissue [3], cord blood [4], banked umbilical cord blood [5], umbilical cord [6], Wharton's jelly [7], placenta [8], and pulp [9]. However, MSCs from bone marrow [10–12] and from adipose tissue [13–15] are two common stem cell sources for treating cartilage degeneration.

Cartilage degeneration or cartilage injury is a common clinical problem and easily leads to osteoarthritis.

Osteoarthritis is a chronic degenerative process characterized by the degeneration of cartilage, bone bud formation, cartilage reorganization, joint erosion, and loss of joint function [16]. Currently, cartilage injury was treated primarily with drugs [17–20] or injection of hyaluronic acid [21, 22] to reduce the symptoms, pain, and inflammation control. However, these therapies's efficiencies were limited and often failed to prevent the degeneration of the joints [23].

MSCs from adipose tissue, also known as stem cells isolated from fat tissue (adipose-derived stem cells—ADSCs), are a suitable source of mesenchymal stem cells for autograft. This stem cell source was used to treat many diseases such as liver fibrosis [24], sciatic nerve defects [25], systemic sclerosis [26], ischemia [27], skeletal muscle injury [28], passive chronic immune thrombocytopenia [29], and infarcted myocardium [30]. Recently, they have been extended to treat cartilage injuries as well as osteoarthritis such as dogs [31–33], rabbits [34], horses [14], rat [35], mice [36], and goats

[37]. These researches demonstrated that neocartilage formed after ADSC transplantation. Some phase I and II clinical trials using ADSCs transplantation are performed to treat osteoarthritis and cartilage degeneration (NCT01300598, NCT01585857, NCT01399749). Pak (2011) showed that all ADSC grafted patients improved the cartilage regeneration [15].

Among all of ADSC transplantation cases, SVF is used as noncultured ADSC (nonexpanded ADSC). SVF transplantation has some advantages such as saving time (from isolation to transplant faster about 2-3 hours), being inexpensive, and reducing the risk of cell culture. Although many studies have demonstrated the benefits of SVF/ADSC transplantation in cartilage injury treatment, especially knee articular cartilage, so far a little comprehensive studies aim to evaluate the safety and efficiency of SVF transplantation for articular cartilage treatment. Therefore, this study aims to evaluate the safety and efficiency of SVF transplantation combined with PRP in the treatment of cartilage injury in the mouse model.

2. Materials and Methods

2.1. SVF and PRP Preparation. Firstly, adipose tissue was collected from abdominal fat tissue of ten consenting healthy donors. About 40–80 mL of fat was collected by syringe and stored in 100 mL sterile bottle. Fat was kept at 2–8°C and then quickly moved to the laboratory. SVF cells are separated from the fat using the extraction kits (Adistem, Australia) according to the manufacturer's guideline. Briefly, fat is washed 3 times with saline solution to eliminate red blood cells. Then, the fat was incubated with a solution AdiExtract (Adistem, Australia). The sample was centrifuged to collect SVF as pellet at the bottom of the tube. To prepare platelet-rich plasma (PRP), 50 mL of peripheral blood was taken from a large vein (arm veins). Blood was centrifuged 1,700 rpm for 10 minutes to get platelet-enriching plasma. This plasma is activated with activator solution (Adistem, Australia). Then, PRP was mixed with SVF to make the cell suspension. Finally, this suspension was stimulated by the LED light (light monochromatic low energy, Adlight, Adistem, Australia) for 30 minutes before using for treatment.

2.2. Quantification of Nucleated Cells from SVF. Cell suspension (SVF and PRP) is used to count the nucleated cells. Cell number and percentage of viable cells were determined by automatically nucleus-based cell counter (NucleoCounter, Chemometec). Total cell numbers were counted after permeabilization of the membrane by Reagent A (lysis buffer, Chemometec) and neutralized with a solution of Reagent B (Neutralized buffer, Chemometec). Cell suspension is loaded into the counting chamber containing Propidium iodide dye. For counting dead cells, suspension cells were mixed with only Reagent B solution and loaded into the counting chamber. Survival rate is calculated as follows: (total cell number – the number of dead cells): the total number of cells × 100%.

2.3. Evaluation of the Existence of ADSC in SVF. The existence of ADSC in SVF determined by flow cytometry. The

process summarized as follows: cells were washed twice in physiological saline of Dulbecco-modified PBS (D-PBS) supplemented with 1% bovine serum albumin (Sigma-Aldrich, St Louis, MO). Cells were stained for 30 min at 4°C with the monoclonal antibody anti-CD44-PE, anti-CD90-PE, and anti-CD105-FITC (BD Biosciences, Franklin Lakes, New Jersey offers). Stained cells were analyzed by flow cytometer FACSCalibur machine (BD Biosciences). Isotype control is used for all analyzes.

2.4. RT-PCR. To evaluate the safety of SVF, we should identify the gene expression levels related to the process of causing tumors and test the ability to form tumors *in vivo.* About gene expression, RNA was isolate by Trizol according to the manufacturer's instructions (Sigma-Aldrich, St Louis, MO). RNA precipitated with isopropanol at room temperature for 10 minutes. ADSC cells analyzed the expression of genes related to markers of cancer cells or embryonic stem cells, Oct-3/4 and Nanog by Real-time kit SYBR RT-PCR one tube-one step (Sigma-Aldrich, St. Louis, MO). The used primers were Oct-3/4, forward primer: F: 5′-GGAGGAAGCTGACAACAATGAAA-3′, reverse primer R: 5′-GGCCTGCACGAGGGTTT-3; Nanog, forward primer: F: 5′-ACAACTGGCCGAAGAATAGCA-3′; reverse primer R: 5′-GGTTCCCAGTCGGGTTCAC-3; GAPDH, forward primer: F: 5′-GGGCTGCTTTTAACTCTGGT-3′; reverse primer: R: 5′-TGGCAGGTTTTTCTAGACGG-3′.

2.5. In Vivo Tumorigenicity Assay. The tumorigenicity of ADSC was evaluated in mice NOD/SCID (NOD.CB17-Prkdcscid/J, Charles River Laboratories). All mice manipulation was according to guideline of laboratory and approved by the Local Ethics Committee of Stem Cell Research and Application, University of Science (VNU-HCM, VN). All mice were kept in clean condition. Mice were injected subcutaneously at a concentration of 10^5, 10^6, and 10^7 cells, respectively in three groups (each group with 3 mice). Control group was injected with PBS. The formation of tumors in mice was followed for 3 months.

2.6. Articular Cartilage Injured Mice Model and Experimental Treatment Schedule. To evaluate the efficiency of SVF transplantation in articular cartilage injury, we used articular cartilage injured mouse model. The NOD mouse/SCID mice were anesthetized with ketamine (40 mg/kg), then joint destruction by fine needle 32.5 G. Normal mice were used as a positive control (uninjured). Nine mice randomly divided into the treatment group (5 mice) and negative control group (4 mice). Six hours after injury, the mice were treated. In the treatment group, 200 μL containing $2 \cdot 10^6$ SVF in PRP (the treatment group) or PBS (the negative control group) was injected into the knee joint via two doses, with a 10 min interval between injections.

Mice were recorded some parameters related to joint regeneration for 45 days. The mice were recorded the movement on the table daily. At the 45th day, all mice were anesthetized, and their hind limbs were cut and used for

Transplantation of Nonexpanded Adipose Stromal Vascular Fraction and Platelet-Rich Plasma for Articular Cartilage Injury Treatment in Mice Model

139

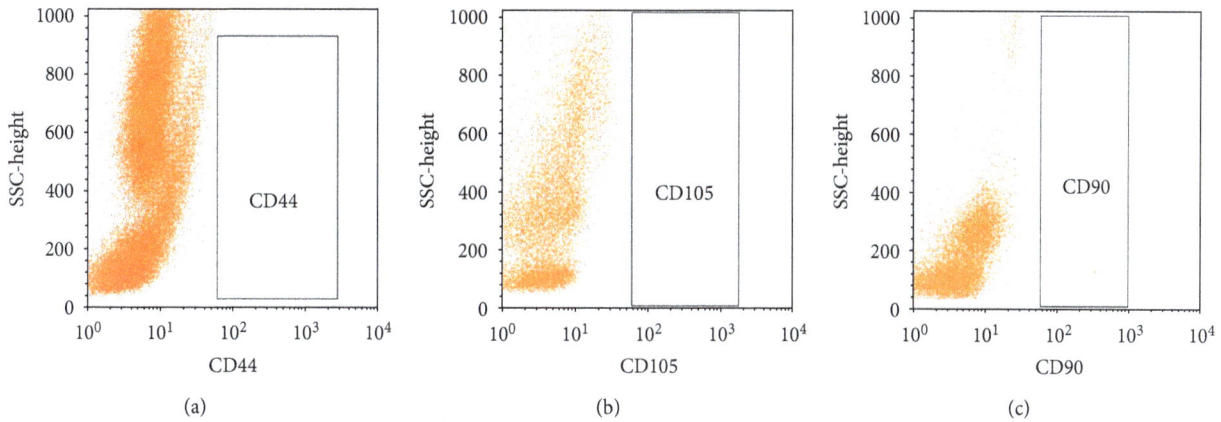

FIGURE 1: Existence of ADSCs in SVF. ADSCs were confirmed based on expression of CD44, CD105 and CD90.

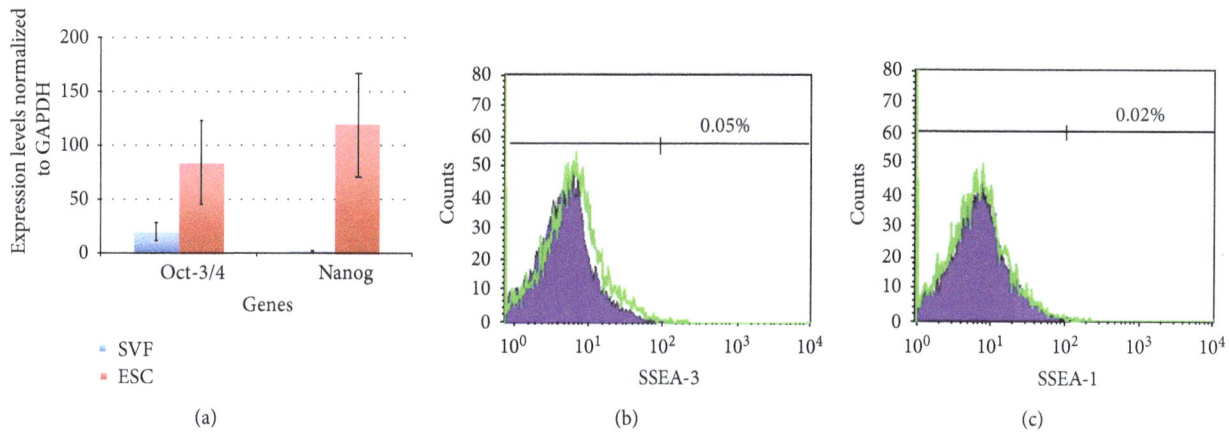

FIGURE 2: Expression of Oct-3/4, Nanog, SSEA-3, and SSEA-1 in SVF. Oct-3/4 and Nanog expressed lower in embryonic stem cell (ESC) (a); while SSEA-3 and SSEA-1 did not express in SVF (b and c).

histological analysis and further experiments. The samples were fixed in 10% formalin, decalcified, sectioned longitudinally, and stained with hematoxylin and eosin (HE) (Sigma-Aldrich, St Louis, MO). Using HE stained slides, three parameters were examined such as the area of injured cartilage (%), the area of neo-cartilage (%), and the number of neocartilage cell layers. The injured cartilage area was determined as the percentage of lost mature cartilage compared to the control. Data was analyzed using Statgraphics software (v7.0; Statgraphics Graphics System, Warrenton, VA).

3. Results and Discussion

MSCs have the large differentiative potential, easily differentiate into bone, cartilage, and adipocyte. Autologous MSC transplantation is considered as a safe and effective therapy in some patients. Recently, adipose tissue was identified as the abundant source of MSCs. ADSCs exist with large amounts of adipose tissue [38]. Similar to MSCs from other sources, ADSC have the ability to differentiate into fat cells, bone, and cartilage and transdifferentiate into neurons and muscle [39–45]. Therefore, ADSC is favored as a source of autologous cell transplantation. However, the isolated ADSC relatively

complex, consuming time, so ADSCs were mainly used as SVF (containing ADSC) without culture. This study aims to evaluate the safety and efficiency of SVF transplantation resuspended in PRP in mouse model.

In the first experiment, we successfully isolated SVF and PRP. Compared to other studies, we have successfully isolated $0.32 \pm 0.15 \times 10^6$ SVF cells from 1 gram of fat with a survival rate of $90.90\% \pm 8.57\%$ ($n = 10$). Next, we assessed the existence of ADSC or MSC in the SVF. Analysis results from 10 samples showed that ADSCs existed in all samples. ADSC counted to $0.89\% \pm 0.11\%$ in the SVF. ADSC populations were identified based on the expression of CD44, CD90, and CD105 of them (Figure 1). These results were similar to many other authors on ADSC markers [43, 46–52]. The markers satisfied the criteria of MSC following to Dominici et al. (2006) [53].

To assess safety, we have evaluated the expression of genes related to cancer. In particular, two genes *Oct-3/4* and *Nanog* were assessed by real-time RT-PCR method and SSEA-3, and SSEA-1 was assessed by flow cytometry. The results showed expression of Oct-3/4, Nanog, SSEA-3, and SSEA-1 much lower than embryonic stem cells (Figure 2). These results demonstrated that the SVF hold low tumorigenicity. In fact,

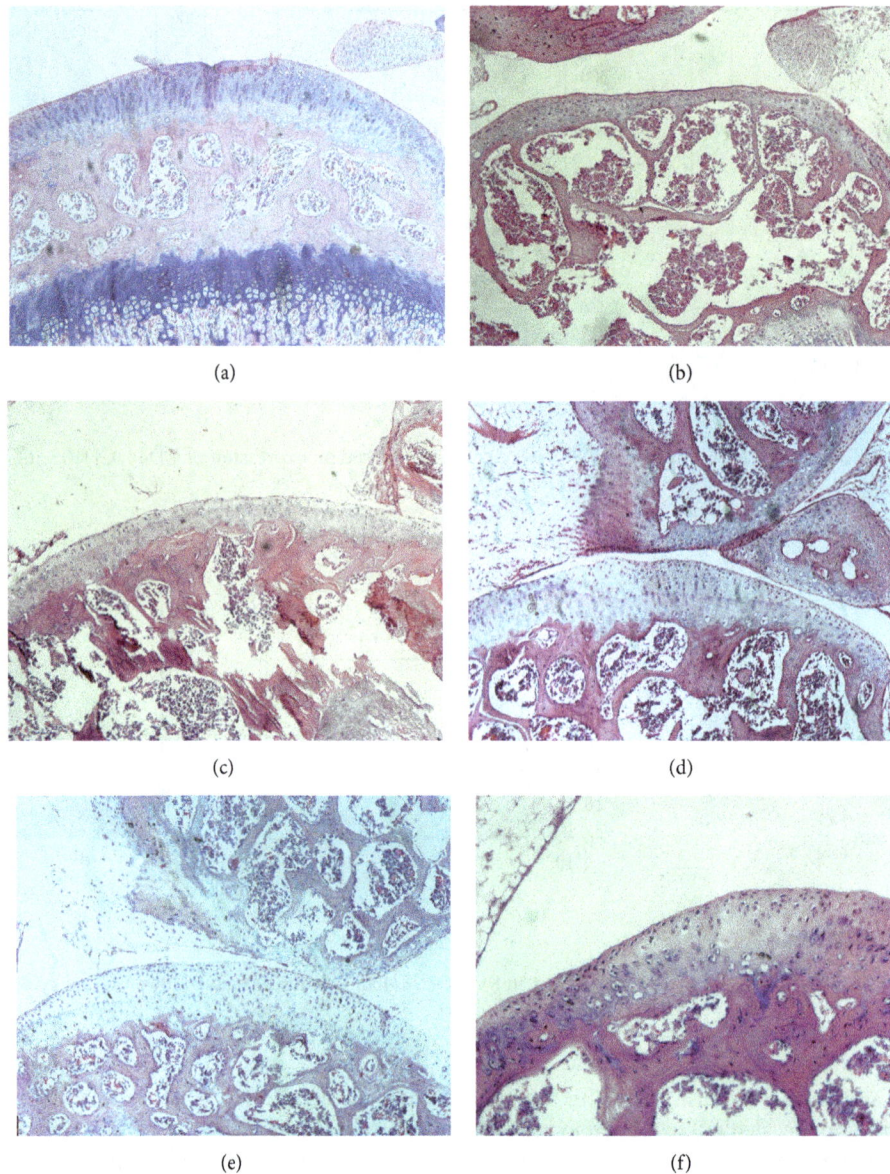

FIGURE 3: HE staining of articular cartilage. Mature cartilage layer was recorded in normal mouse (a). Mature cartilage was thinned by needle (b). Injured cartilage was regenerated in negative control group (c, e) and treated group (d, f). However, the neocartilage in treated group was thicker than in negative control group.

Nanog and Oct-3/4 participate in the process of self-renewal of embryonic stem cells [54, 55]. Moreover, these proteins related to the tumorigenicity process in mature germ cells [56], carcinoma oral squamous cell [57], lung cancer [58], breast cancer [59], and gliomas [60]. The SVF and PRP injection under the skin mouse NOD/SCID could not form teratomas. With these experiments, we concluded that the SVF plus PRP has a promising therapy with a high safety for transplantation experiments.

In the next experiment, we evaluated the efficiency of SVF transplantation in articular cartilage injury. The results showed that the SVF plus PRP transplantation significantly improved the articular cartilage injury compared to control. In the treated group, mice exhibited a reduction of the time required that mice could move on the table by injured hind

limb compared to control. In the control group, mice can move by injured legs after 38.5 ± 4.30 days, while in the treated group, mice could move by injured legs after 29.4 ± 4.32 days.

About histological analysis, in the treated group, an average area of the cartilage damage was 62.60%, and there was 35.5% of neocartilage formation after 45 days ($n = 5$). While in the control group, average area of cartilage lesions was 53.13%, but only 15.5% of neocartilage formation after 45 days (Figure 3).

The grade of cartilage injury between two experimental groups was different due to the effects of the dissimilar force from needle. After 45 days, results showed that 35.5% of neocartilage formed in treated group, while only 15.5% of neocartilage formed in the negative control group. This

Transplantation of Nonexpanded Adipose Stromal Vascular Fraction and Platelet-Rich Plasma for Articular Cartilage Injury
Treatment in Mice Model

141

suggested that SVF and PRP gave benefit effects on the enhancement as well as trigger the neocartilage forming. More importantly, the articular cartilage in both of groups completed at the same level after 45 days with 12 cell layers. These results demonstrated that the SVF and PRP could participate in the process of self-renewal of joint cartilage at the joint microenvironment. Especially, there were no scar tissues or tumors forming at the graft sites. This result was similar to the previous publications about SVF plus PRP transplantation in the treatment of cartilage injury in dog [31–33], rabbits [34, 61], horses [14, 61], rat [35], mice [36], and goats [37]. For example, in joint injured mice model by collagenase, Ter Huurne et al. (2011) showed that the level of damage nearly 50% reduction in ADSC transplanted mice compared to control after 42 days [36]. Specifically, knee injury went down to 25% in treated mice compare to 88% in controls. They suggested that the transplanted ADSC protected and healed of injured cartilage [37]. The findings of Dragoo et al. (2007) showed that autologous ADSCs could reestablish the joint surface in rabbits, in which 100% of rabbits (12/12) had the occurrence of neocartilage, while only 8% rabbit (1/12) in the control had the appearance of neocartilage ($P < 0.001$) [62].

Roles of SVF or ADSCs in cartilage regeneration were recorded with many different effects. In fact, similar to MSCs derived from bone marrow, ADSC had anti-inflammatory properties [63, 64] and inhibition of graft versus host disease (GVHD) [65]. The transplantation of ADSC could successfully treat graft versus host disease with steroid-resistant form [66, 67]. All of these roles could add more effects to trigger rapid cartilage regeneration in this study.

Besides, in this study, ingredients from PRP also had important roles in stimulated grafted cells as well as endogenous cells growth and differentiation. There are at least six known growth factors such as platelet-derived growth factor (PDGF) that promotes blood vessel growth, cell division, and forming the skin; transforming growth factor-beta (TGF-β) that promotes cell division mitosis and bone metabolism; vascular endothelial growth factor (VEGF) that promotes the blood vessel formation; epidermal growth factor (EGF) that promotes cell growth and differentiation, angiogenesis, and collagen formation; fibroblast growth factor-2 (FGF-2) that promotes the growth of cell differentiation and angiogenesis; and insulin-like growth factor (IGF) that is a regulator in all cell types of the body [68, 69]. PRP injection also showed that improvements in knee injury and osteoarthritis score, including pain and symptom relief [70, 71].

Combining effect of SVF and PRP has a positive effect on the stimulation of proliferation, differentiation, and regeneration of cartilage in a mouse model. However, SVF also has a few limitations, notably the relatively low presence of ADSCs in the SVF. Therefore, SVF cultured to enrich ADSC before being transplanted may be essential, especially a little obtained fat cases.

4. Conclusion

Adipose tissue provides a rich source of MSCs. The SVF and PRP injection are a promising therapy in injured articular cartilage regeneration. This therapy significantly improved the injured articular cartilage. However, this study only assesses the ability of tumorigenicity and efficiency in mouse. Some side effects such as fever and muscle pain as well as the tumorigenicity in human being when using SVF and PRP could not be checked in this research.

Abbreviations

SVF: Stromal vascular fraction
PRP: Platelet-rich plasma
RT-PCR: Reverse transcription polymerase chain reaction
NOD/SCID: Nonobese diabetic/severe combined immunodeficient
SSEA: Stage-specific embryonic antigen
MSC: Mesenchymal stem cell
ADSC: Adipose-derived stem cell
D-PBS: Dulbecco-modified phosphate buffered saline
HE: Hematoxylin and eosin
ESC: Embryonic stem cell
GVHD: Graft versus host disease.

References

[1] D. J. Prockop, "Marrow stromal cells as stem cells for non-hematopoietic tissues," *Science*, vol. 276, pp. 71–74, 1997.

[2] S. M. Phadnis, M. V. Joglekar, M. P. Dalvi et al., "Human bone marrow-derived mesenchymal cells differentiate and mature into endocrine pancreatic lineage in vivo," *Cytotherapy*, vol. 13, no. 3, pp. 279–293, 2011.

[3] B. T. Estes, B. O. Diekman, J. M. Gimble, and F. Guilak, "Isolation of adipose-derived stem cells and their induction to a chondrogenic phenotype," *Nature Protocols*, vol. 5, no. 7, pp. 1294–1311, 2010.

[4] A. Reinisch, C. Bartmann, E. Rohde et al., "Humanized system to propagate cord blood-derived multipotent mesenchymal stromal cells for clinical application," *Regenerative Medicine*, vol. 2, no. 4, pp. 371–382, 2007.

[5] P. V. Phuc, T. H. Nhung, D. T. T. Loan, D. C. Chung, and P. K. Ngoc, "Differentiating of banked human umbilical cord blood-derived mesenchymal stem cells into insulin-secreting cells," *In Vitro Cellular and Developmental Biology—Animal*, vol. 47, no. 1, pp. 54–63, 2011.

[6] V. A. Farias, J. L. Linares-Fernández, J. L. Peñalver et al., "Human umbilical cord stromal stem cell express CD10 and exert contractile properties," *Placenta*, vol. 32, pp. 86–95, 2011.

[7] J. Peng, Y. Wang, L. Zhang et al., "Humanumbilical cord Wharton's jelly-derived mesenchymal stem cells differentiate into a Schwann-cell phenotype and promote neurite outgrowth in vitro," *Brain Research Bulletin*, vol. 84, pp. 235–243, 2011.

[8] G. A. Pilz, C. Ulrich, M. Ruh et al., "Human term placenta-derived mesenchymal stromal cells are less prone to osteogenic differentiation than bone marrow-derived mesenchymal stromal cells," *Stem Cells and Development*, vol. 20, no. 4, pp. 635–646, 2011.

[9] L. Spath, V. Rotilio, M. Alessandrini et al., "Explant-derived human dental pulp stem cells enhance differentiation and proliferation potentials," *Journal of Cellular and Molecular Medicine*, vol. 14, pp. 1635–1644, 2010.

[10] A. M. Lubis and V. K. Lubis, "Adult bone marrow stem cells in cartilage therapy," *Acta Medica Indonesiana*, vol. 44, pp. 62–68, 2012.

[11] C. Kasemkijwattana, S. Hongeng, S. Kesprayura, V. Rungsina-porn, K. Chaipinyo, and K. Chansiri, "Autologous bone marrow mesenchymal stem cells implantation for cartilage defects: two cases report," *Journal of the Medical Association of Thailand*, vol. 94, no. 3, pp. 395–400, 2011.

[12] F. Davatchi, B. S. Abdollahi, M. Mohyeddin, F. Shahram, and B. Nikbin, "Mesenchymal stem cell therapy for knee osteoarthritis. Preliminary report of four patients," *International Journal of Rheumatic Diseases*, vol. 14, no. 2, pp. 211–215, 2011.

[13] D. Minteer, K. G. Marra, and J. P. Rubin, "Adipose-derived mesenchymal stem cells: biology and potential applications," *Advances in Biochemical Engineering/Biotechnology*. In press.

[14] D. D. Frisbie, J. D. Kisiday, C. E. Kawcak, N. M. Werpy, and C. W. McIlwraith, "Evaluation of adipose-derived stromal vascular fraction or bone marrow-derived mesenchymal stem cells for treatment of osteoarthritis," *Journal of Orthopaedic Research*, vol. 27, no. 12, pp. 1675–1680, 2009.

[15] J. Pak, "Regeneration of human bones in hip osteonecrosis and human cartilage in knee osteoarthritis with autologous adipose-tissue-derived stem cells: a case series," *Journal of Medical Case Reports*, vol. 5, p. 296, 2011.

[16] H. A. Wieland, M. Michaelis, B. J. Kirschbaum, and K. A. Rudolphi, "Osteoarthritis—an untreatable disease?" *Nature Reviews Drug Discovery*, vol. 4, pp. 331–344, 2005.

[17] J. A. Buckwalter, C. Saltzman, and T. Brown, "The impact of osteoarthritis: implications for research," *Clinical Orthopaedics and Related Research*, vol. 427, pp. S6–S15, 2004.

[18] M. Dougados, "The role of anti-inflammatory drugs in the treatment of osteoarthritis: a European viewpoint," *Clinical and Experimental Rheumatology*, vol. 19, pp. S9–S14, 2001.

[19] T. Pincus, G. G. Koch, T. Sokka et al., "A randomized, double-blind, crossover clinical trial of diclofenac plus misoprostol versus acetaminophen in patients with osteoarthritis of the hip or knee," *Arthritis & Rheumatism*, vol. 44, pp. 1587–1598, 2001.

[20] S. Eyigor, S. Hepguler, M. Sezak, F. Öztop, and K. Capaci, "Effects of intra-articular hyaluronic acid and corticosteroid therapies on articular cartilage in experimental severe osteoarthritis," *Clinical and Experimental Rheumatology*, vol. 24, no. 6, p. 724, 2006.

[21] T. Spaková, J. Rosocha, M. Lacko, D. Harvanová, and A. Gharaibeh, "Treatment of knee joint osteoarthritis with autologous platelet-rich plasma in comparison with hyaluronic acid," *American Journal of Physical Medicine and Rehabilitation*, vol. 91, no. 5, pp. 411–417, 2012.

[22] V. Karatosun, B. Unver, A. Ozden, Z. Ozay, and I. Gunal, "Intra-articular hyaluronic acid compared to exercise therapy in osteoarthritis of the ankle. A prospective randomized trial with long-term follow-up," *Clinical and Experimental Rheumatology*, vol. 26, no. 2, pp. 288–294, 2008.

[23] J. P. Schroeppel, J. D. Crist, H. C. Anderson, and J. Wang, "Molecular regulation of articular chondrocyte function and its significance in osteoarthritis," *Histology and Histopathology*, vol. 26, pp. 377–394, 2011.

[24] H. J. Harn, S. Z. Lin, S. H. Hung et al., "Adipose-derived stem cells can abrogate chemical-induced liver fibrosis and facilitate recovery of liver function," *Cell Transplantation*. In press.

[25] J. H. Gu, Y. H. Ji, E. S. Dhong, D. H. Kim, and E. S. Yoon, "Transplantation of adipose derived stem cells for peripheral nerve regeneration in sciatic nerve defects of the rat," *Current Stem Cell Research & Therapy*, vol. 7, no. 5, pp. 347–355, 2012.

[26] N. Scuderi, S. Ceccarelli, M. G. Onesti et al., "Human adipose derived stem cells for cell based therapies in the treatment of systemic sclerosis," *Cell Transplantation*. In press.

[27] M. Mazo, S. Hernández, J. J. Gavira et al., "Treatment of reperfused ischemia with adipose-derived stem cells in a preclinical swine model of myocardial infarction," *Cell Transplantation*. In press.

[28] R. Peçanha, L. L. Bagno, M. B. Ribeiro et al., "Adipose-derived stem-cell treatment of skeletal muscle injury," *The Journal of Bone & Joint Surgery*, vol. 94, pp. 609–617, 2012.

[29] J. Xiao, C. Zhang, Y. Zhang et al., "Transplantation of adipose-derived mesenchymal stem cells into a murine model of passive chronic immune thrombocytopenia," *Transfusion*, vol. 52, no. 12, pp. 2551–2558, 2012.

[30] J. J. Yang, X. Yang, Z. Q. Liu et al., "Transplantation of adipose tissue-derived stem cells overexpressing heme oxygenase-1 improves functions and remodeling of infarcted myocardium in rabbits," *The Tohoku Journal of Experimental Medicine*, vol. 226, pp. 231–241, 2012.

[31] L. L. Black, J. Gaynor, C. Adams et al., "Effect of intraarticular injection of autologous adipose-derived mesenchymal stem and regenerative cells on clinical signs of chronic osteoarthritis of the elbow joint in dogs," *Veterinary Therapeutics*, vol. 9, no. 3, pp. 192–200, 2008.

[32] L. L. Black, J. Gaynor, D. Gahring et al., "Effect of adipose-derived mesenchymal stem and regenerative cells on lameness in dogs with chronic osteoarthritis of the coxofemoral joints: a randomized, double-blinded, multicenter, controlled trial," *Veterinary Therapeutics*, vol. 8, no. 4, pp. 272–284, 2007.

[33] A. Guercio, P. Di Marco, S. Casella et al., "Production of canine mesenchymal stem cells from adipose tissue and their application in dogs with chronic osteoarthritis of the humeroradial joints," *Cell Biology International*, vol. 36, pp. 189–194, 2012.

[34] F. S. Toghraie, N. Chenari, M. A. Gholipour et al., "Treatment of osteoarthritis with infrapatellar fat pad derived mesenchymal stem cells in Rabbit," *Knee*, vol. 18, no. 2, pp. 71–75, 2011.

[35] J. M. Lee and G. I. Im, "SOX trio-co-transduced adipose stem cells in fibrin gel to enhance cartilage repair and delay the progression of osteoarthritis in the rat," *Biomaterials*, vol. 33, pp. 2016–2024, 2012.

[36] M. C. ter Huurne, P. L. E. M. van Lent, A. B. Blom et al., "A single injection of adipose-derived stem cells protects against cartilage damage and lowers synovial activation in experimental osteoarthritis," *Arthritis & Rheumatism*, vol. 63, p. 1784, 2011.

[37] J. M. Gimble and F. Guilak, "Differentiation potential of adipose derived adult stem cell (ADAS) cells," *Current Topics in Developmental Biology*, vol. 58, pp. 137–160, 2003.

[38] J. M. Murphy, D. J. Fink, E. B. Hunziker, and F. P. Barry, "Stem cell therapy in a caprine model of osteoarthritis," *Arthritis & Rheumatism*, vol. 48, pp. 3464–3474, 2003.

[39] P. A. Zuk, M. Zhu, H. Mizuno et al., "Multilineage cells from human adipose tissue: implications for cell-based therapies," *Tissue Engineering*, vol. 7, no. 2, pp. 211–228, 2001.

[40] Y. D. C. Halvorsen, D. Franklin, A. L. Bond et al., "Extracellular matrix mineralization and osteoblast gene expression by human adipose tissue-derived stromal cells," *Tissue Engineering*, vol. 7, no. 6, pp. 729–741, 2001.

[41] H. Mizuno, P. A. Zuk, M. Zhu, H. P. Lorenz, P. Benhaim, and M. H. Hedrick, "Myogenic differentiation by human processed

Transplantation of Nonexpanded Adipose Stromal Vascular Fraction and Platelet-Rich Plasma for Articular Cartilage Injury Treatment in Mice Model

143

lipoaspirate cells," *Plastic and Reconstructive Surgery*, vol. 109, pp. 199–209, 2002.

[42] W. Wagner, F. Wein, A. Seckinger et al., "Comparative characteristics of mesenchymal stem cells from human bone marrow, adipose tissue, and umbilical cord blood," *Experimental Hematology*, vol. 33, no. 11, pp. 1402–1416, 2005.

[43] S. Kern, H. Eichler, J. Stoeve, H. Klüter, and K. Bieback, "Comparative analysis of mesenchymal stem cells from bone marrow, umbilical cord blood, or adipose tissue," *Stem Cells*, vol. 24, no. 5, pp. 1294–1301, 2006.

[44] K. M. Safford, K. C. Hicok, S. D. Safford et al., "Neurogenic differentiation of murine and human adipose-derived stromal cells," *Biochemical and Biophysical Research Communications*, vol. 294, no. 2, pp. 371–379, 2002.

[45] R. Izadpanah, C. Trygg, B. Patel et al., "Biologic properties of mesenchymal stem cells derived from bone marrow and adipose tissue," *Journal of Cellular Biochemistry*, vol. 99, no. 5, pp. 1285–1297, 2006.

[46] A. N. Patel, J. Yochman, V. Vargas, and D. A. Bull, "Putative population of adipose derived stem cells isolated from mediastinal tissue during cardiac surgery," *Cell Transplantation*. In press.

[47] G. Musumeci, D. Lo Furno, C. Loreto et al., "Mesenchymal stem cells from adipose tissue which have been differentiated into chondrocytes in three-dimensional culture express lubricin," *Experimental Biology and Medicine*, vol. 236, pp. 1333–1341, 2011.

[48] V. Zachar, J. G. Rasmussen, and T. Fink, "Isolation and growth of adipose tissue-derived stem cells," *Methods in Molecular Biology*, vol. 698, pp. 37–49, 2011.

[49] C. K. Rebelatto, A. M. Aguiar, M. P. Moretão et al., "Dissimilar differentiation of mesenchymal stem cells from bone marrow, umbilical cord blood, and adipose tissue," *Experimental Biology and Medicine*, vol. 233, no. 7, pp. 901–913, 2008.

[50] I. S. Blande, V. Bassaneze, C. Lavini-Ramos et al., "Adipose tissue mesenchymal stem cell expansion in animal serum-free medium supplemented with autologous human platelet lysate," *Transfusion*, vol. 49, no. 12, pp. 2680–2685, 2009.

[51] C. Dromard, P. Bourin, M. André, S. De Barros, L. Casteilla, and V. Planat-Benard, "Human adipose derived stroma/stem cells grow in serum-free medium as floating spheres," *Experimental Cell Research*, vol. 317, no. 6, pp. 770–780, 2011.

[52] D. T. B. Shih, J. C. Chen, W. Y. Chen, Y. P. Kuo, C. Y. Su, and T. Burnouf, "Expansion of adipose tissue mesenchymal stromal progenitors in serum-free medium supplemented with virally inactivated allogeneic human platelet lysate," *Transfusion*, vol. 51, no. 4, pp. 770–778, 2011.

[53] M. Dominici, K. Le Blanc, I. Mueller et al., "Minimal criteria for defining multipotent mesenchymal stromal cells. The International Society for Cellular Therapy position statement," *Cytotherapy*, vol. 8, no. 4, pp. 315–317, 2006.

[54] H. Niwa, J. Miyazaki, and A. G. Smith, "Quantitative expression of Oct-3/4 defines differentiation, dedifferentiation or self-renewal of ES cells," *Nature Genetics*, vol. 24, pp. 372–376, 2000.

[55] K. Mitsui, Y. Tokuzawa, H. Itoh et al., "The homeoprotein nanog is required for maintenance of pluripotency in mouse epiblast and ES cells," *Cell*, vol. 113, no. 5, pp. 631–642, 2003.

[56] K. Hochedlinger, Y. Yamada, C. Beard, and R. Jaenisch, "Ectopic expression of Oct-4 blocks progenitor-cell differentiation and causes dysplasia in epithelial tissues," *Cell*, vol. 121, no. 3, pp. 465–477, 2005.

[57] S. H. Chiou, C. C. Yu, C. Y. Huang et al., "Positive correlations of Oct-4 and Nanog in oral cancer stem-like cells and high-grade oral squamous cell carcinoma," *Clinical Cancer Research*, vol. 14, no. 13, pp. 4085–4095, 2008.

[58] Y. C. Chen, H. S. Hsu, Y. W. Chen et al., "Oct-4 expression maintained cancer stem-like properties in lung cancer-derived CD133-positive cells," *PLoS One*, vol. 3, article e2637, 2008.

[59] C. Liu, X. Cao, Y. Zhang et al., "Co-expression of Oct-4 and Nestin in human breast cancers," *Molecular Biology Reports*, vol. 39, pp. 5875–5881, 2012.

[60] Y. Guo, S. Liu, P. Wang et al., "Expression profile of embryonic stem cell-associated genes Oct4, Sox2 and Nanog in human gliomas," *Histopathology*, vol. 59, pp. 763–775, 2011.

[61] J. T. Oliveira, L. S. Gardel, T. Rada, L. Martins, M. E. Gomes, and R. L. Reis, "Injectable gellan gum hydrogels with autologous cells for the treatment of rabbit articular cartilage defects," *Journal of Orthopaedic Research*, vol. 28, no. 9, pp. 1193–1199, 2010.

[62] J. L. Dragoo, G. Carlson, F. McCormick et al., "Healing full-thickness cartilage defects using adipose-derived stem cells," *Tissue Engineering*, vol. 13, no. 7, pp. 1615–1621, 2007.

[63] P. F. Caimi, J. Reese, Z. Lee, and H. M. Lazarus, "Emerging therapeutic approaches for multipotent mesenchymal stromal cells," *Current Opinion in Hematology*, vol. 17, no. 6, pp. 505–513, 2010.

[64] N. G. Singer and A. I. Caplan, "Mesenchymal stem cells: mechanisms of inflammation," *Annual Review of Pathology*, vol. 6, pp. 457–478, 2011.

[65] O. Ringdén, M. Uzunel, I. Rasmusson et al., "Mesenchymal stem cells for treatment of therapy-resistant graft-versus-host disease," *Transplantation*, vol. 81, no. 10, pp. 1390–1397, 2006.

[66] B. Fang, Y. Song, L. Liao, Y. Zhang, and R. C. Zhao, "Favorable response to human adipose tissue-derived mesenchymal stem cells in steroid-refractory acute graft-versus-host disease," *Transplantation Proceedings*, vol. 39, no. 10, pp. 3358–3362, 2007.

[67] B. Fang, Y. Song, R. C. Zhao, Q. Han, and Q. Lin, "Using human adipose tissue-derived mesenchymal stem cells as salvage therapy for hepatic graft-versus-host disease resembling acute hepatitis," *Transplantation Proceedings*, vol. 39, no. 5, pp. 1710–1713, 2007.

[68] P. Borrione, A. D. Gianfrancesco, M. T. Pereira, and F. Pigozzi, "Platelet-rich plasma in muscle healing," *American Journal of Physical Medicine & Rehabilitation*, vol. 89, pp. 854–861, 2010.

[69] W. Yu, J. Wang, and J. Yin, "Platelet-rich plasma: a promising product for treatment of peripheral nerve regeneration after nerve injury," *International Journal of Neuroscience*, vol. 121, pp. 176–180, 2011.

[70] S. Sampson, M. Reed, H. Silvers, M. Meng, and B. Mandelbaum, "Injection of platelet-rich plasma in patients with primary and secondary knee osteoarthritis: a pilot study," *American Journal of Physical Medicine & Rehabilitation*, vol. 89, no. 12, pp. 961–969, 2010.

[71] G. Filardo, E. Kon, R. Buda et al., "Platelet-rich plasma intra-articular knee injections for the treatment of degenerative cartilage lesions and osteoarthritis," *Knee Surgery, Sports Traumatology, Arthroscopy*, vol. 19, no. 4, pp. 528–535, 2011.

16

Using Design of Experiments Methods for Assessing Peak Contact Pressure to Material Properties of Soft Tissue in Human Knee

Marjan Bahraminasab,[1] Ali Jahan,[2] Barkawi Sahari,[1,3] Manohar Arumugam,[4] Mahmoud Shamsborhan,[5] and Mohd Roshdi Hassan[1]

[1] Department of Mechanical and Manufacturing Engineering, Universiti Putra Malaysia, 43400 Selangor, Malaysia
[2] Faculty of Engineering, Semnan Branch, Islamic Azad University, Semnan, Iran
[3] Institute of Advanced Technology, ITMA, Universiti Putra Malaysia, Selangor, Malaysia
[4] Department of Orthopedic Surgery, Faculty of Medicine & Health Science, Universiti Putra Malaysia, Selangor, Malaysia
[5] Department of Mechanical Engineering, K.N. Toosi University of Technology, Tehran, Iran

Correspondence should be addressed to Marjan Bahraminasab; m.bahraminasab@yahoo.com

Academic Editor: Kaisar Alam

Contact pressure in the knee joint is a key element in the mechanisms of knee pain and osteoarthritis. Assessing the contact pressure in tibiofemoral joint is a challenging mechanical problem due to uncertainty in material properties. In this study, a sensitivity analysis of tibiofemoral peak contact pressure to the material properties of the soft tissue was carried out through fractional factorial and Box-Behnken designs. The cartilage was modeled as linear elastic material, and in addition to its elastic modulus, interaction effects of soft tissue material properties were added compared to previous research. The results indicated that elastic modulus of the cartilage is the most effective factor. Interaction effects of axial/radial modulus with elastic modulus of cartilage, circumferential and axial/radial moduli of meniscus were other influential factors. Furthermore this study showed how design of experiment methods can help designers to reduce the number of finite element analyses and to better interpret the results.

1. Introduction

Knee joint contact pressure is of critical importance in the mechanisms of knee pain and osteoarthritis [1, 2]. Computational models and finite element analyses (FEA) have been utilized to study contact characteristics of normal and injured knees, as well as total knee replacements (TKR) [3–8]. The purpose of these studies was to determine peak contact pressure in order to predict either tissue degradation of the knee or wear of ultra-high molecular weight polyethylene (UHMWPE) in TKR. Some biomechanical factors, such as material properties and geometries of tissues [9, 10], and knee kinematic [11] can affect the contact behavior of the knee and consequently the design of TKR. Impacts of horn attachments stiffness and meniscal material properties on tibiofemoral contact pressure using "semiautomatic" optimization method were investigated by Haut Donahue et al.

[9], who set tolerances on the variables to restore the contact pressure to within a specified error. The authors, however, performed more than 60 analyses to determine whether an individual factor is of importance. Meanwhile, interaction effects between different factors were not considered in their study. In order to better interpret the effects of variations in the material properties of soft tissue, a powerful statistical approach is required to design computational experiments.

Design of experiments (DOE) is a formal mathematical method that helps to solve complicated problems and to save time and resources (cost) by reducing the number of required experiments (runs) while obtaining all the necessary information. However, reducing runs associate with decrease in resolution. Usually in an experiment, one or more factors are deliberately changed in order to observe the effect of these changes on one or more response variables. The statistical design of experiments is an efficient method for planning

Using Design of Experiments Methods for Assessing Peak Contact Pressure to Material Properties of
Soft Tissue in Human Knee

145

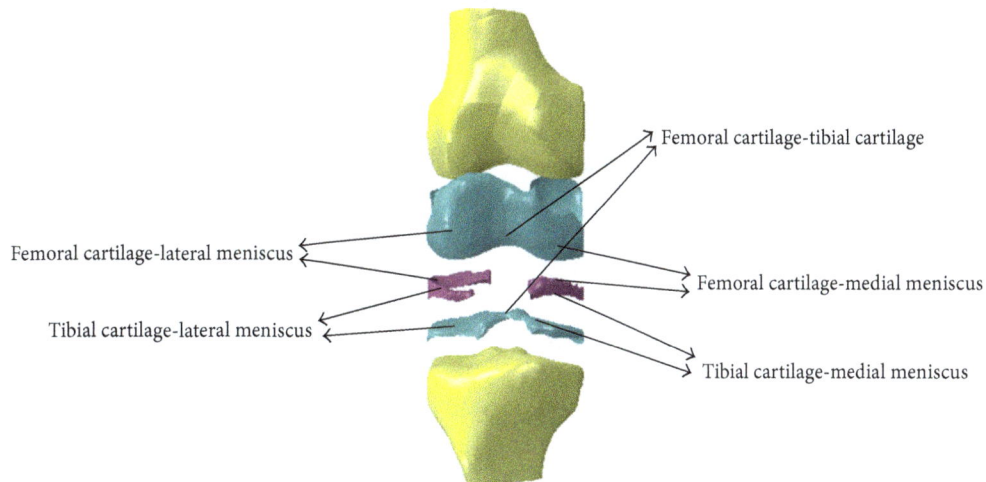

FIGURE 1: Different parts of FE model and contact pairs.

experiments so that it can be analyzed to yield valid and objective conclusions that can be obtained for a given amount of experimental efforts. Recently, there has been an increasing interest in use of DOE for sensitivity analysis based on FEA in biomedical applications [11–15]. So far, this method has only been applied in few studies related to the human knee joint [11, 14, 15]. Yao et al. [15] focused on the medial compartment of the knee and investigated the sensitivities of medial meniscal motion and deformation to material properties of soft tissues. They used Taguchi approach and central composite design to fit the finite element model (FEM) to the experimental data in the anterior cruciate ligament-deficient knee. Furthermore, Julkunen et al. [14] used a three-level fractional factorial design in combination with composition-based finite element model to determine the effect of different cartilage constituents on the mechanical response of the tissue. Due to the uncertainty in material properties [16], finite element analysis of the tibiofemoral joint becomes a very challenging mechanical problem. Therefore, the aims of this paper are to explore the most important parameters related to the material properties of meniscus and cartilage affecting the tibiofemoral joint contact pressure and to make a regression model based on main interaction and quadratic effects of variables to understand how they influence and minimize the error of FEA output. In this regard, fractional factorial design was applied in screening step and Box-Behnken method was used in response to surface method and optimization process.

2. Methods

2.1. Creation of Finite Element Analysis. Geometries of bony structures and soft tissues were taken from a healthy human knee of a 24-year-old man. Solid models of the femur and tibia and geometries of soft tissues, including articular cartilages and menisci, were developed from the magnetic resonance images (MRI). Each image was taken at 3.2 mm interval in a sagittal plane. The obtained data, subsequently, was used to create a three dimensional computer aided design (3D CAD) model in order to import into ABAQUS 6.8

software (Dassault Systèmes Simulia Corp., Providence, RI, USA). The model consisted of two bony structures (femur and tibia), both the femoral and tibial articular cartilages, and both the medial and lateral menisci. Figure 1 shows the generated 3D model in details. The model did not include ligaments. The finite element mesh generation was performed leading to 41709 linear 4-noded tetrahedron elements for articular cartilage and menisci (25293 for femoral cartilage, 9130 for tibial cartilage, 3866 for medial meniscus, and 3420 for lateral meniscus). Contact was defined between the femoral cartilage and meniscus, the meniscus and tibial cartilage, and femoral cartilage and tibial cartilage for both lateral and medial compartments, resulting in six contact-surface pairs. Completely general contact condition involving small sliding of pairs was applied on the model and all contact surfaces were modeled as frictionless. The cartilage in the knee is a complex structure, composed mainly of networks of collagen fibrils that embed water and a non-fibrillar matrix. The cartilage is known to be inhomogeneous and anisotropic material, but considering that the loading time of interest is related to a single leg stance and that the viscoelastic time constant for cartilage is approximately 1500 seconds from biphasic theory [3, 9], the elastic solution does not diverge from the biphasic solution [17]. The cartilage, therefore, was assumed to behave as a homogeneous linearly isotropic elastic material for contact pressure computations, similar to the previous studies [18, 19]. The meniscus, also, has similar structure to that of cartilage and it is also known to be inhomogeneous and anisotropic material, but various material property definitions can be found in the literature for this component [20–22]. Furthermore, the meniscus has a time constant, as large as 3300 seconds [9], and can also be considered as an elastic material for compression of the joint during the short loading times (single leg stance). In this study, the menisci were treated as linearly elastic, transversely isotropic material to represent the circumferential fiber arrangement. Femur and tibia were represented by rigid bodies because this is time efficient in a nonlinear analysis and accurate due to their much larger stiffness compared to that of soft tissues. Meanwhile, the previous study [3] confirmed that this simplification has no substantive effect on contact variables. Horn attachments, in the current model, were defined by 10 linear springs. For boundary conditions, the tibia was fully constrained and femur was constrained from rotation and free to translate in anterior-posterior, medial-lateral, and inferior-superior axes.

For validation of the model, static loads equivalent to 0, 500, 734, 800, 1000, 1500, 2000, and 2500 N were applied on the model at 0° flexion angle in order to compare with previously reported measurements and predictions [3, 29–31]. In this regard, the initial cartilage elastic modulus and Poisson's ratio were considered as 15 MPa and 0.475, respectively [25], and for menisci the primary values were moduli of 20 MPa in axial/radial directions and 140 MPa in circumferential direction. The values used for in-plane and out-of-plane Poisson's ratios and shear modulus were 0.2, 0.3, and 50 MPa, respectively [20, 21, 23, 24]. The stiffness of horn attachments was considered 200 N/mm, which resulted in 2000 N/mm total stiffness. Figure 2 shows the finite element representation of the joint.

2.2. Verifying the Results of FEA. The results of peak contact pressure for different magnitudes of force are shown in Figure 3. The applied force is transferred through the femur-meniscus, femur-tibia, and meniscus-tibia at the contact regions. The stresses were computed and it was seen that the total stress multiples by area equilibrate the total load in the knee joint. The predicted reaction forces at each loading condition, also, were in equilibrium with the applied load. Although, the finite element solution may have satisfied the equilibrium, indicating that the finite element solution was accurate to some extent, confidence in the validity of the model itself were obtained by comparing the computed values of the peak contact pressure with the previously reported measurements and predictions. Among the various researches that have measured the peak contact pressure on the tibial plateau [3, 4, 8–10, 29–31], studies of Brown and Shaw [31], Ahmed et al. [29], Fukubayashi and Kurosawa [30] and Donahue et al. [3] were chosen because they used a load application system with various compressive loads (734, 800, 1000, 1500, and 2500 N) at 0 flexion angle and computed the peak contact pressure on the tibial cartilage. It can be seen that, the results of present study fall well within the ranges provided by the literature. Hence the present results are verified.

2.3. Design of Experiments. DOE starts with determining the objectives of an experiment. These objectives are as follows: comparative, screening, and modeling [32–34]. Objective of comparative designs is to find a suitable method for an initial comparison. Screening designs identify which factors are important and help to screen out unimportant factors. Response surface modeling seeks for one or more of the following objectives: hit a target, maximize or minimize a response or make it robust.

In this research, seven factors including axial/radial and circumferential elastic moduli of meniscus ($E_{2,3}$ and E_1),

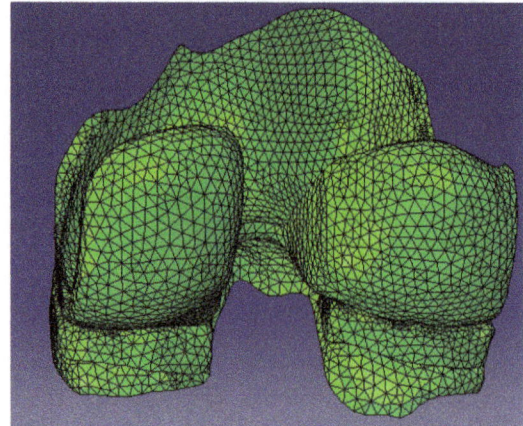

FIGURE 2: The finite element representation of the joint.

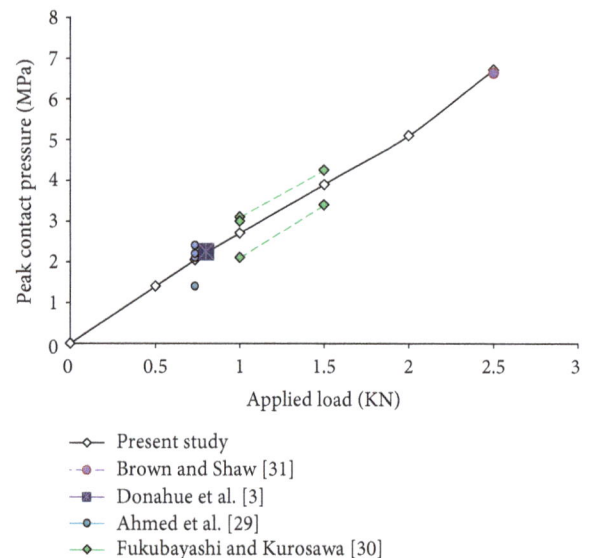

Present study
Brown and Shaw [31]
Donahue et al. [3]
Ahmed et al. [29]
Fukubayashi and Kurosawa [30]

FIGURE 3: Comparison of the results of peak contact pressure on the tibial plateau.

stiffness of meniscus horn attachment (K), in-plane and out-of-plane Poisson's ratios (v_{12}, v_{23}), shear modulus (G_{12}), and elastic modulus of cartilage (E) were considered as initial variables for sensitivity analysis. Due to the large number of factors and levels, in the first step fractional factorial design was applied to screen out less significant factors. It is useful and efficient when full factorial design becomes unpractical [35]. Two levels for each factor impose ($2^7 = 128$) treatment combinations for full factorial design, but the 1/8 fractional factorial design suggested 16, so it can be used as a rational way for choosing the treatment combinations of experiments. The 1/8 fractional design corresponds to resolution IV in which the main effects are not confounded with two-way interactions. However, a limitation of fractional factorial design is the use of only two levels for each factor and the responses are assumed to be approximately linear over the range of the factor levels chosen. More detailed discussions can be found in Montgomery [36]. In the next step, after

Using Design of Experiments Methods for Assessing Peak Contact Pressure to Material Properties of
Soft Tissue in Human Knee

147

TABLE 1: Name and variation range of factors.

Factor	Name	Range of variation	References
A	E_1 (Mpa): circumferential modulus of meniscus	100–200	[20, 21, 23]
B	$E_{2,3}$ (Mpa): axial/radial modulus of meniscus	15–60	[20, 23, 24]
C	v_{12}: in-plane Poisson's ratio	0.1–0.4	[9]
D	v_{23}: out-of-plane Poisson's ratio	0.1–0.35	[9]
E	G_{12} (Mpa): shear modulus of meniscus	27.7–77.7	[9]
F	K (N/mm): stiffness of meniscus horn attachment	500–30,000	[9]
G	E (Mpa): elastic modulus of cartilage	5–20	[25–28]

TABLE 2: Actual values of 2^{7-3} screening design and response.

Run no.	E_1 (Mpa)	$E_{2,3}$ (Mpa)	v_{12}	v_{23}	G_{12} (Mpa)	K (N/mm)	E (Mpa)	Peak contact pressure (Mpa)
1	180	50	0.2	0.3	30	1500	12	7.396
2	120	50	0.3	0.2	30	1500	20	6.381
3	120	15	0.2	0.2	30	1500	12	7.546
4	120	15	0.2	0.3	30	6000	20	6.361
5	180	15	0.3	0.2	30	6000	12	7.542
6	180	50	0.3	0.3	60	6000	20	6.389
7	120	15	0.3	0.3	60	1500	12	7.537
8	180	50	0.2	0.2	30	6000	20	6.386
9	180	15	0.3	0.3	30	1500	20	6.364
10	180	15	0.2	0.2	60	1500	20	6.366
11	120	50	0.3	0.3	30	6000	12	7.400
12	180	15	0.2	0.3	60	6000	12	7.533
13	120	50	0.2	0.2	60	6000	12	7.492
14	180	50	0.3	0.2	60	1500	12	7.386
15	120	50	0.2	0.3	60	1500	20	6.384
16	120	15	0.3	0.2	60	6000	20	6.362

screening out the less significant factors, Box-Behnken design was applied to do more investigations and hit the value of experiment in the FE model. The Box-Behnken is a good design in response surface methodology due to estimation of the parameters in the quadratic model. Furthermore, it is slightly more efficient than the central composite design [37], which was used by Yao et al. [15] in the sensitivity analysis of the knee joint.

In this study, sensitivity analysis was performed under 2500 (N) static load at full extension and the reference value of the peak contact pressure was taken from the experimental work of Brown and Shaw [31], which was equivalent to 6.5 (MPa). At this load, the optimum values of parameters were obtained by the estimated model based on Box-Behnken design. The predicted optimum values were subsequently tested for other applied loads at 0 flexion angle. The study of Brown and Shaw [31] was chosen because it measured the peak contact pressure on tibial cartilage under the same loading condition (static load of 2500 N at 0 flexion angle). The considered factors and their investigated ranges based on the literature are demonstrated in Table 1. The combination of parameters was generated and analyzed using Minitab software [38], and the design generators were E = ABC, F = BCD, and G = ACD.

3. Result and Discussion

3.1. Screening Analysis. Table 2 shows the treatment combinations and the results of peak contact pressure, according to fractional factorial design. The ranges in this step were chosen according to the most prevalent values used in the literature. For example, however the given range for elastic modulus of cartilage is 5–20 MPa, the majority of studies have considered values equal or more than 12 MPa [3, 4, 8, 9, 11, 39, 40].

The location of peak contact pressure was on the lateral compartment of the tibial cartilage in all FE analyses. Figure 4 shows the distribution of contact pressure for the first experiment; the red region represents the maximum contact pressure. The maximum variations in the location of peak contact pressure were 0.10 mm in anterior-posterior direction to the anterior and 1.74 mm in medial-lateral direction to the lateral side.

Table 3 demonstrates both the magnitude and the importance of the parameters effects. Any absolute value of the effect greater than $\alpha = 0.05$ is potentially important. Hence, factor E which represents the elastic modulus of cartilage has the most effect on peak contact pressure, followed by $E_1 * K$ (representing both circumferential and horn stiffness) and $E_{2,3}$ (axial/radial modulus of meniscus), respectively.

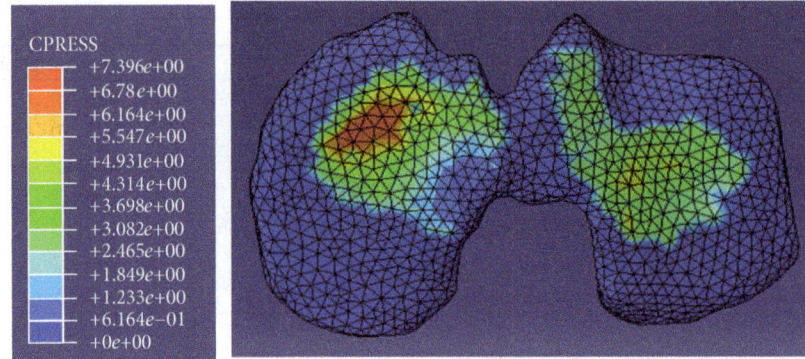

FIGURE 4: Contact pressure distribution in the tibial cartilage in first experiment.

TABLE 3: Considered terms, effects, and alias structure in 1/8 fractional design.

Term	Effect	Alias structure (up to order 3)
E_1	−0.0126	$E_1 + E_{2,3} * V_{12} * G_{12} + E_{2,3} * K * E + V_{12} * V_{23} * E + V_{23} * G_{12} * K$
$E_{2,3}$	−0.0496	$E_{2,3} + E_1 * V_{12} * G_{12} + E_1 * K * E + V_{12} * V_{23} * K + V_{23} * G_{12} * E$
V_{12}	−0.0129	$V_{12} + E_1 * E_{2,3} * G_{12} + E_1 * V_{23} * E + E_{2,3} * V_{23} * K + G_{12} * K * E$
V_{23}	−0.0121	$V_{23} + E_1 * V_{12} * E + E_1 * G_{12} * K + E_{2,3} * V_{12} * K + E_{2,3} * G_{12} * E$
G_{12}	0.0091	$G_{12} + E_1 * E_{2,3} * V_{12} + E_1 * V_{23} * K + E_{2,3} * V_{23} * E + V_{12} * K * E$
K	0.0131	$K + E_1 * E_{2,3} * E + E_1 * V_{23} * G_{12} + E_{2,3} * V_{12} * V_{23} + V_{12} * G_{12} * E$
E	−1.1049	$E + E_1 * E_{2,3} * K + E_1 * V_{12} * V_{23} + E_{2,3} * V_{23} * G_{12} + V_{12} * G_{12} * K$
$E_1 * E_{2,3}$	−0.0124	$E_1 * E_{2,3} + V_{12} * G_{12} + K * E$
$E_1 * V_{12}$	0.0129	$E_1 * V_{12} + E_{2,3} * G_{12} + V_{23} * E$
$E_1 * V_{23}$	0.0126	$E_1 * V_{23} + V_{12} * E + G_{12} * K$
$E_1 * G_{12}$	−0.0126	$E_1 * G_{12} + E_{2,3} * V_{12} + V_{23} * K$
$E_1 * K$	**0.0714**	**$E_1 * K + E_{2,3} * E + V_{23} * G_{12}$**
$E_1 * E$	0.0169	$E_1 * E + E_{2,3} * K + V_{12} * V_{23}$
$E_{2,3} * V_{23}$	−0.0069	$E_{2,3} * V_{23} + V_{12} * K + G_{12} * E$
$E_1 * E_{2,3} * V_{23}$	0.0129	$E_1 * E_{2,3} * V_{23} + E_1 * V_{12} * K + E_1 * G_{12} * E + E_{2,3} * V_{12} * E + E_{2,3} * G_{12} * K + V_{12} * V_{23} * G_{12} + V_{23} * K * E$

The bold item shows the most important term.

The other parameters are not significant at the 5% level. It can be seen that, however, other factors including E_1 and K are not significant, interaction of E_1 and K is significant.

Interaction is the variation among the differences between means for different levels of one factor over different levels of the other factor; and since in resolution IV designs, two-factor interaction effects may be confounded with other two-factor interactions, screening out of factors should be done carefully, because confounding pattern makes it difficult to determine which factors are the most important ones. In this regard, Table 3 also shows alias structure up to order 3. The alias structure indicates which effects are confounded with each other. As shown in Table 3, effect of $E_1 * K$ aliased to $E_{2,3} * E$ and $v_{23} * G_{12}$. Since factors $E_{2,3}$ and E have been significant, so probably the main reason for significance of $E_1 * K$ is due to the interaction of these two significant factors ($E_{2,3} * E$); but for more confidence, factors E_1 and K are kept for further investigation due to their higher effect (0.0126, 0.0131, resp.) comparing to v_{23} and G_{12} (0.0121, 0.0091, resp.). Moreover, this result is in agreement with research of Haut Donahue et al. [9] that showed the importance of E_1 and K for contact variables of the tibial plateau. Furthermore, the importance of axial/radial modulus ($E_{2,3}$) and circumferential moduli of meniscus (E_1), as a result of the present study, is consistent with the findings of Yao et al. [15], who revealed that the meniscal motion and deformation are most sensitive to the circumferential and radial/axial moduli of menisci.

According to screening experiments, the following results can be estimated. (1) Peak contact pressure for $E = 12$ (Mpa) is, on average, 1.1049 (Mpa) more than that for $E = 20$ (Mpa). (2) Peak contact pressure at $E_{2,3} = 15$ (Mpa) is, on average, 0.0496 (Mpa) more than that at $E_{2,3} = 50$ (Mpa). (3) The effect of 0.0714 for $E_1 * K$ can be interpreted to mean that the effect of combining the high level factor E_1 with K ($E_1 = 180$ Mpa, $K = 1500$, and 6000 N/mm) is, on average, 0.0714 more than the effect of low level of factor E_1 with K ($E_1 = 120$ Mpa, $K = 1500$, and 6000 N/mm). (4) However, the effect of other factors and their interaction are not significant. Figure 5 shows the main effect of factors. The contact pressure decreases by increasing the E, E_1, $E_{2,3}$, v_{12}, and v_{23} and increases by increasing the G_{12} and K. It is

Using Design of Experiments Methods for Assessing Peak Contact Pressure to Material Properties of Soft Tissue in Human Knee

149

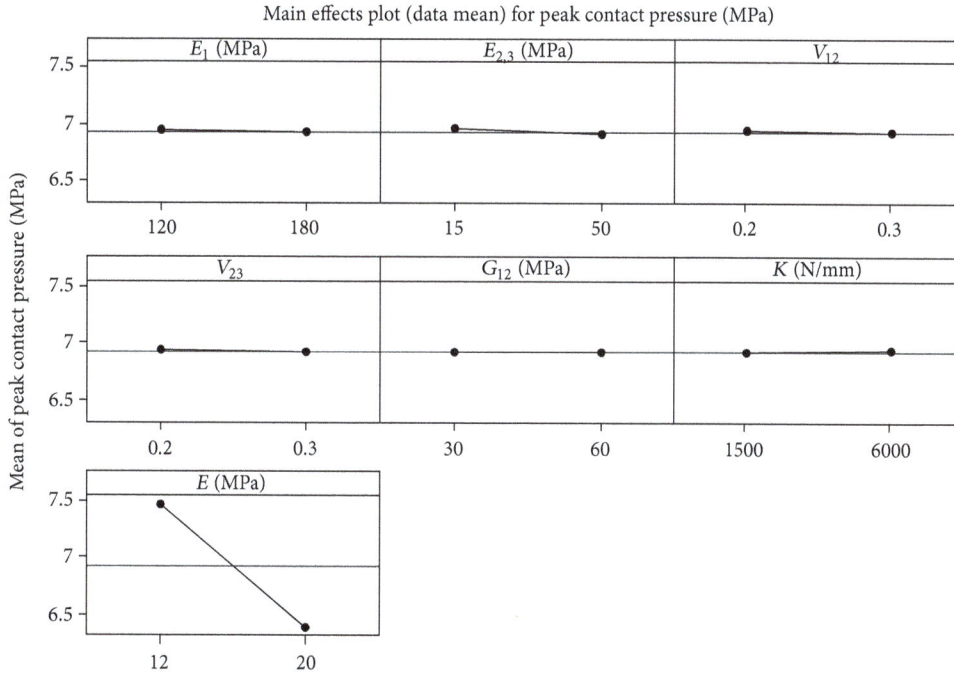

FIGURE 5: The main effects plot (data means) of peak contact pressure.

obvious that the main effect of factor E is much larger than the other factors and overshadows them. Interaction effect of E_1 and K is shown in Figure 6.

From the results of the screening experiment, factors of E, E_1, K, and $E_{2,3}$ were collected for more investigation by response surface method (RSM).

3.2. Response Surface Method . In this section, the response surface method was used as a statistical design of experiment tool, in order to produce precise maps based on mathematical models leading to optimum performance [41]. The regression model was built in two phases. First started with linear model but due to the lack of linear fit, quadratic model was applied subsequently. For choosing the level of factors which screened out in the last section, less interactions with other factors were considered ($v_{23} = 0.2$, $G_{12} = 60$ MPa, and $v_{12} = 0.2$). The Box-Behnken method was used at three levels of each factor and a single center point was considered because the FEM experiments include no actual experimental error; thus, duplication of center point was not necessary [42, 43]. The results of simulation runs and Box-Behnken design in RSM are given in Table 4. In this step, in order to assess more levels of each factor, other ranges of variables were chosen.

Estimated linear regression for peak contact pressure (PCP) is as follows:

$$PCP = 8.80182 - 0.13871 * E^{***} \quad \left(R_{adj}^2 = 95.8\%\right). \quad (1)$$

Results of ANOVA revealed that with 99% confidence, increase of one unit of E (P value <0.01) will result in decrease of peak contact pressure by 0.13871 MPa.

Although, the adjusted R^2 demonstrates that 95.8% of variation in peak contact pressure can be explained by

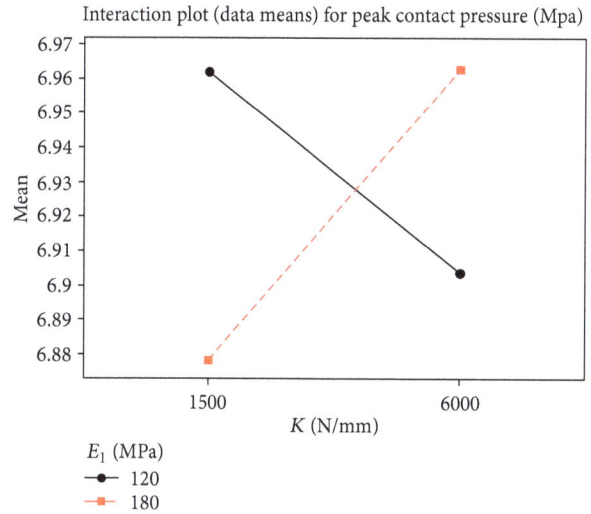

FIGURE 6: Interaction plot of peak contact pressure for E_1 and K.

variation in E, Figure 7(a) shows that the residuals of linear regression model for peak contact pressure are not normal; hence, the regression model should be revised. The full quadratic regression coefficients for peak contact pressure are estimated as follows:

$$PCP = 13.4733 - 0.7265 * E^{***} + 0.0182 * E^{2***}$$
$$+ 0.0001 * E * E_{2,3}^{**} \quad \left(R_{adj}^2 = 100\%\right). \quad (2)$$

According to the outputs of ANOVA the following can be concluded. (1) E affects peak contact pressure with 99%

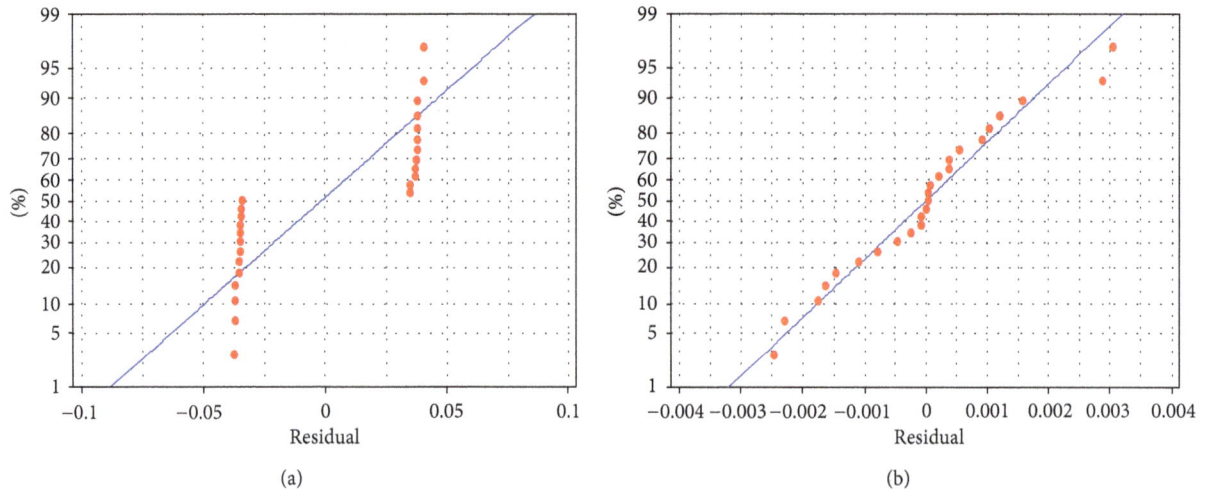

FIGURE 7: Normal probability plot of residuals. (a) Linear and (b) full quadratic regression model (response is peak contact pressure).

TABLE 4: Response, factors, and levels for the Box-Behnken experimental design.

Run no.	E (Mpa)	E_1 (Mpa)	K (N/mm)	$E_{2,3}$ (Mpa)	Peak contact pressure (Mpa)
1	16	150	2000	40	6.572
2	14	130	4000	30	6.917
3	16	150	2000	20	6.562
4	16	150	6000	40	6.572
5	16	130	4000	40	6.571
6	14	150	2000	30	6.918
7	18	150	6000	30	6.363
8	14	150	4000	40	6.916
9	16	150	6000	20	6.562
10	18	150	4000	40	6.367
11	16	150	4000	30	6.568
12	16	170	6000	30	6.569
13	18	150	2000	30	6.363
14	14	150	6000	30	6.918
15	18	130	4000	30	6.362
16	18	170	4000	30	6.364
17	16	130	4000	20	6.560
18	18	150	4000	20	6.359
19	16	170	2000	30	6.569
20	14	150	4000	20	6.920
21	16	130	6000	30	6.567
22	14	170	4000	30	6.918
23	16	130	2000	30	6.566
24	16	170	4000	40	6.573
25	16	170	4000	20	6.564

confidence (***means P value <0.01), as $-0.7265 * E + 0.0182 * E^2$, and with 95% confidence (**means P value <0.05) as $0.0001\ E * E_{2,3}$, if the effects of factor $E_{2,3}$ are held

constant. (2) $E_{2,3}$ affects the peak contact pressure with 95% confidence (P value <0.05) as $0.0001 * E * E_{2,3}$, if the effects of factor E are held constant. (3) There is no reason to believe the importance, of other factors, interactions and any other quadratic effects. Residuals of full quadratic regression model are shown in Figure 7(b). According to the results of RSM, surface plot of peak contact pressure versus $E_{2,3}$ and E is shown in Figure 8. It demonstrates the negative effect of E on peak contact pressure. Contour plot of peak contact pressure in Figure 9 shows that $E_{2,3}$ has more effect on contact pressure than E_1. Figure 10 represents that the peak contact pressure does not change when K is greater than 4000 (N/mm).

3.3. Optimization. The optimal region to run a process is typically determined after a sequence of experiments and developing empirical models. From a mathematical viewpoint, the objective is to find the operating conditions that maximize, minimize, or close the system's response to the true one. Therefore, the goal of this section is minimizing difference between estimated quadratic model obtained in the last section and experimental data of Brown and Shaw [31]. By considering 99% confidence, the estimated quadratic model will only include factor E. Figure 11 shows the behavior of peak contact pressure with respect to E. In regression analysis, usually developing a model includes the fewest numbers of explanatory variables which permit an adequate interpretation:

$$\text{Min}\ f = \text{PCP} - 6.5$$
$$= \left(13.4733 - 0.7265 * E + 0.0182 * E^2\right) - 6.5 \quad (3)$$
$$\text{s.t}\ 14 \le E \le 18.$$

The value of E was obtained to be 16.059 (Mpa) by solving the above nonlinear problem using generalized reduced gradient (GRG) algorithm on Microsoft Excel software. Furthermore, according to Figure 11, it is obvious that the minimum contact pressure is at $E = 20$ (MPa).

Using Design of Experiments Methods for Assessing Peak Contact Pressure to Material Properties of
Soft Tissue in Human Knee

151

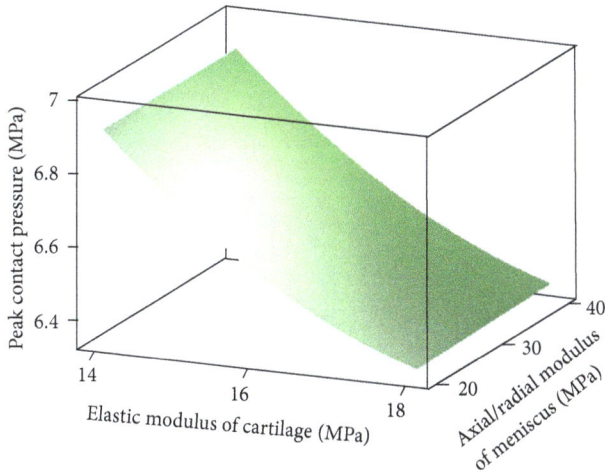

FIGURE 8: Surface plot of peak contact pressure versus $E_{2,3}$ and E (at $E_1 = 1500$ (Mpa), $K = 4000$ (N/mm)).

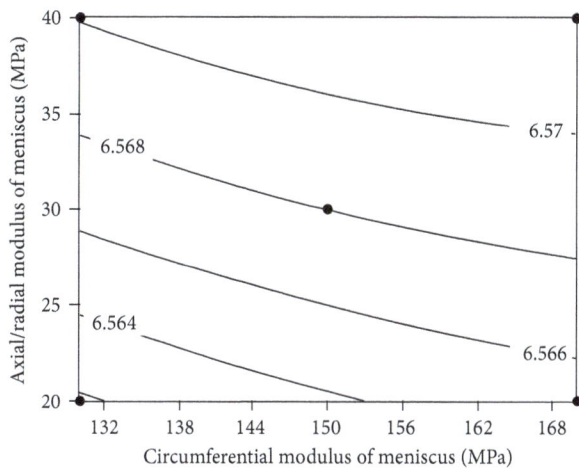

FIGURE 9: Contour plot of peak contact pressure (Mpa) versus $E_{2,3}$, E_1 (at $E = 16$ (Mpa), $K = 4000$ (N/mm)).

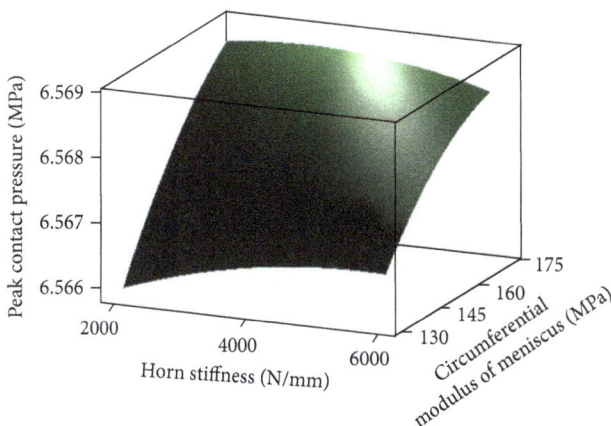

FIGURE 10: Surface plot of peak contact pressure versus E_1 and K (at $E = 16$ (Mpa), $E_{2,3} = 30$ (N/mm)).

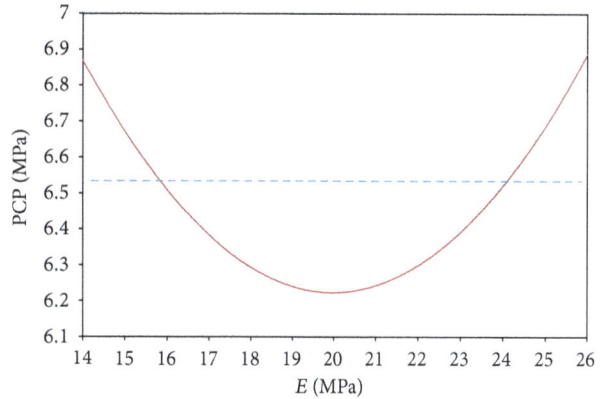

FIGURE 11: Behavior of quadratic estimated model for peak contact pressure with respect to E.

TABLE 5: Box-Behnken experimental design after optimizing the E.

Run no.	E_1 (Mpa)	$E_{2,3}$ (Mpa)	K (N/mm)	Peak contact pressure (Mpa)
1	100	60	6000	6.566
2	200	40	10000	6.567
3	150	20	2000	6.554
4	150	40	6000	6.565
5	150	60	2000	6.569
6	200	20	6000	6.558
7	100	40	10000	6.561
8	200	60	6000	6.572
9	150	60	10000	6.570
10	200	40	2000	6.567
11	100	40	2000	6.560
12	150	20	10000	6.554
13	100	20	6000	6.550

$G_{12} = 60$ (MPa), $v_{23} = 0.2$, $v_{12} = 0.2$, and $E = 16.059$ (MPa).

Further analyses were carried out after optimizing the E and removing its strong shadow on the other factors. Table 5 shows the Box-Behnken design with one center point, three factors, and peak contact pressure. In this stage, wider ranges of parameters were considered to investigate the maximum effects of factors.

Estimated full quadratic regression coefficients for peak contact pressure by considering optimized value of E, factors E_1, $E_{2,3}$, and K are as follows:

$$\text{PCP} = 6.517 + 0.00099 * E_{2,3} + 0.00019 * E_1$$
$$- 0.00001 * E_{2,3}^2 \quad \left(R_{\text{adj}}^2 = 99.8\%\right). \tag{4}$$

According to the outputs of ANOVA, it can be concluded that with 99% confidence, $E_{2,3}$ affects peak contact pressure as $0.00099 * E_{2,3} - 0.00001 * E_{2,3}^2$, E_1 as $0.00019 * E_1$ and there is no reason to believe the importance of factor K, other interactions, and any other quadratic effect. Figure 12 shows normality of residual in the above estimated regression model

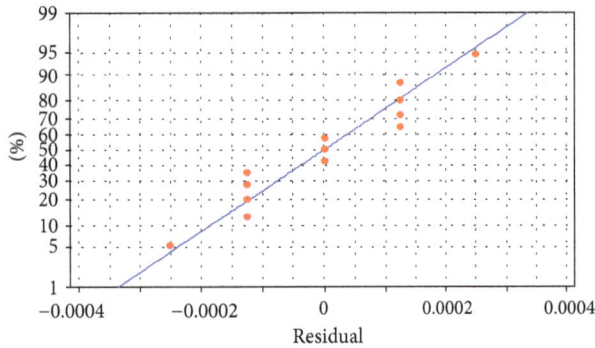

FIGURE 12: Normal probability plot of residuals for regression model using optimized value of E.

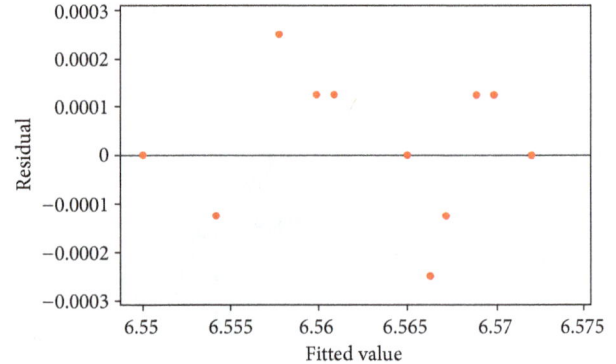

FIGURE 13: Residuals versus the fitted values in estimated regression model using optimized value of E.

and Figure 13 demonstrates random distribution of residuals versus the fitted values.

In order to find the optimum value for E_1 and $E_{2,3}$, it was tried to minimize the difference between the above estimated quadratic model and experimental data again:

$$\text{Min} \ f = \text{PCP} - 6.5$$

$$= \left(-0.00001 * E_{2,3}^2 6.517 + 0.00099 * E_{2,3} \right.$$

$$\left. + 0.00019 * E_1 - 0.00001 * E_{2,3}^2 \right) - 6.5 \quad (5)$$

$$\text{s.t} \ 100 \leq E_1 \leq 200$$

$$20 \leq E_{2,3} \leq 60.$$

The optimum values for E_1 and $E_{2,3}$ were 100 and 20 MPa, respectively. The suitable value for K can be considered ≥ 2000 N/mm. This value supports the idea that meniscal replacement surgery should attach the horns through a technique providing high stiffness. Using the values obtained from the optimization process, finally the results of the proposed model were tested by running another two simulations with two compressive loads of 1000 N [30] and 3000 N [31] at 0 degrees of flexion. The errors between FEA and experimental data of peak contact pressure decreased to be less than 10% and 5% for 1000 and 3000 N, respectively.

It should be pointed out that in the FEM, some anatomical geometries are missing or simplified depending on the complexity of the problem. Ligaments were not included in our FE model. The posterior cruciate ligament (PCL) and lateral collateral ligament (LCL) are slack under axial compressive loading at 0 flexion angle [8], but the anterior cruciate ligament (ACL) and the medial collateral ligament (MCL) both contribute in the axial compression experienced by the joint. Under no external load, the joint is primarily compressed due to the prestress in these two ligaments and the axial compression sustained by the joint is, thus, greater than the applied external load. Therefore, the influence of missing these two ligaments in this FEA is that the results of peak contact pressure only correspond to the external load. Future studies will consider including ACL and MCL. However, according to Haut Donahue et al. [9], contact characteristics are not sensitive to the nonlinear material properties of

MCL during axial compression. Furthermore, cartilage can be assumed to have linear elastic material property in contact analyses from the biphasic solution, but the huge influence of its elastic modulus on contact pressure might indicate the requirement for more precise material model in this component. This is supported by the study of Julkunen et al. [14] which showed that the mechanical responses of the cartilage under different loading conditions are dependent on tissue composition and structure. Therefore, future investigations will focus on the effect of anisotropic nonlinear behavior of cartilage on contact outputs. Moreover, it will be interesting to investigate the sensitivity of contact pressure to material properties under different degrees of flexion, which was not considered in this research.

4. Conclusion

A sensitivity analysis of tibiofemoral peak contact pressure to the material properties of soft tissue was performed and design of experiments methods was used to reduce the number of program runs and to minimize the contact pressure error. The present study evaluated the effect of cartilage elastic modulus and interaction effects of the parameters in addition to previous research. It was demonstrated that elastic modulus of the cartilage is the most influential factor. Another important finding was that after cartilage elastic modulus, interaction of axial/radial modulus with elastic modulus of cartilage, circumferential and axial/radial moduli of meniscus are significant factors. The importance of circumferential and axial/radial moduli of meniscus as a result of this study is in agreement with the past predictions. Furthermore, this research demonstrated the complex relations between material properties of tissue and contact pressure of tibiofemoral joint. The result of sensitivity analyses can be used as a guideline for experimental efforts intended at determining material properties of soft tissue, because estimating the most sensitive parameters should be done precisely. However, this analysis is only valid under full extension loading mode and with elastic assumptions of soft tissues. Further biomaterial studies may reveal more factors or more realistic form of material properties of human tissues.

Using Design of Experiments Methods for Assessing Peak Contact Pressure to Material Properties of Soft Tissue in Human Knee

153

However, more investigation in this regard based on DOE techniques will provide a remarkably versatile strategy for analysis of knee joint biomechanics and help researchers with faster and more reliable analysis.

References

[1] T. P. Andriacchi and A. Mündermann, "The role of ambulatory mechanics in the initiation and progression of knee osteoarthritis," *Current Opinion in Rheumatology*, vol. 18, no. 5, pp. 514–518, 2006.

[2] J. J. Elias, D. R. Wilson, R. Adamson, and A. J. Cosgarea, "Evaluation of a computational model used to predict the patellofemoral contact pressure distribution," *Journal of Biomechanics*, vol. 37, no. 3, pp. 295–302, 2004.

[3] T. L. H. Donahue, M. L. Hull, M. M. Rashid, and C. R. Jacobs, "A finite element model of the human knee joint for the study of tibio-femoral contact," *Journal of Biomechanical Engineering*, vol. 124, no. 3, pp. 273–280, 2002.

[4] P. Beillas, G. Papaioannou, S. Tashman, and K. H. Yang, "A new method to investigate in vivo knee behavior using a finite element model of the lower limb," *Journal of Biomechanics*, vol. 37, no. 7, pp. 1019–1030, 2004.

[5] A. C. Godest, M. Beaugonin, E. Haug, M. Taylor, and P. J. Gregson, "Simulation of a knee joint replacement during a gait cycle using explicit finite element analysis," *Journal of Biomechanics*, vol. 35, no. 2, pp. 267–275, 2002.

[6] J. P. Halloran, A. J. Petrella, and P. J. Rullkoetter, "Explicit finite element modeling of total knee replacement mechanics," *Journal of Biomechanics*, vol. 38, no. 2, pp. 323–331, 2005.

[7] T. Villa, F. Migliavacca, D. Gastaldi, M. Colombo, and R. Pietrabissa, "Contact stresses and fatigue life in a knee prosthesis: comparison between in vitro measurements and computational simulations," *Journal of Biomechanics*, vol. 37, no. 1, pp. 45–53, 2004.

[8] M. Bendjaballah, A. Shirazi-Adl, and D. Zukor, "Biomechanics of the human knee joint in compression: reconstruction, mesh generation and finite element analysis," *The Knee*, vol. 2, no. 2, pp. 69–79, 1995.

[9] T. L. Haut Donahue, M. L. Hull, M. M. Rashid, and C. R. Jacobs, "How the stiffness of meniscal attachments and meniscal material properties affect tibio-femoral contact pressure computed using a validated finite element model of the human knee joint," *Journal of Biomechanics*, vol. 36, no. 1, pp. 19–34, 2003.

[10] T. L. Haut Donahue, M. L. Hull, M. M. Rashid, and C. R. Jacobs, "The sensitivity of tibiofemoral contact pressure to the size and shape of the lateral and medial menisci," *Journal of Orthopaedic Research*, vol. 22, no. 4, pp. 807–814, 2004.

[11] J. Yao, A. D. Salo, J. Lee, and A. L. Lerner, "Sensitivity of tibio-menisco-femoral joint contact behavior to variations in knee kinematics," *Journal of Biomechanics*, vol. 41, no. 2, pp. 390–398, 2008.

[12] H. Isaksson, C. C. van Donkelaar, and K. Ito, "Sensitivity of tissue differentiation and bone healing predictions to tissue properties," *Journal of Biomechanics*, vol. 42, no. 5, pp. 555–564, 2009.

[13] H. Isaksson, C. C. van Donkelaar, R. Huiskes, J. Yao, and K. Ito, "Determining the most important cellular characteristics for fracture healing using design of experiments methods," *Journal of Theoretical Biology*, vol. 255, no. 1, pp. 26–39, 2008.

[14] P. Julkunen, J. S. Jurvelin, and H. Isaksson, "Contribution of tissue composition and structure to mechanical response of articular cartilage under different loading geometries and strain rates," *Biomechanics and Modeling in Mechanobiology*, vol. 9, no. 2, pp. 237–245, 2010.

[15] J. Yao, P. D. Funkenbusch, J. Snibbe, M. Maloney, and A. L. Lerner, "Sensitivities of medial meniscal motion and deformation to material properties of articular cartilage, meniscus and meniscal attachments using design of experiments methods," *Journal of Biomechanical Engineering*, vol. 128, no. 3, pp. 399–408, 2006.

[16] Y. Y. Dhaher, T.-H. Kwon, and M. Barry, "The effect of connective tissue material uncertainties on knee joint mechanics under isolated loading conditions," *Journal of Biomechanics*, vol. 43, no. 16, pp. 3118–3125, 2010.

[17] J. J. Garcia, N. J. Altiero, and R. C. Haut, "An approach for the stress analysis of transversely isotropic biphasic cartilage under impact load," *Journal of Biomechanical Engineering*, vol. 120, no. 5, pp. 608–613, 1998.

[18] P. S. Donzelli, R. L. Spilker, G. A. Ateshian, and V. C. Mow, "Contact analysis of biphasic transversely isotropic cartilage layers and correlations with tissue failure," *Journal of Biomechanics*, vol. 32, no. 10, pp. 1037–1047, 1999.

[19] A. W. Eberhardt, L. M. Keer, J. L. Lewis, and V. Vithoontien, "An analytical model of joint contact," *Journal of Biomechanical Engineering*, vol. 112, no. 4, pp. 407–413, 1990.

[20] M. Tissakht and A. M. Ahmed, "Tensile stress-strain characteristics of the human meniscal material," *Journal of Biomechanics*, vol. 28, no. 4, pp. 411–422, 1995.

[21] D. C. Fithian, M. A. Kelly, and V. C. Mow, "Material properties and structure-function relationships in the menisci," *Clinical Orthopaedics and Related Research*, no. 252, pp. 19–31, 1990.

[22] D. J. Goertzen, D. R. Budney, and J. G. Cinats, "Methodology and apparatus to determine material properties of the knee joint meniscus," *Medical Engineering and Physics*, vol. 19, no. 5, pp. 412–419, 1997.

[23] R. Whipple, "Advances in bioengineering," in *Advances in Bioengineering*, ASME, New Orleans, La, USA, 1984.

[24] D. L. Skaggs, W. H. Warden, and V. C. Mow, "Radial tie fibers influence the tensile properties of the bovine medial meniscus," *Journal of Orthopaedic Research*, vol. 12, no. 2, pp. 176–185, 1994.

[25] D. E. T. Shepherd and B. B. Seedhom, "The "instantaneous" compressive modulus of human articular cartilage in joints of the lower limb," *Rheumatology*, vol. 38, no. 2, pp. 124–132, 1999.

[26] L. Blankevoort and R. Huiskes, "Ligament-bone interaction in a three-dimensional model of the knee," *Journal of Biomechanical Engineering*, vol. 113, no. 3, pp. 263–269, 1991.

[27] A. Oloyede, R. Flachsmann, and N. D. Broom, "The dramatic influence of loading velocity on the compressive response of articular cartilage," *Connective Tissue Research*, vol. 27, no. 4, pp. 211–224, 1992.

[28] R. U. Repo and J. B. Finlay, "Survival of articular cartilage after controlled impact," *Journal of Bone and Joint Surgery A*, vol. 59, no. 8, pp. 1068–1076, 1977.

[29] A. M. Ahmed, D. L. Burke, and A. Yu, "In-vitro measurement of static pressure distribution in synovial joints—part II: retropatellar surface," *Journal of Biomechanical Engineering*, vol. 105, no. 3, pp. 226–236, 1983.

[30] T. Fukubayashi and H. Kurosawa, "The contact area and pressure distribution pattern of the knee. A study of normal and osteoarthrotic knee joints," *Acta Orthopaedica Scandinavica*, vol. 51, no. 6, pp. 871–879, 1980.

[31] T. D. Brown and D. T. Shaw, "In vitro contact stress distribution on the femoral condyles," *Journal of Orthopaedic Research*, vol. 2, no. 2, pp. 190–199, 1984.

[32] N. R. P. Costa, "Multiple response optimisation: methods and results," *International Journal of Industrial and Systems Engineering*, vol. 5, no. 4, pp. 442–459, 2010.

[33] C. M. Anderson-Cook, C. M. Borror, and D. C. Montgomery, "Response surface design evaluation and comparison," *Journal of Statistical Planning and Inference*, vol. 139, no. 2, pp. 629–641, 2009.

[34] R. Bailey, *Design of Comparative Experiments*, Cambridge University Press, Cambridge, UK, 2008.

[35] L. Trutna and J. J. Filliben, "Process improvement," in *Engineering Statistics Handbook*, chapter 5, National Institute of Standards and Technology, 2000.

[36] D. C. Montgomery, *Design and Analysis of Experiments*, John Wiley & Sons, New York, NY, USA, 3rd edition, 1991.

[37] S. L. C. Ferreira, R. E. Bruns, H. S. Ferreira et al., "Box-Behnken design: an alternative for the optimization of analytical methods," *Analytica Chimica Acta*, vol. 597, no. 2, pp. 179–186, 2007.

[38] *Minitab 14.1 Statistical Software*, Computer Software, Minitab, State College, Pa, USA, 2003.

[39] W. Mesfar and A. Shirazi-Adl, "Biomechanics of changes in ACL and PCL material properties or prestrains in flexion under muscle force-implications in ligament reconstruction," *Computer Methods in Biomechanics and Biomedical Engineering*, vol. 9, no. 4, pp. 201–209, 2006.

[40] K. E. Moglo and A. Shirazi-Adl, "On the coupling between anterior and posterior cruciate ligaments, and knee joint response under anterior femoral drawer in flexion: a finite element study," *Clinical Biomechanics*, vol. 18, no. 8, pp. 751–759, 2003.

[41] R. H. Myers, D. C. Montgomery, and C. M. Anderson-Cook, *Response Surface Methodology: Process and Product Optimization Using Designed Experiments*, John Wiley & Sons, New York, NY, USA, 2009.

[42] G. J. Besseris, "Analysis of an unreplicated fractional-factorial design using nonparametric tests," *Quality Engineering*, vol. 20, no. 1, pp. 96–112, 2008.

[43] N. Costa and Z. L. Pereira, "Decision-making in the analysis of unreplicated factorial designs," *Quality Engineering*, vol. 19, no. 3, pp. 215–225, 2007.

17

Muscle Contributions to $L_{4\text{-}5}$ Joint Rotational Stiffness following Sudden Trunk Flexion and Extension Perturbations

Joel A. Cort,[1] **James P. Dickey,**[2] **and Jim R. Potvin**[3]

[1] *Department of Kinesiology, University of Windsor, 401 Sunset Avenue, Windsor, ON, Canada N9B 3P4*
[2] *School of Kinesiology, The University of Western Ontario, 1151 Richmond Street, London, ON, Canada N6A 3K7*
[3] *Department of Kinesiology, McMaster University, 1280 Main Street West, Hamilton, ON, Canada L8S 4L8*

Correspondence should be addressed to Joel A. Cort; cortj@uwindsor.ca

Academic Editor: Ayako Oyane

The purpose of this study was to investigate the contribution of individual muscles ($MJRS_m$) to total joint rotational stiffness ($MJRS_T$) about the lumbar spine's $L_{4\text{-}5}$ joint prior to, and following, sudden dynamic flexion or extension perturbations to the trunk. We collected kinematic and surface electromyography (sEMG) data while subjects maintained a kneeling posture on a parallel robotic platform, with their pelvis constrained by a harness. The parallel robotic platform caused sudden inertial trunk flexion or extension perturbations, with and without the subjects being aware of the timing and direction. Prevoluntary muscle forces incorporating both short and medium latency neuromuscular responses contributed significantly to joint rotational stiffness, following both sudden trunk flexion and extension motions. $MJRS_T$ did not change with perturbation direction awareness. The lumbar erector spinae were always the greatest contributor to $MJRS_T$. This indicates that the neuromuscular feedback system significantly contributed to $MJRS_T$, and this behaviour likely enhances joint stability following sudden trunk flexion and extension perturbations.

1. Introduction

There is a complex arrangement of bones, ligaments, muscle, and nervous tissue which combine to maintain the structural integrity of the spine, thus reducing the potential for system buckling. For stability maintenance, Bergmark [1] identified the importance of the force distribution of the lumbar musculature. Other research has shown that passive tissues of the lumbar spine can only provide minimal resistance to compressive loads (up to 90 N), thus the majority of stiffness is provided by the muscles, demonstrating the importance of muscles for joint safety [2]. Moorhouse and Granata [3] and Sinkjaer et al. [4] stated that involuntary muscle force contributions account for 35 to 42% of the total joint stiffness following a perturbation. Although muscles are vital for joint safety, their force distribution relies on the careful control of the nervous system to properly coordinate the required joint stiffness. Poor neuromuscular coordination has been suggested to be a risk factor for mechanical failure following kinematic disturbances [5–9]. Granata and England [10] were among the first to characterize the neuromuscular control of stability during *dynamic* trunk flexion/extension movements. However, that research did not account for scenarios where the *timing* of trunk disturbances was unknown and, thus, the results cannot be used to explain the implications of the common scenario of an unexpected kinematic disturbance, such as a slip or shift in load, where involuntary muscle force contributions are crucial.

Numerous studies have contributed to our understanding of lumbar spine stability; however, there are limits to the conclusions about stability due to the majority of these studies either quantifying joint stability during static conditions [11–15], using theoretical and mathematical concepts [16–19], utilizing in vitro techniques [20–25] or approximated joint stability using electrophysiology combined with joint kinematics [26–30]. Furthermore, of the studies that calculated stability, only net joint stability throughout the motion was reported without information detailing the individual muscle contribution to stability [11–15]. In must be noted that Brown and Potvin [17] calculated individual muscle contributions to joint rotational stiffness (MJRS); however, since empirical-based data were not used in this work, only theorically based results were provided. Thus, there is a need for further research of the role that the neuromuscular system plays in maintaining stability in response to a sudden perturbation, through the control of individual muscles. However, in order to understand these roles, it is imperative that the complexities caused by the interaction between the skeletal and neuromuscular systems are minimized. Specific to the lumbar spine, to limit such interactions the sudden perturbations should cause joint motion about the flexion/extension axis given that rotation about this axis presents less of a challenge to the neuromuscular system based on the symmetrical design of the bilateral flexor and extensor musculature. This type of study design will provide for an initial and basic understanding of how the neuromuscular system aids in joint stability of the lumbar spine. Detail at this level can contribute to furthering our understanding of how various modes of joint instability can ultimately contribute to injury risk [17].

The purpose of this research was to investigate the contribution of the trunk muscles to joint rotational stiffness about the lumbar spine's L_{4-5} joint prior to, and following, sudden dynamic flexion and extension perturbations to the trunk. In particular, this project examined the sum of all muscles contributing to the total MJRS ($MJRS_T$), as well as the contribution of individual muscles to $MJRS_T$ ($MJRS_m$). It was hypothesized that prior knowledge of both perturbation timing and direction would be accompanied by increased $MJRS_T$ prior to the perturbation, resulting in decreased trunk motion. In addition, it was hypothesized that prior knowledge of the perturbation direction would cause a neuromuscular strategy such that individual muscle contributions to $MJRS_T$ would be dependent upon the forced direction.

2. Methods

2.1. Subjects. This study included 7 male subjects with a mean age of 24.7 ± 2.4 years, height of 178.5 ± 4.6 cm, and mass of 77.0 ± 8.5 kg. All subjects were free of musculoskeletal injury to the trunk, neck, and upper limbs. The University's Research Ethics Board approved all aspects of the study.

2.2. Instrumentation and Data Acquisition. We collected fourteen channels of surface electromyography (sEMG), using the placement protocol outlined in Cholewicki and McGill [31], bilaterally for the following muscles: rectus abdominis (RA), external oblique (EO), internal oblique (IO), lumbar erector spinae (LES), thoracic erector spinae (TES), multifidus (MULT), and latissimus dorsi (LD). We positioned disposable bipolar Ag-AgCl surface electrodes (Medi-trace disposable electrodes, Kendall, Mansfield, MA) in an-orientation parallel to each muscle's line of action, between the myotendinous junctions and innervation zones as per Shiraishi et al. [32]. The interelectrode distance was 2.5 cm. We collected and amplified the sEMG signals using two Bortec AMT-8 systems (Bortec Biomedical, Calgary, Canada, 10–1000 Hz, CMMR = 115 dB, gain = 500–1000, input impedance = 10 GΩ). We A/D converted these signals

at a sample rate of 2000 Hz using a 16-bit A/D converter (ODAU II, Northern Digital Inc., Waterloo, Canada).

We collected kinematic data using an active marker system (Optotrak 3020, Northern Digital Inc., Waterloo, Canada) sampling at 100 Hz. We placed two marker arrays on rigid fins, each with four infrared emitting diodes, and rigidly secured them to the midline of the body at the pelvis (middle of sacrum), representing the lumbar region, and rib cage (approximately at T9 level), representing the thoracic region. We used a parallel robotic platform (R2000 Rotopod, PRSCo, NH, USA) to apply the sudden inertial trunk flexion or extension perturbations. Finally, to measure acceleration and timing of the platform perturbations, we attached a triaxial accelerometer (Crossbow CXL75M3, Crossbow Technology Inc., Milpitas, CA) to the robotic platform and sampled the data at 2000 Hz.

2.3. Experimental Procedures and Protocol. Prior to the experimental trials, subjects performed isometric maximal voluntary exertions (MVEs) for each muscle to be later used to normalize the sEMG data collected during experimental trials. To obtain the MVE of the abdominals (RA, IO, and EO), subjects laid in a supine position, replicating a "sit-up" position with the feet braced to ground, and performed a sequence of isometric maximal trunk flexion efforts that also included twist and lateral bend efforts, against the resistance of the researchers. The subjects performed the MVEs for the trunk extensor muscles (LES, TES, LATS, MULT) while lying in a prone position with the feet braced, and subjects executed a sequence of maximal trunk extension efforts, against resistance manually applied by the researchers. Each of the abdominal and back muscle efforts were isometrically held for 2-3 seconds and 30 second rests were provided in between each of the efforts.

After this, we positioned the subjects in a kneeling posture on a robotic platform and harnessed them into an apparatus that minimized motion below the pelvis, but allowed for unconstrained motion of the trunk and head. Also, subjects crossed their arms in front of their chest to minimize motion of the upper limbs and to maintain an erect trunk posture (Figure 1). The parallel robotic platform applied the sudden inertial trunk flexion or extension perturbations, through rapid linear anterior or posterior 4 cm displacements of the platform (peak accelerations = 4 m/s/s). Preexperimental testing showed that the perturbation profiles were sufficient to elicit an electromyographic response.

We exposed each subject to 16 perturbation conditions, which included two timing-knowledge conditions and two direction-knowledge conditions in four perturbation directions, assigned in a random order. The timing knowledge conditions were (1) known timing (KT) and (2) unknown timing (UT). The perturbation device was equipped with dual controls such that it could be engaged manually by the subject during the KT conditions, via an electronic trigger button, or through computer activation using a digital trigger signal for UT conditions. During UT conditions, we informed the subjects of the start of the trial; however the computer randomly assigned a time to engage the perturbation device within a 15-second period after the informed start. The directional

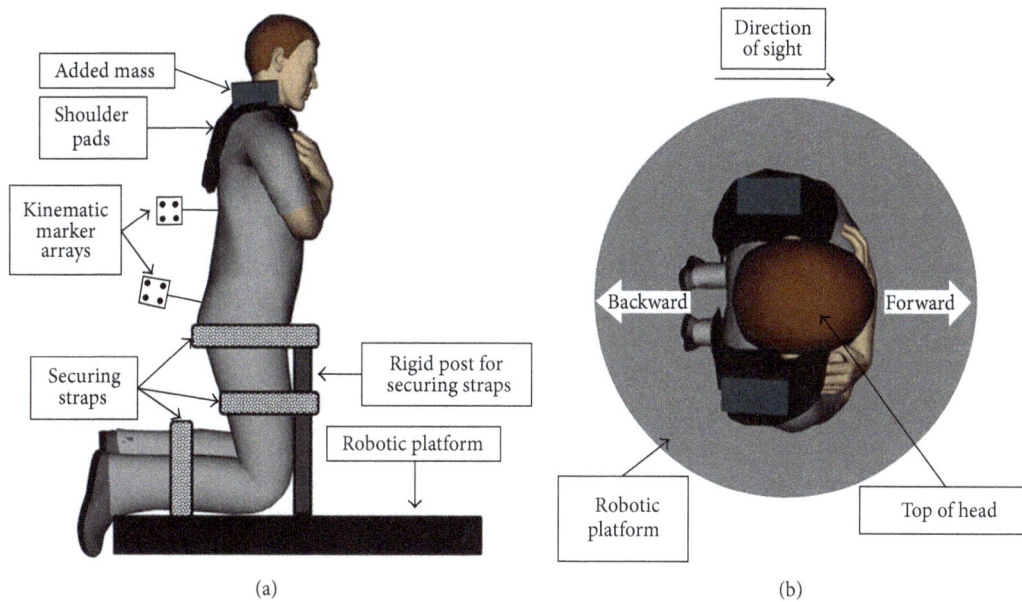

FIGURE 1: An illustration of the experimental device in a sagittal (a) and a coronal view (b). Subjects knelt on the robotic platform and legs (below the pelvis) were secured to framing that was attached to the platform. Subjects wore modified shoulder pads and maintained an upright neutral trunk posture with both arms crossed in front of the chest.

knowledge conditions were (1) known direction (KD) and (2) unknown direction (UD). The different perturbation directions were forced trunk: (1) flexion via posterior linear platform displacements (P_{FLEX}), (2) extension via anterior platform displacements (P_{EXT}), (3) left lateral bend via right platform displacements, and (4) right lateral bend via left platform displacements. Only data from the forced flexion and extension trials will be discussed in this paper. To enhance the effect of the perturbations, we rigidly attached modified football shoulder pads to the trunk that allowed us to add mass to the trunk via evenly distributed fixed weights to each shoulder (15% of each subject's upper body mass, including head, trunk, and upper extremities taken from [33]).

2.4. Data Analysis. We conditioned all sEMG data by removing the DC bias, high pass filtering at 140 Hz (Butterworth, 6th order) [34, 35], rectifying, low-pass filtering at 2.5 Hz (Butterworth, 2nd order) and normalizing to the MVE. In addition, we used the thoracic and lumbar kinematic marker arrays to determine the relative angle of the trunk. Specifically, the thoracic segment was defined by the marker array that was fixed to the spinous process at T9 and the lumbar segment was defined by the marker array attached to the sacrum (described in Section 2.2). Using this method the trunk angle was calculated as the intersection of the line connecting the thoracic and lumbar marker arrays [36]. The lumbar angle was represented as a fraction of the total trunk angle. For each of the orthogonal axes, the following percentages represent the lumbar component of the overall angle: flexion = 72.2%, extension = 43.5%, lateral bend = 49.1%, and axial twist = 5.6% [37–39]. Furthermore, the L$_{4-5}$

joint angle was represented as a fraction of the total lumbar angle. The L$_{4-5}$ component of the overall lumbar angle for each axis are as follows: flexion = 22.4%, extension = 9.5%, lateral bend = 16.2%, and axial twist = 13.3% [37–39]. We processed the joint angles with a critically damped dual-pass Butterworth filter with a final cut-off of 5 Hz (2nd order). The trunk angles were reported as the calculated displacement from the resting sitting angle to the peak angle following the perturbation. Also, we dual lowpass Butterworth filtered the tri-axial accelerometer data using a 50 Hz cutoff. Following conditioning, we downsampled all signals to 100 Hz.

We utilized the normalized and conditioned instantaneous bilateral sEMG and joint angle data as inputs to a biomechanical trunk model developed by Cholewicki and McGill [31], to determine muscle forces and moments. These data were used to calculate MJRS$_T$ about L$_{4-5}$ about the flexion/extension, lateral bend, and axial twist axes. Specifically, the Cholewicki and McGill [31] kinematic lumbar spine model was utilized in this study to determine the kinematics of each muscle's instantaneous length, velocity, and moment arm. We used the normalized and conditioned instantaneous sEMG data as input into this model to provide a first approximation of instantaneous muscle force based on each muscle's sEMG (normalized to MVE), instantaneous muscle length (as per [40]), velocity (as per [41]), and maximal muscle stress set at 1 N/cm^2. While common estimates of muscle stress typically fall within the range of 30–100 N/cm^2, the actual magnitude of this variable was not a critical component of the current calculation since the focus of this study was to examine the contribution of individual muscles as percentage of a theoretical maximum MJRS$_T$, which is described in more detail in a later paragraph. Thus, the maximum muscle

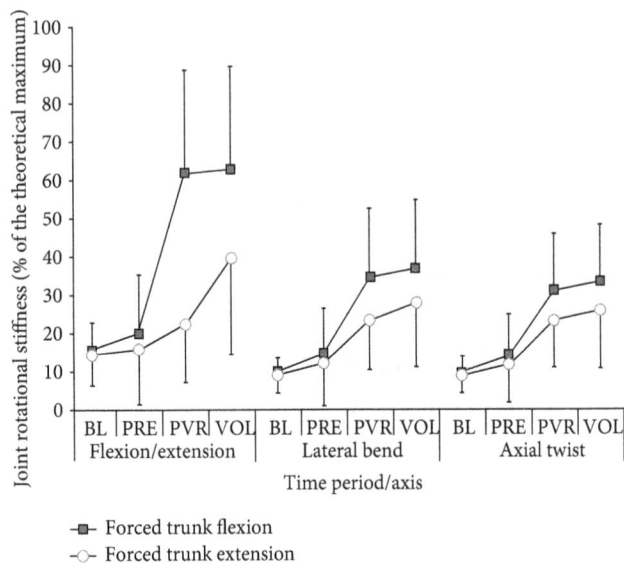

FIGURE 2: The MJRS$_T$ (as a percentage of the theoretical MJRS maximum) is shown by time period for each axis of the three axes. Displayed is the MJRS$_T$ for each axis for both the forced trunk flexion and forced trunk extension. Included in the graph are the standard deviations for each of the data points.

stress value was arbitrary as it was held constant (value of 1) during the sEMG-muscle force modelling between the theoretical maximum and the experimental conditions.

We utilized the equation of Potvin and Brown [19] to calculate the MJRS$_m$ about the three orthopaedic axes of the L$_{4-5}$ joint. In this study, a constant relating muscle force to muscle stiffness (q) was set to 10 as recommended by Potvin and Brown [19]. The q value was further corrected to account for muscle contraction velocity, as Cholewicki and McGill [42] found that muscle stiffness decreases as muscle contraction velocity increases (both concentrically and eccentrically). We developed regression equations ($r^2 = 0.99$) based on the stiffness curve in Figure 2 of Cholewicki and McGill [42], such that outputs from these equations modulated each muscles q value to accommodate the effects of contraction velocity. The muscle stiffness corrections were then multiplied by the constant q value for each muscle's instantaneous contraction velocity. For each muscle, the MJRS equation then used the estimated muscle forces, described above, and the geometric orientation of the muscles and their nodes, to calculate MJRS$_m$ values about each of the three axes.

The summation of all individual MJRS$_m$ contributions within each respective axis, at each instant in time, allowed us to determine the MJRS$_T$. Rather than reports the actual estimated MJRS$_m$ and MJRS$_T$ values, we normalized these values as a percentage of the theoretical maximum MJRS$_T$ when the trunk was presumed to have maximal stiffness in the upright neutral posture (0 degree trunk flexion angle). Specifically, we calculated muscle kinetics using the previously described modelling methodology; however, we used the theoretical sEMG values in place of experimentally recorded data. We assigned an activation of 100% MVE to the RA, IO, and

EO muscles, of the weaker trunk flexor muscle group, and then we calculated the activation of the stronger trunk extensor group (LES, TES, MULT, and LATS), necessary to balance the moment about the flexion/extension axis to zero. We used these theoretical activations to calculate the individual muscle forces, assuming a maximal muscle stress of 1 N/cm^2, and subsequent MJRS$_m$ and MJRS$_T$ values about each of the three axes. We considered these MJRS$_T$ values as the maximum theoretical magnitudes about each axis and used them normalize all previously estimated experimental MJRS$_m$ values as a percentage of maximum theoretical value within each axis.

We windowed the MJRS$_T$ MJRS$_T$ and MJRS$_m$ data into four time periods based on Stokes et al. [29]: (1) baseline (BL) from 500 to 450 ms prior to the perturbation, (2) preperturbation (PRE) from the 50 ms prior to the perturbation, (3) prevoluntary response period (PVR) from 25–150 ms after perturbation (incorporating both short and medium latency neuromuscular responses), and (4) voluntary response period (VOL) from 150 to 300 ms after perturbation. We calculated the mean and standard deviations for MJRS$_T$ and MJRS$_m$ during BL and PRE. To ensure that the full response of the system was captured following the perturbation, we determined the individual peak MJRS$_T$ values within each of the PVR and VOL time periods.

Finally, the sEMG onset was used to estimate the timing of each muscle amplitude change following the perturbations [29, 43]. For each trial and muscle, sEMG onset was determined using the integration method of Santello and McDonagh [44] and manually confirmed based on the threshold method described by Hodges and Bui [43]. We removed any onset timing data from the analysis if the detected onset occurred 400 ms after the perturbation, based on work by Wilder et al. [45], who found that muscular responses that occurred 400 ms or more after a perturbation are not a direct result of the perturbation.

2.5. Statistical Analysis. For all 8 conditions, within each subject, we calculated means and standard deviations for each dependent variable across the five repeated trials. We used these mean values to represent each subject's response to that condition within the subsequent statistical analysis. A $2 \times 2 \times 2 \times 2 \times 4$ analysis of variance (ANOVA), with repeated measures, was used to determine the influence of each of the five independent variables: muscle side location (left and right), time knowledge (KT and UT), perturbation direction (P_{EXT} and P_{FLEX}), and direction knowledge (KD and UD), as well as time period (BL, PRE, PVR, and VOL). The significance level for each ANOVA was set at $P < 0.05$. The dependent variables for this analysis included MJRS$_T$ and MJRS$_m$ for each muscle. For the significant main and interaction effects, we compared means with a Tukey's HSD post hoc test. We also used an ω^2 analysis on each statistical interaction to calculate the percentage of the total variance explained by the interaction. To be considered for discussion, we required all interactions to account for at least 1% of the total variance [46, 47]. In addition, a 2×2 ANOVA, with repeated measures, was used to determine the effect of

TABLE 1: Summary of the mean and standard deviations of the joint angle and acceleration magnitudes prior to (BL time period) and following the perturbations (VOL time periods). The BL angles and accelerations were calculated as the average magnitudes during that time period, whereas the peak magnitudes found during the VOL time period are reported.

Measure	Axis	Trunk		L_{4-5}	
		BL	VOL	BL	VOL
Joint angle (degs)	Flex/Ext	3.7 ± 2.3	5.6 ± 2.5	0.7 ± 0.8	1.1 ± 0.9
	Lat. bend	1.8 ± 1.1	2.3 ± 1.0	0.2 ± 0.1	0.2 ± 0.1
	Twist	1.3 ± 1.1	2.0 ± 1.1	0.1 ± 0.1	0.2 ± 0.1
Joint acceleration (degs/s/s)	Flex/Ext	7.1 ± 27.8	336.3 ± 122.7	1.2 ± 5.8	51.7 ± 31.8
	Lat. bend	3.8 ± 4.5	70.3 ± 25.0	0.4 ± 0.5	7.5 ± 2.7
	Twist	5.8 ± 6.2	66.8 ± 23.4	0.6 ± 0.7	7.4 ± 2.6

perturbation direction and direction knowledge on the sEMG onset timing (excluding KT data) dependent measure. We used the same post hoc test and ω^2 analysis as described above on the statistical analysis for this dependent measure.

3. Results

The results of the dependent measures from this study are detailed within this section. To better understand the magnitude of the perturbations, we have included the calculated joint angles and accelerations for the trunk and L_{4-5} for each axis in Table 1.

3.1. Total L_{4-5} Joint Rotational Stiffness. The total theoretical maximum $MJRS_T$ was 412, 419, and 241 Nm/rad for the FE, lateral bend, and axial twist axes, respectively (Figure 2). For all 3 axes, there was a significant interaction between time period and perturbation direction (F/E $P < 0.001$, lateral bend $P < 0.01$, and axial twist $P < 0.01$). Post-hoc analysis showed that, for the F/E axis during the forced flexion, the $MJRS_T$ increased as the time period progressed from BL to PVR, BL to VOL, PRE to PVR, and PRE to VOL. Also the post-hoc analysis revealed that, during the forced extension, $MJRS_T$ increased from BL to VOL, PRE to VOL, and PVR to VOL. For both the lateral bend and axial twist axes, in both the forced flexion and extension conditions, $MJRS_T$ increased from BL to PVR, BL to VOL, PRE to PVR and PRE to VOL. Interestingly, the direction knowledge variable did not significantly influence $MJRS_T$ for any of the 3 axes.

3.2. Individual Relative Muscle Contributions of Total Joint Rotational Stiffness. We calculated the muscle contributions to $MJRS_T$ about each orthogonal axis; however, only contributions about the F/E axis will be presented, as it is the primary axis about which the perturbation acted (Figure 3). There was no significant effect of muscle side, indicating symmetrical trunk motion, so we averaged data from the left and right sides for each muscle. Also, we assumed that changes of less than 2% of $MJRS_T$ were not functionally relevant and, thus, only significant ($P < 0.05$) effects, with average differences greater than 2% of $MJRS_T$, are presented. The RA and LATS were the only muscles that did not ever meet this requirement.

There was a significant three-way interaction between time period, perturbation direction, and timing knowledge

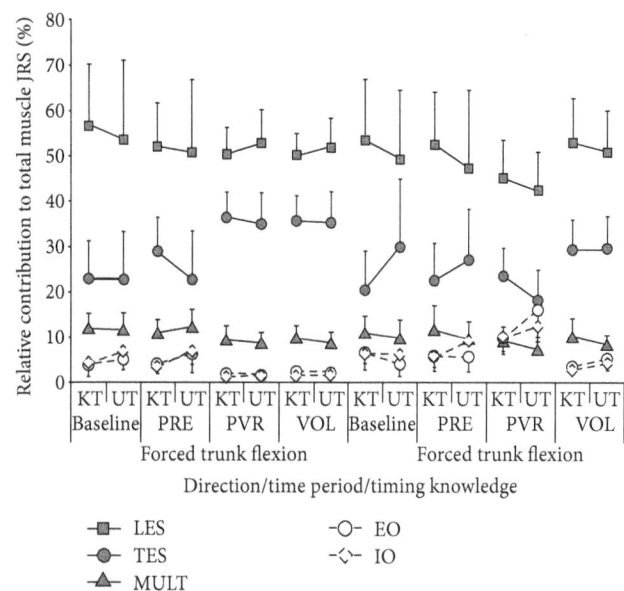

FIGURE 3: The $MJRS_m$ data is shown for each orthogonal axis when subjects both possessed and did not possess perturbation timing awareness. In addition, these data are also separated into each of the experiment time period classification. Included in the graph are the standard deviations for each of the data points.

for the EO muscle ($P < 0.001$). Further post-hoc analyses revealed no differences between the known and unknown timing within any of the time periods during the forced flexion. However, during the forced extension trial, KT was higher than UT at PRE and UT was higher than KT at PVR. There also was a significant interaction between time period and perturbation direction ($P < 0.0001$). During the P_{FLEX} condition, we found a significant decrease in IO's relative contribution to $MJRS_T$ from both BL and PRE to both PVR and VOL. During the P_{EXT} conditions, there was an increase as time periods advanced from BL and PRE to PVR and significantly lower values at VOL than at both PRE and PVR. Finally, direction knowledge did not significantly influence the response of any of the trunk flexor muscles.

The relative contribution of LES to $MJRS_T$ had a significant interaction between time period and perturbation direction ($P < 0.05$). There were no differences between time

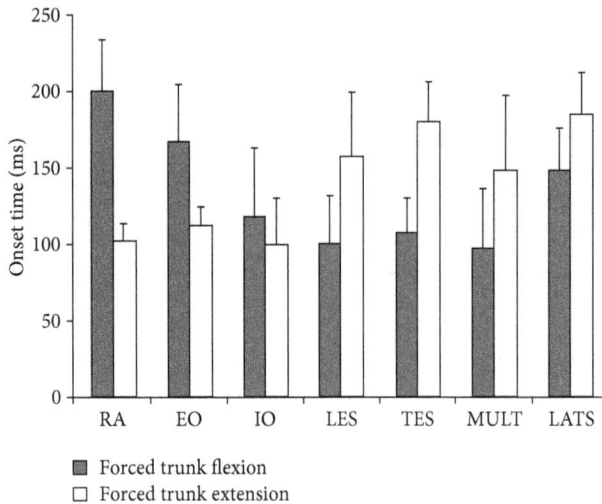

FIGURE 4: The mean and standard deviations of the sEMG onset timings for each recorded muscle (ms).

periods for the P_{FLEX} condition. However, for the P_{EXT} condition, the PVR values were lower than those at BL, PRE and VOL. The TES contribution to $MJRS_T$ had a 3-way interaction between time period, perturbation direction, and timing knowledge ($P < 0.05$). Although there were no differences found in the P_{FLEX} data, UT was higher than KT at BL for the P_{EXT} condition. Also, there was a main effect of time period for the MULT $MJRS_T$ contribution ($P < 0.05$). Post-hoc analyses showed a 27% decrease in contribution as time period advanced from BL to PVR and PRE to PVR. Lastly, direction knowledge did not significantly influence the response of any of the trunk extensor muscles.

3.3. sEMG Onset Timings. Main effects of perturbation direction for sEMG onset timing were found for all muscles, except for IO (Figure 4). Specifically, the onset times for EO and RA were higher in the P_{EXT} compared to the P_{FLEX} condition ($P < 0.01$ and $P < 0.001$ resp.), and both the RA ($P < 0.001$) and EO ($P < 0.01$) had later onset times. The LES, TES, MULT, and LATS showed a main effect of perturbation direction ($P < 0.001$, $P < 0.0001$, $P < 0.01$, $P < 0.05$, resp.), and post-hoc analyses showed that onset times were higher for these muscles in the P_{EXT} compared to the P_{FLEX} condition. In addition, for the MULT muscle, we found the UD onset times to be 10% higher than for KD ($P < 0.05$).

4. Discussion

The purpose of this research was to investigate trunk muscle contributions to joint rotational stiffness about the lumbar L_{4-5} joint prior to, and following, sudden inertial flexion and extension perturbations to the trunk. Our unique perturbation methodology allowed for us to determine that possessing the knowledge of perturbation direction does not affect $MJRS_T$, whereas awareness of the perturbation timing does cause an increase in $MJRS_T$ magnitude. In addition, based on our knowledge this is the first work that determined

individual muscle contributions to joint rotational stiffness, prior to and following sudden trunk perturbations. Based on our work we found that the LES was the greatest contributor to $MJRS_T$, followed in order by the TES, MULT, EO, and IO.

We also found that the response of the neuromuscular system, immediately following forced trunk flexion and extension, was a significant contributor to $MJRS_T$, which supports previous research findings. In our work the greatest $MJRS_T$ magnitude was always about the flexion/extension axis, followed by the lateral bend and axial twist axes. Since the F/E axis was the primary contributor to $MJRS_T$ in the current study, the remainder of this discussion will focus on that axis.

Our work suggests that it is most likely that the prevoluntary response, incorporating both short and medium latency neuromuscular responses, was an attempt to limit the perturbation motion. It served as a first responder, initially providing stiffness until the voluntary component began its contribution. Albeit smaller in magnitude, this prevoluntary response likely plays a critical role in injury avoidance, given that the voluntary response may not occur early enough after the perturbation.

4.1. $MJRS_T$: Timing Knowledge. The $MJRS_T$ increased when the subjects knew the perturbation timing, demonstrating that timing awareness promoted increased joint rotational stiffness. This finding is consistent with previous studies that identified that subjects increased muscle activation and, thus joint stiffness, prior to the perturbation [27–30, 48, 49].

A deeper investigation of our data showed that, with timing knowledge, most subjects tended to increase $MJRS_T$ from the baseline measure to just prior to the perturbation (PRE). This suggests an anticipatory adjustment in preparation for the forced motion. However, there were two subjects who, during each of the known timing-trunk extension trials, showed increased $MJRS_T$ magnitudes during the PRE and PVR time periods with respect to the values calculated during the baseline periods. While this approach may provide maximum safety against the expected perturbation, it is also metabolically inefficient to maintain elevated muscle activity for unnecessarily long-time periods.

4.2. $MJRS_T$: Direction Knowledge. The robotic device allowed for multidirectional forced motion. This device enabled a unique inertial perturbation approach, compared to most previous experimental protocols used for sudden loading studies where a harness-cable system has been used to perturb subjects. Given that the required cable used in such a system to pull the body segment to produce the perturbation provided subjects with knowledge of the perturbation direction, only timing knowledge could be manipulated. Our robotic platform also allowed for increased uncertainty with regard to the direction of the perturbation. Nevertheless, the results revealed that direction knowledge did not affect the neuromuscular response to trunk perturbations. This was unexpected as we had hypothesized that the awareness of direction, like that seen for timing knowledge, would offer assistance to the neuromuscular system for coordinating the recruitment of muscle forces for increased $MJRS_T$.

To the best of our knowledge, this is the first published sudden trunk loading research that incorporated conditions where the perturbation direction was completely unknown to the subject. Masani et al. [50] completed a multidirectional perturbation study of the trunk and found that muscle responses were dependent upon the forced direction; however, their subjects were always aware of the perturbation direction. Cholewicki and VanVliet [14] showed that loading direction affects the contribution of individual muscles to joint stability during isometric trunk exertions; however, the preexisting data does not provide details on whether such coordination occurs in preparation for an unexpected disturbances. It is possible that it may be difficult to prioritize specific individual muscle recruitment for optimal joint rotational stiffness, in preparation for sudden motion. Brown et al. [51] found that cocontraction (abdominal muscle force during forced trunk extension) increased trunk stiffness prior to a sudden perturbation; however, their subjects lacked the ability to selectively increase abdominal muscle force without a subsequent increase in back muscle activity, which potentially increases the risk of injury given the subsequent increase in trunk compressive forces.

4.3. MJRS$_m$ General Considerations. Of the seven bilateral muscles recorded and modeled, the RA and LATS did not meet the statistical requirements, discussed previously, to be considered significant contributors to MJRS$_T$ in the context of this research. However, the IO, EO, MULT, LES, and TES all contributed to MJRS$_T$, albeit at various levels.

A qualitative comparison of each muscle's contribution showed that the LES was the greatest contributor followed, in order, by the TES, MULT, EO, and IO (see Figure 3). This order of muscle contribution is reflected in other similar studies, such as Chiang and Potvin [27], Krajcarski et al. [28], and Thomas et al. [30]. These findings demonstrate that no one muscle is exclusively responsible for generating joint rotational stiffness, but that it is a collection of muscles acting together to generate the required resistance. Furthermore, both Brown and Potvin [17] and Crisco and Panjabi [52] suggest that the "global" multisegmental muscles, which possess larger moment arms, are the main contributor to joint rotational stiffness. This concept is supported by the current work where the primary contributors to MJRS$_T$, LES, and TES have the longest moment arms.

During the forced extension conditions, we expected that the IO and EO muscles would be the main contributors to MJRS$_T$, since they acted as antagonists during the motion. However, this was not the case and may be a result of the relatively small trunk extension motion that was caused by the perturbation. This is a limitation in our study. The magnitude of the extension perturbation was set to a level that would have minimal risk of injury; however, this may have been insufficient to elicit substantial length changes for the abdominal muscles and cause them to activate.

4.4. MJRS$_m$: Timing Knowledge, Direction, and Direction Knowledge Interaction. The TES and EO were unique in that their contributions were dependent on all of the experimental variables (timing knowledge, direction of the forced motion, and time period). During the unexpected timing conditions, when forced into trunk extension, there was a greater relative contribution from the EO just prior to the perturbation. In the same experimental condition, the EO greatly increased its relative contribution to MJRS$_T$ during the prevoluntary time period, when timing knowledge was not provided. Vera-Garcia et al. [53] found similar EO response patterns during unanticipated trunk extension perturbations; however, when subjects anticipated the perturbation, as seen through increased voluntary contraction of the other monitored muscles, the EO response was significantly reduced. For the TES, timing knowledge only impacted the baseline time period, with no muscle contribution changes observed just prior to, or following, the perturbation. As such, these results are considered to be functionally irrelevant and are likely due to slight adjustments in trunk posture at the start of the trials.

The behaviour of the EO is likely the result of increased magnitudes of MJRS$_T$ associated with the anticipation of the perturbation. Specifically, in the presence of timing awareness, the anticipatory activity of this muscle raised the magnitude of its MJRS$_T$. Accordingly the joint became stiffer prior to, and throughout, the forced motion. This ultimately allowed for less dependence on the prevoluntary contribution. Thus, in order to obtain the necessary levels of stiffness, a feed-forward neuromuscular strategy was utilized reducing the dependency on the involuntary muscle response as seen during the unexpected timing conditions.

Qualitative examination of the individual muscle contributions to MJRS$_T$ revealed that the antagonist muscles (those muscles not involved in arresting the forced motion) were active both prior to (PRE), and following (PVR and VOL), the perturbation. Rather than aiding in arresting the forced motion, it is likely that these muscles are utilized to increase L$_{4-5}$ joint's overall rotational stiffness, and thus joint safety, at the expense of greater moment in the direction caused by the perturbation. However, this increase in joint moment caused by the cocontracting muscles may be a necessary "tradeoff" to ensure adequate joint stiffness. Increased muscle forces of the trunk through cocontraction are thought to be important for stiffness of the spine, which ultimately aids in stabilizing the joint [54, 55].

As mentioned earlier, reliance on the feedback mechanism, when timing awareness is not available, may be intended to optimize the balance between tissue loading and joint stiffness. Granata and Marras [54] noted that there is a "tradeoff" between tissue loading and spine stability; a balance is needed in order for lumbar spine motions to occur with minimal risk of injury. A strategy of muscle preactivation, in anticipation of a kinematic disturbance, results in greater muscle forces (although not calculated in this study), and may cause higher compressive loads on the spine [27, 51, 53, 54]. These higher compressive loads are important since high compressive forces are a risk factor for low back injury [56].

It must be noted that only the EO and TES were affected by the relationship between timing awareness and time period, whereas the remaining muscles were not affected by this relationship. Similar to the findings for MJRS$_T$, we have concluded that some subjects tended to increase their levels of muscle activation right from the beginning of the trial (starting at BL) through to the end. However, not all subjects employed this approach and due to this, we have hypothesized that those having timing awareness, that showed increased responses following the perturbation (and thus minimal pre-perturbation muscle anticipation), were exhibiting physiologically efficiency, as they would have been required to maintain higher levels of muscle activation for extended periods of time. Therefore, those subjects showed that it is more physiologically economical, in cases where timing was unknown, to begin activation just prior to the perturbation, while maintaining joint rotational stiffness.

5. Conclusions

Although the magnitudes of the prevoluntary muscle forces are smaller than those produced voluntarily, our data suggests that subjects adopted a response strategy that relies on prevoluntary (reflex) muscle forces to produce rapid increases in joint rotational stiffness following a perturbation. Findings from this study support those of Moorhouse and Granata [3], Granata and England [10] and Sinkjaer et al. [4], as these authors observed that prevoluntary muscle force contributions are important to joint integrity during either simple voluntary trunk motion or following sudden trunk perturbations. Our work shows that a strategy that includes MJRS from the reflex response could be considered superior since an immediate but lower magnitude response allows the system to safely increase joint stiffness, rather than deferring the full responsibility later in time to the voluntary response. Based on this work, it is apparent that the early muscle response plays a vital role in joint safety during sudden kinematic disturbances. These findings can be used to better understand the role of the neuromuscular system during sudden trunk perturbations, both when timing and direction knowledge are varied.

Acknowledgment

This project was funded by the Natural Sciences and Engineering Research Council (NSERC) of Canada.

References

[1] A. Bergmark, "Stability of the lumbar spine. A study in mechanical engineering," *Acta Orthopaedica Scandinavica*, vol. 60, no. 230, pp. 2–54, 1989.

[2] J. J. Crisco, M. M. Panjabi, I. Yamamoto, and T. R. Oxland, "Euler stability of the human ligamentous lumbar spine—Part II: experiment," *Clinical Biomechanics*, vol. 7, no. 1, pp. 27–32, 1992.

[3] K. M. Moorhouse and K. P. Granata, "Role of reflex dynamics in spinal stability: intrinsic muscle stiffness alone is insufficient for stability," *Journal of Biomechanics*, vol. 40, no. 5, pp. 1058–1065, 2007.

[4] T. Sinkjaer, E. Toft, S. Andreassen, and B. C. Hornemann, "Muscle stiffness in human ankle dorsiflexors: intrinsic and reflex components," *Journal of Neurophysiology*, vol. 60, no. 3, pp. 1110–1121, 1988.

[5] K. P. Granata, K. F. Orishimo, and A. H. Sanford, "Trunk muscle coactivation in preparation for sudden load," *Journal of Electromyography and Kinesiology*, vol. 11, no. 4, pp. 247–254, 2001.

[6] M. M. Panjabi, "The stabilizing system of the spine. Part I. Function, dysfunction, adaptation, and enhancement," *Journal of Spinal Disorders*, vol. 5, no. 4, pp. 383–389, 1992.

[7] A. Radebold, J. Cholewicki, M. M. Panjabi, and T. C. Patel, "Muscle response pattern to sudden trunk loading in healthy individuals and in patients with chronic low back pain," *Spine*, vol. 25, no. 8, pp. 947–954, 2000.

[8] N. P. Reeves and J. Cholewicki, "Modeling the human lumbar spine for assessing spinal loads, stability, and risk of injury," *Critical Reviews in Biomedical Engineering*, vol. 21, no. 1, pp. 73–139, 2003.

[9] N. Peter Reeves, K. S. Narendra, and J. Cholewicki, "Spine stability: the six blind men and the elephant," *Clinical Biomechanics*, vol. 22, no. 3, pp. 266–274, 2007.

[10] K. P. Granata and S. A. England, "Stability of dynamic trunk movement," *Spine*, vol. 31, no. 10, pp. E271–E276, 2006.

[11] S. H. M. Brown and J. R. Potvin, "Constraining spine stability levels in an optimization model leads to the prediction of trunk muscle cocontraction and improved spine compression force estimates," *Journal of Biomechanics*, vol. 38, no. 4, pp. 745–754, 2005.

[12] J. Cholewicki, "The effects of lumbosacral orthoses on spine stability: what changes in EMG can be expected?" *Journal of Orthopaedic Research*, vol. 22, no. 5, pp. 1150–1155, 2004.

[13] J. Cholewicki, A. P. D. Simons, and A. Radebold, "Effects of external trunk loads on lumbar spine stability," *Journal of Biomechanics*, vol. 33, no. 11, pp. 1377–1385, 2000.

[14] J. Cholewicki and J. J. VanVliet, "Relative contribution of trunk muscles to the stability of the lumbar spine during isometric exertions," *Clinical Biomechanics*, vol. 17, no. 2, pp. 99–105, 2002.

[15] K. P. Granata and S. E. Wilson, "Trunk posture and spinal stability," *Clinical Biomechanics*, vol. 16, no. 8, pp. 650–659, 2001.

[16] S. J. Howarth, A. E. Allison, S. G. Grenier, J. Cholewicki, and S. M. McGill, "On the implications of interpreting the stability index: a spine example," *Journal of Biomechanics*, vol. 37, no. 8, pp. 1147–1154, 2004.

[17] S. H. M. Brown and J. R. Potvin, "Exploring the geometric and mechanical characteristics of the spine musculature to provide rotational stiffness to two spine joints in the neutral posture," *Human Movement Science*, vol. 26, no. 1, pp. 113–123, 2007.

[18] M. Gardner-Morse, I. A. F. Stokes, and J. P. Laible, "Role of muscles in lumbar spine stability in maximum extension efforts," *Journal of Orthopaedic Research*, vol. 13, no. 5, pp. 802–808, 1995.

[19] J. R. Potvin and S. H. M. Brown, "An equation to calculate individual muscle contributions to joint stability," *Journal of Biomechanics*, vol. 38, no. 5, pp. 973–980, 2005.

[20] M. M. Panjabi, "Clinical spinal instability and low back pain," *Journal of Electromyography and Kinesiology*, vol. 13, no. 4, pp. 371–379, 2003.

[21] M. Panjabi, K. Abumi, J. Duranceau, and T. Oxland, "Spinal stability and intersegmental muscle forces. A biomechanical model," *Spine*, vol. 14, no. 2, pp. 194–200, 1989.

[22] I. A. Stokes, M. Gardner-Morse, D. Churchill, and J. P. Laible, "Measurement of a spinal motion segment stiffness matrix," *Journal of Biomechanics*, vol. 35, no. 4, pp. 517–521, 2002.

[23] I. A. F. Stokes and M. Gardner-Morse, "Spinal stiffness increases with axial load: another stabilizing consequence of muscle action," *Journal of Electromyography and Kinesiology*, vol. 13, no. 4, pp. 397–402, 2003.

[24] K. M. Tesh, J. S. Dunn, and J. H. Evans, "The abdominal muscles and vertebral stability," *Spine*, vol. 12, no. 5, pp. 501–508, 1987.

[25] M. M. Panjabi, "The stabilizing system of the spine. Part II. Neutral zone and instability hypothesis," *Journal of Spinal Disorders*, vol. 5, no. 4, pp. 390–396, 1992.

[26] S. H. M. Brown, M. L. Haumann, and J. R. Potvin, "The responses of leg and trunk muscles to sudden unloading of the hands: Implications for balance and spine stability," *Clinical Biomechanics*, vol. 18, no. 9, pp. 812–820, 2003.

[27] J. Chiang and J. R. Potvin, "The in vivo dynamic response of the human spine to rapid lateral bend perturbation: effects of preload and step input magnitude," *Spine*, vol. 26, no. 13, pp. 1457–1464, 2001.

[28] S. R. Krajcarski, J. R. Potvin, and J. Chiang, "The in vivo dynamic response of the spine to perturbations causing rapid flexion: effects of pre-load and step input magnitude," *Clinical Biomechanics*, vol. 14, no. 1, pp. 54–62, 1999.

[29] I. A. F. Stokes, M. Gardner-Morse, S. M. Henry, and G. J. Badger, "Decrease in trunk muscular response to perturbation with preactivation of lumbar spinal musculature," *Spine*, vol. 25, no. 15, pp. 1957–1964, 2000.

[30] J. S. Thomas, S. A. Lavender, D. M. Corcos, and G. B. J. Andersson, "Trunk kinematics and trunk muscle activity during a rapidly applied load," *Journal of Electromyography and Kinesiology*, vol. 8, no. 4, pp. 215–225, 1998.

[31] J. Cholewicki and S. M. McGill, "Mechanical stability of the in vivo lumbar spine: implications for injury and chronic low back pain," *Clinical Biomechanics*, vol. 11, no. 1, pp. 1–15, 1996.

[32] M. Shiraishi, T. Masuda, T. Sadoyama, and M. Okada, "Innervation zones in the back muscles investigated by multichannel surface EMG," *Journal of Electromyography and Kinesiology*, vol. 5, no. 3, pp. 161–167, 1995.

[33] P. De Leva, "Adjustments to zatsiorsky-seluyanov's segment inertia parameters," *Journal of Biomechanics*, vol. 29, no. 9, pp. 1223–1230, 1996.

[34] J. R. Potvin and S. H. M. Brown, "Less is more: high pass filtering, to remove up to 99% of the surface EMG signal power, improves EMG-based biceps brachii muscle force estimates," *Journal of Electromyography and Kinesiology*, vol. 14, no. 3, pp. 389–399, 2004.

[35] D. Staudenmann, J. R. Potvin, I. Kingma, D. F. Stegeman, and J. H. van Dieën, "Effects of EMG processing on biomechanical models of muscle joint systems: sensitivity of trunk muscle moments, spinal forces, and stability," *Journal of Biomechanics*, vol. 40, no. 4, pp. 900–909, 2007.

[36] I. Kingma, H. M. Toussaint, M. P. De Looze, and J. H. Van Dieen, "Segment inertial parameter evaluation in two anthropometric models by application of a dynamic linked segment model," *Journal of Biomechanics*, vol. 29, no. 5, pp. 693–704, 1996.

[37] M. Pearcy, I. Portek, and J. Shepherd, "Three-dimensional X-ray analysis of normal movement in the lumbar spine," *Spine*, vol. 9, no. 3, pp. 294–297, 1984.

[38] M. J. Pearcy and S. B. Tibrewal, "Axial rotation and lateral bending in the normal lumbar spine measured by three-dimensional radiography," *Spine*, vol. 9, no. 6, pp. 582–587, 1984.

[39] A. A. White and M. M. Panjabi, *Clinical Biomechanics of the Spine*, Lippincott, Philadelphia, Pa, USA, 1990.

[40] S. L. Delp, J. P. Loan, M. G. Hoy, F. E. Zajac, E. L. Topp, and J. M. Rosen, "An interactive graphics-based model of the lower extremity to study orthopaedic surgical procedures," *IEEE Transactions on Biomedical Engineering*, vol. 37, no. 8, pp. 757–767, 1990.

[41] S. M. McGill and R. W. Norman, "1986 Volvo award in biomechanics: partitioning of the L4-L5 dynamic moment into disc, ligamentous, and muscular components during lifting," *Spine*, vol. 11, no. 7, pp. 666–678, 1986.

[42] J. Cholewicki and S. M. McGill, "Relationship between muscle force and stiffness in the whole mammalian muscle: a simulation study," *Journal of Biomechanical Engineering*, vol. 117, no. 3, pp. 339–342, 1995.

[43] P. W. Hodges and B. H. Bui, "A comparison of computer-based methods for the determination of onset of muscle contraction using electromyography," *Electroencephalography and Clinical Neurophysiology*, vol. 101, no. 6, pp. 511–519, 1996.

[44] M. Santello and M. J. N. Mcdonagh, "The control of timing and amplitude of EMG activity in landing movements in humans," *Experimental Physiology*, vol. 83, no. 6, pp. 857–874, 1998.

[45] D. G. Wilder, A. R. Aleksiev, M. L. Magnusson, M. H. Pope, K. F. Spratt, and V. K. Goel, "Muscular response to sudden load: a tool to evaluate fatigue and rehabilitation," *Spine*, vol. 21, no. 22, pp. 2628–2639, 1996.

[46] P. L. Weir, A. M. Holmes, D. Andrews, W. J. Albert, N. R. Azar, and J. P. Callaghan, "Determination of the just noticeable difference (JND) in trunk posture perception," *Theoretical Issues in Ergonomics Science*, vol. 8, no. 3, pp. 185–199, 2007.

[47] G. Keppel and T. D. Wickens, *Design and Analysis: A Researcher's Handbook*, Pearson Prentice Hall, 2004.

[48] S. A. Lavender and W. S. Marras, "The effects of a temporal warning signal on the biomechanical preparations for sudden loading," *Journal of Electromyography and Kinesiology*, vol. 5, no. 1, pp. 45–56, 1995.

[49] S. A. Lavender, G. A. Mirka, R. W. Schoenmarklin, C. M. Sommerich, L. R. Sudhakar, and W. S. Marras, "The effects of preview and task symmetry on trunk muscle response to sudden loading," *Human Factors*, vol. 31, no. 1, pp. 101–115, 1989.

[50] K. Masani, V. W. Sin, A. H. Vette et al., "Postural reactions of the trunk muscles to multi-directional perturbations in sitting," *Clinical Biomechanics*, vol. 24, no. 2, pp. 176–182, 2009.

[51] S. H. M. Brown, F. J. Vera-Garcia, and S. M. McGill, "Effects of abdominal muscle coactivation on the externally preloaded trunk: variations in motor control and its effect on spine stability," *Spine*, vol. 31, no. 13, pp. E387–E393, 2006.

[52] J. J. Crisco and M. M. Panjabi, "Euler stability of the human ligamentous lumbar spine—part I: theory," *Clinical Biomechanics*, vol. 7, no. 1, pp. 19–26, 1992.

[53] F. J. Vera-Garcia, S. H. M. Brown, J. R. Gray, and S. M. McGill, "Effects of different levels of torso coactivation on trunk muscular and kinematic responses to posteriorly applied sudden loads," *Clinical Biomechanics*, vol. 21, no. 5, pp. 443–455, 2006.

[54] K. P. Granata and W. S. Marras, "Cost-benefit of muscle cocontraction in protecting against spinal instability," *Spine*, vol. 25, no. 11, pp. 1398–1404, 2000.

[55] J. Cholewicki, M. M. Panjabi, and A. Khachatryan, "Stabilizing function of trunk flexor-extensor muscles around a neutral spine posture," *Spine*, vol. 22, no. 19, pp. 2207–2212, 1997.

[56] R. Norman, R. Wells, P. Neumann et al., "A comparison of peak vs cumulative physical work exposure risk factors for the reporting of low back pain in the automotive industry," *Clinical Biomechanics*, vol. 13, no. 8, pp. 561–573, 1998.

18

Development of an Anatomically Realistic Forward Solver for Thoracic Electrical Impedance Tomography

Fei Yang,[1] Jie Zhang,[2] and Robert Patterson[3]

[1] Washington University School of Medicine, Saint Louis, MO 63110, USA
[2] Division of Radiological Medical Physics, University of Kentucky, Lexington, KY 40536, USA
[3] Institute of Engineering in Medicine, University of Minnesota, Minneapolis, MN 55455, USA

Correspondence should be addressed to Fei Yang; fei.yang@wustl.edu

Academic Editor: Hengyong Yu

Electrical impedance tomography (EIT) has the potential to provide a low cost and safe imaging modality for clinically monitoring patients being treated with mechanical ventilation. Variations in reconstruction algorithms at different clinical settings, however, make interpretation of regional ventilation across institutions difficult, presenting the need for a unified algorithm for thoracic EIT reconstruction. Development of such a consensual reconstruction algorithm necessitates a forward model capable of predicting surface impedance measurements as well as electric fields in the interior of the modeled thoracic volume. In this paper, we present an anatomically realistic forward solver for thoracic EIT that was built based on high resolution MR image data of a representative adult. Accuracy assessment of the developed forward solver in predicting surface impedance measurements by comparing the predicted and observed impedance measurements shows that the relative error is within the order of 5%, demonstrating the ability of the presented forward solver in generating high-fidelity surface thoracic impedance data for thoracic EIT algorithm development and evaluation.

1. Introduction

Electrical impedance tomography (EIT) is a medical imaging technique in which an image of the conductivity distribution in a part of the body is inferred from surface electrical potentials resulting from application of a number of current patterns through the body. Because of its noninvasiveness, portability, and low cost, EIT has been actively investigated since the 1970s [1] and finds potential applications in a wide variety of clinical areas including monitoring of lung problems such as pulmonary edema [2] or pneumothorax [3], non-invasive monitoring of heart function and blood flow [4], localization of epileptic foci [5], investigating gastric emptying [6], and measuring local internal temperature increases associated with hyperthermia therapy [7]. Lately, similar technique has also been proposed for stenotic plaque detection [8].

Among those, one of the most promising applications of EIT is continuous regional pulmonary monitoring, especially for monitoring patients being treated with mechanical ventilation. Mechanical ventilation is indicated when the patient's spontaneous ventilation is inadequate and is one of the most common interventions administered in intensive care. Mechanical ventilation can improve the prognosis for acute phase patients; however, it also often leads to potential complications such as ventilator-associated lung injury (VALI) and ventilator-induced lung injury (VILI). It thus poses an urgent demand on continuous and noninvasive monitoring of regional ventilation at the bedside of patients with respiratory failure. The feasibility of EIT for monitoring regional ventilation as demonstrated by previous studies on the one hand and the capability of EIT to resolve the volume changes between dependent and nondependent lung regions as ventilator parameters change thus allowing patient-specified ventilator settings during lung protective ventilation on the other have fostered a growing interest in using EIT for monitoring patients under mechanical ventilation [9].

However, in most clinical or research settings images are reconstructed using a variety of algorithms and thus

TABLE 1: Conductivity values of the major tissues.

Tissue	Conductivity (S/m)
Air	10^{-18}
Lung	0.0714
Blood	0.6667
Heart muscle	0.4
Liver and kidney	0.1667
Skeletal muscle	0.4444
Fat, bone, and cartilage	0.05

show spatial nonuniformity in image amplitude, position, and resolution, thereby making interpretation of regional ventilation difficult or even error prone [10]. In addition, these algorithms have poorly understood behaviors in real patients, which makes it difficult to interpret whether a certain behavior is real or artefact due to the algorithms. As an initiative to address this issue, GREIT (Graz consensus Reconstruction algorithm for EIT) was proposed in an attempt to develop a consensus framework for a reconstruction algorithm [10]. One of the major challenges for developing such a reconstruction algorithm is the dearth of well-accepted standard thoracic impedance datasets. This is due to the costs of acquisition and the difficulties of *in vivo* quantifying the severity of lung lesions as well as the ethical issues in sharing patient data amongst others. In view of this need, this paper presents an anatomically detailed forward solver for thoracic EIT built based on ECG-gated MR images from a representative adult male aiming at generating well-characterized thoracic impedance datasets for thoracic EIT reconstruction algorithm development and evaluation.

2. Methods

2.1. The Electrical Model of the Thorax. The subject imaged as the anatomical source for the model was a 63-year-old male, weighing approximately 100 kg and approximately 180 cm in height. MR scans were performed with a 1.5 T Siemens Sonata instrument. Forty-three breath-held thoracic transverse images, gated to coincide with end diastole, were obtained from abdomen to neck. The images were digitized with transverse resolution of 1.5 mm × 1.5 mm and axial resolution of 5 mm (equal to the MR slice thickness). Organs and tissues were segmented manually and confirmed by a pathologist. Upon segmentation, a total of thirty-six tissue types and blood-containing regions were obtained. For each of the identified components, a 3D volumetric mask was created in the interest of automatic electrical conductivity assignment. Conductivities for the major tissues used in the model are listed in Table 1 [11]. A 3D electrical model of the thorax at the end of diastole was thus created, with a resolution of $1.5 \times 1.5 \times 5$ mm^3 and 3.8 million elements. As a sample, one segmented slice is shown in Figure 1.

2.2. The Forward Problem. In the electrical frequency range of 1 kHz to 1 MHz, the human body can be conceived of as a

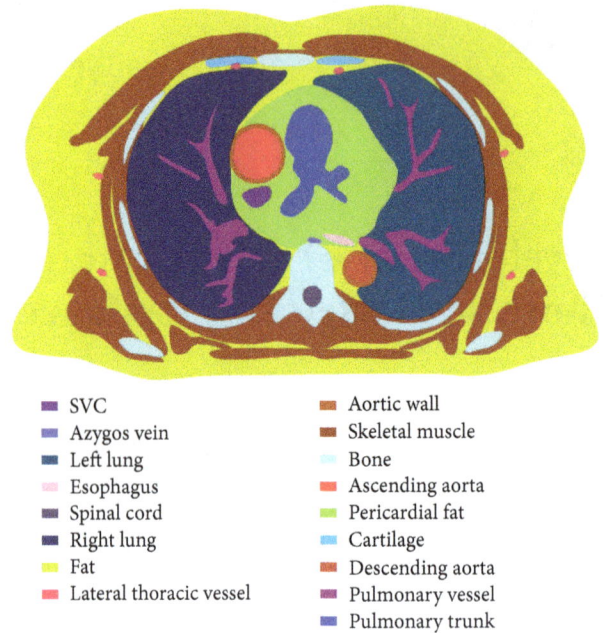

SVC
Azygos vein
Left lung
Esophagus
Spinal cord
Right lung
Fat
Lateral thoracic vessel
Aortic wall
Skeletal muscle
Bone
Ascending aorta
Pericardial fat
Cartilage
Descending aorta
Pulmonary vessel
Pulmonary trunk

FIGURE 1: Sample of a segmented axial image. Inferior view at T6 vertebral level.

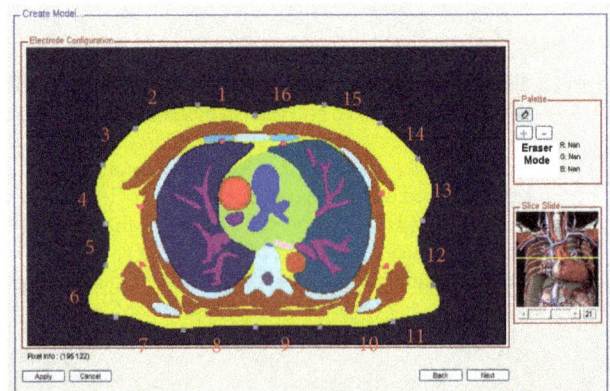

FIGURE 2: A screen shot of the developed GUI. Electrodes are paired and numbered to show the default adjacent excitation scheme.

piecewise, homogeneous, and resistive system with neglecting the reactive effects of the body [12]. If it is also assumed that the thorax is a source-free region and that the current flux normal to the body surface is zero except under the current electrodes, then for a given conductivity distribution in the defined volume conductor, the potential distribution induced by current sources obeys the generalized Laplace's equation subject to the integrated Neumann boundary conditions on the electrodes and the Dirichlet boundary conditions on other body surface. Due to its complex geometry and inhomogeneities in electrical conductivity, the modeled thoracic volume was discretized into hexahedral elements and the finite difference method was employed for the numerical solution of the governing equation. By applying Ohm's and Kirchoff's laws on the discretized nodes, equations describing

Table 2: Percentage of difference between the predicted and observed impedance measurements.

Current excitation electrode pair	Voltage pick-up electrode pair															
---	1	2	3	4	5	6	7	8	9	10	11	12	13	14	15	16
1			2.2	-2.1	-3.9	1.5	-0.7	2.7	2.4	1.0	-2.8	-1.5	0.1	-1.2	-4.0	
2				-2.5	-4.6	-3.3	0.3	-4.1	-3.0	4.2	1.1	-2.3	0.6	-0.9	-4.8	2.4
3	3.7				-0.5	-0.9	4.9	3.5	0.7	2.6	2.6	-1.5	0.5	2.9	0.6	-3.0
4	-4.1	-4.6				-0.5	1.6	-3.3	-4.1	3.8	1.0	0.5	1.9	2.0	3.9	4.7
5	-1.3	-1.6	-1.9				3.4	-0.4	3.2	3.2	4.0	-0.5	-0.1	0.3	2.4	-4.3
6	4.0	4.6	1.8	3.4				-5.0	2.4	2.6	3.7	2.0	-0.1	2.7	0.9	-2.3
7	2.2	-3.5	-4.5	3.8	-3.4				3.5	0.8	-4.4	-2.4	-3.2	-3.4	-0.9	0.1
8	-0.5	-4.3	4.7	0.0	2.5	-2.5				4.5	-4.6	3.1	-2.9	3.1	2.4	-0.8
9	-4.6	-1.8	0.6	-4.6	-3.1	1.2	3.3				-0.8	-1.9	-1.4	2.7	-1.0	2.6
10	-4.6	-2.6	1.2	1.6	2.6	-0.7	-1.0	-4.6				3.2	1.2	3.3	-2.1	0.4
11	3.4	-2.4	-2.7	-0.9	-4.3	4.2	2.4	5.0	-0.5				-3.1	2.7	2.8	-4.6
12	-4.7	1.1	-3.0	2.8	1.5	4.5	-1.5	-2.7	3.6	0.1				0.6	3.8	-0.5
13	-4.6	-1.6	3.1	-2.5	3.0	-0.3	-1.9	-3.2	-3.5	1.6	-3.2				2.0	-0.2
14	0.1	3.3	0.1	2.4	2.5	-2.8	-2.5	-3.7	-1.2	-4.0	4.8	4.2				2.7
15	-3.0	-2.1	0.5	-0.1	-1.2	-4.3	0.5	4.2	1.5	1.3	4.6	-4.3	-3.8			
16		4.7	-1.5	-2.0	0.3	3.7	4.2	1.0	-0.1	3.9	3.3	-2.8	-0.5	3.2		

the potential at one node as a function of the potentials at the adjacent nodes were established, from which a large sparse system of linear equations accounting for the potentials throughout the modeled conductive volume was able to be assembled and solved consequently.

2.3. The Graphic User Interface Environment. The above-mentioned modeling approach requires many steps and involves multiple applications. To achieve ease of use, model data and simulation programs were further integrated in a MATLAB-based Graphic User Interface (GUI) environment that guides users through the simulation process in a step-by-step manner. The following lists some important features implemented in the environment.

(1) Ability to modify the accompanying thoracic model, if necessary. The segmented anatomical geometry can be changed in a pixelwise fashion and/or as a whole be linearly scaled up or scaled down up to 20%. Furthermore, electrical conductivities are tabularized, allowing adjustment of the conductivity for any given tissue.

(2) Interactive electrode placement. The GUI allows users to interactively insert electrodes into the rendered images of the segmented volume. By default, there are 16 electrodes evenly spaced around the body periphery at a mid-thoracic level and the adjacent excitation scheme is used. Figure 2 shows a screen capture shot of the interface screen for electrode placement and the assumed electrode pairs for the adjacent excitation strategy.

(3) Automated electrode arrangement recognition. The developed software environment is able to automatically cluster the placed dot electrodes into distinct groups based on spatial adjacency. Upon recognition, the electrodes are paired and numbered anticlockwise starting with electrode pair 1 at the 12 o'clock position.

Upon completion of simulations, the software generates two output files. One is a picture file which gives the location of each electrode pair, and the other gives the transfer impedance data which can be readily used or can be easily adapted to the expected form of a specific algorithm, for image reconstructions.

2.4. Image Reconstruction. The reconstruction of conductivity distribution from surface voltage measurements is complicated by the fact that electric current is distributed over the entire volume being modeled. Therefore, reconstructing the electrical conductivity distribution of the volume conductor is significantly more difficult than that of other medical imaging modalities such as CT where the photon travels essentially in straight lines. Besides that, the ill-conditioned nature of problems of this kind imposes further difficulties. Forty years after the first impedance image was published, EIT reconstruction continues to be an area of active research. Among various proposed reconstruction algorithms [13], two most representative algorithms, the Sheffield filtered backprojection algorithm and the GREIT algorithm [10], were taken as examples in the current study to show the utility of the developed software environment as a forward solver for thoracic EIT. The Sheffield filtered backprojection algorithm, distributed in the commonly used Sheffield DAS-01 P EIT system, aims at projecting changes of the surface impedance measurements along the equipotential lines calculated from the homogeneous medium. The GREIT algorithm, developed by a consensus of a large group of experts in EIT algorithm design and clinical applications for pulmonary

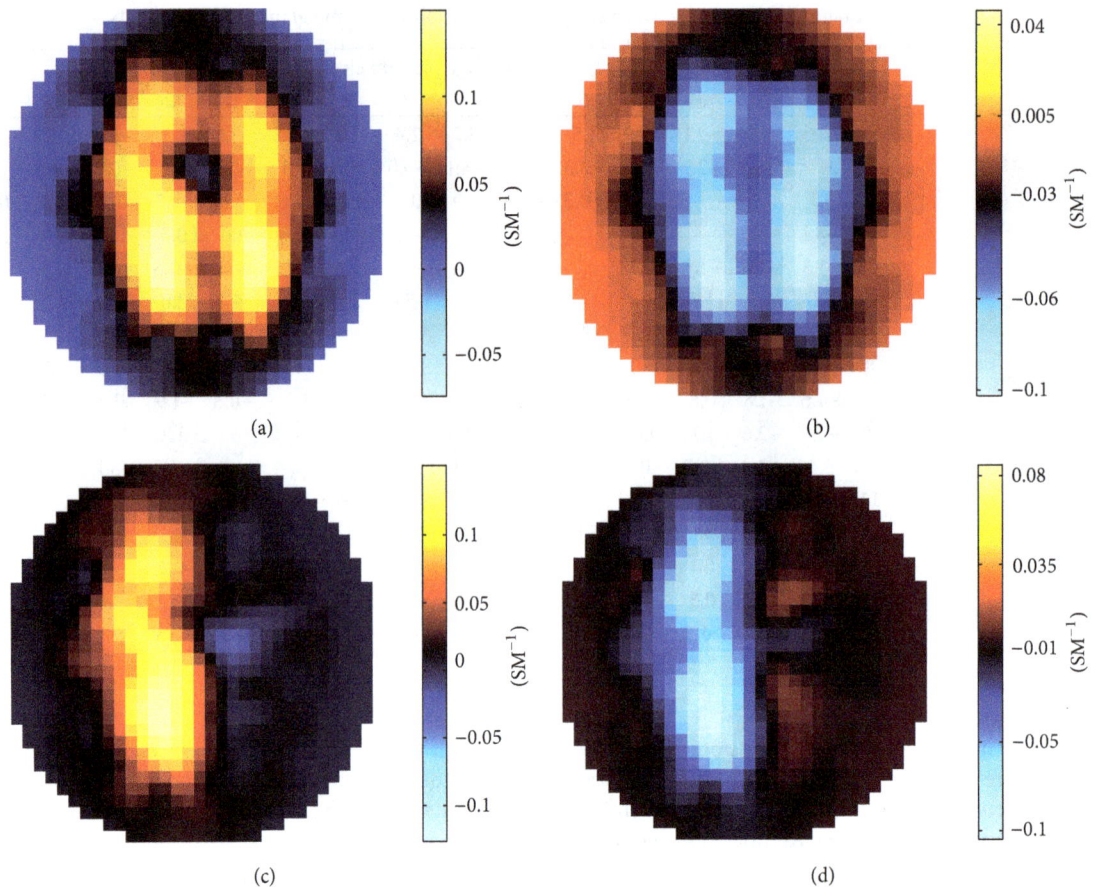

Figure 3: Conductivity differential images reconstructed with the Sheffield filtered backprojection algorithm. (a) Conductivities of both lungs were changed from 0.0714 S/m to 0.1000 S/m; (b) conductivities of both lungs were changed from 0.0714 S/m to 0.0556 S/m; (c) conductivity of the right lung was changed from 0.0714 S/m to 0.1000 S/m; (d) conductivity of the right lung was changed from 0.0714 S/m to 0.0556 S/m.

monitoring attempting to provide a unified approach for real-time thoracic EIT reconstruction, derives the conductivity distribution based on a matrix pretrained with various performance requirement criteria such as uniform amplitude and uniform resolution. Current implementations of both algorithms reconstructed conductivity differential images onto a circular field of 32×32 pixel.

3. Results

3.1. Accuracy of the Forward Solution. Accuracy of the developed forward solver in predicting surface impedance measurements was appraised by comparing the predicted impedance measurements with the impedance measurements made on the same subject whose MR data was used as the anatomical source of the model. To obtain impedance measurements from the subject, 16 electrodes were placed evenly around the body plane of the subject as shown in Figure 2 and the impedance measurements were recorded with a BIOPAC system. For 16 electrodes with the adjacent excitation scheme, a total of 208 impedance measurements were obtained. Table 2 presents the percentage of differences between the predicted impedance measurements and the observed impedance measurements. It shows that the relative

error is within the order of 5%, demonstrating the ability of the presented forward solver in generating high-fidelity surface thoracic impedance data.

3.2. Reconstruction Demonstration. Impedance measurements generated from the developed forward solver for image reconstruction demonstration consisted of the following conductivity change scenarios of either or both lungs:

(1) changing the conductivity of both lungs from the standard value of 0.0714 S/m up to 0.1000 S/m,

(2) changing the conductivity of both lungs from 0.0714 S/m down to 0.0556 S/m,

(3) changing the conductivity of the right lung from 0.0714 S/m up to 0.1000 S/m,

(4) changing the conductivity of the right lung from 0.0714 S/m down to 0.0556 S/m.

All the simulations were carried out on a Dell Precision T7400 workstation with a 2 GB memory and for each case it took about 4 hours. The differential images reconstructed with using the Sheffield filtered backprojection algorithm and the GREIT algorithm are presented in Figures 3 and

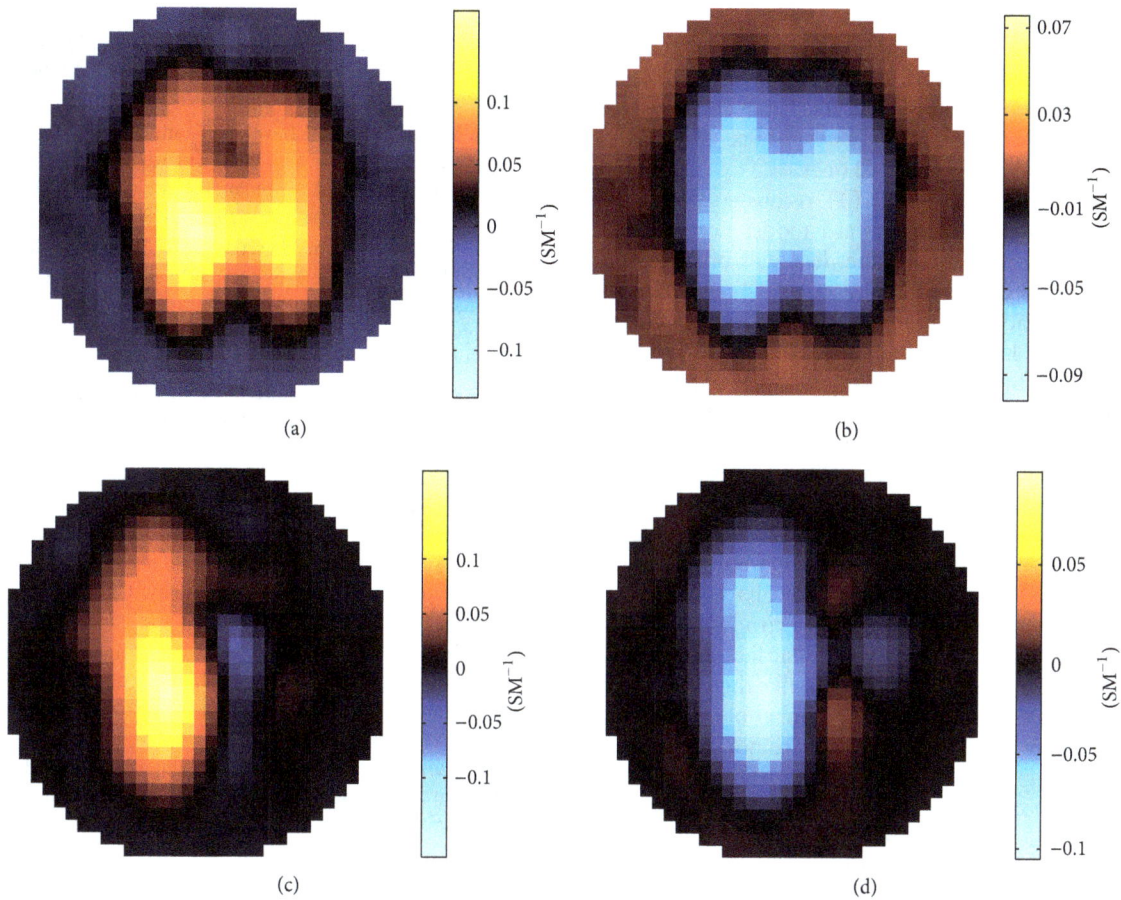

FIGURE 4: Conductivity differential images reconstructed with the GREIT algorithm. (a) Conductivities of both lungs were changed from 0.0714 S/m to 0.1000 S/m; (b) conductivities of both lungs were changed from 0.0714 S/m to 0.0556 S/m; (c) conductivity of the right lung was changed from 0.0714 S/m to 0.1000 S/m; (d) conductivity of the right lung was changed from 0.0714 S/m to 0.0556 S/m.

4, respectively. The reconstructed conductivity differences shown in these images exhibit location agreements with the lungs underwent conductivity change, demonstrating the utility of the developed forward solver in generating surface impedance measurements for thoracic EIT.

4. Discussion

The forward problem of EIT involves building a model to determine the surface impedance measurements when current excitation is applied on the boundary of the volume being modeled. Constructing such a model generally assumes the geometry and conductivity values are known and entails solving the governing Laplace's equation in the interior of the modeled volume with proper boundary conditions. In order to accurately determine surface voltage measurements as well as the interior electric fields, models with realistic representation of anatomical variability inside the thorax are highly desirable. To date, a number of electrical models of the human thorax have been developed. Based on the US National Library of Medicines Visible Human Man data, Kauppinen et al. [14] built a finite difference thoracic model that has a total number of control elements of 477,611

with size ranging from 0.011 to 0.78 cm^3 and 28 identified tissue types. de Jongh et al. [15] developed a finite element thoracic model comprising 214,957 control elements and 5 identified tissue types. Aguel et al. [16] constructed a finite element model of the human thorax consisting of 933,409 control elements and 4 identified tissue types. Wtorek [17] developed a finite element model of the thorax that consisted of 18,165 control elements with size ranging from 0.053 to 20.83 cm^3. Mocanu et al. [18] developed a thoracic model containing approximately 400,000 control elements with size of 0.054 cm^3.

Compared to these existing thoracic electrical models, the thoracic model accompanying the developed forward solver comprising 36 different types of tissues and consisting of 3.8 million control elements—about three times more than the best model among those aforementioned—with a uniform element size of 0.011 cm^3 represents the human thorax with a much higher level of anatomical accuracy, thus allowing more accurate prediction of surface impedance measurements and electric fields in the interior of the thorax. Furthermore, it is built based on MR scans of a living subject, thus offering a more authentic representation of thoracic anatomy in comparison with the majority of the existing models where cadaveric anatomy was employed in that most organs inside the thorax undergo structural and functional changes after death. Moreover, the presented forward solver, coming equipped with a user-friendly GUI and being independent from commercial software, achieves ease of use to a greater degree than the previously reported ones.

Given the ill-posed nature of EIT reconstructions, one may wonder how detailed a forward model needs to be, or even whether a model with fine details as presented in this work is needed. In our opinion, this is the line of reasoning which leads us to propose the use of highly accurate models. EIT is a low resolution modality, but for many applications, it is important to study whether a particular small feature can be seen in the images, and if so, whether its location in the images will be perturbed by nearby anatomical structures. In order to study this effect, it is important to have a thoracic model featured with a high level of anatomical accuracy. One limitation of the developed forward solver is the inaccurate representation of the electrode-body interface. For reconstruction approaches employing voltages measured from current injection electrodes, the electrode impedance would have an evident impact on the surface impedance measurements. However, for applications like the GREIT and Sheffield backprojection along with a number of other methods where current injection electrodes are not included in the voltage pick-up schemes, its effect on surface impedance measurements would be expected to be very limited as having been demonstrated here in the current study. Use of the presented thoracic forward solver would thus be restricted to reconstruction strategies leaving out injection electrodes for impedance measuring. Incorporation of more realistic electrode models into the developed forward solver is currently under investigation.

5. Conclusion

In summary, we have developed an anatomically detailed forward solver for thoracic EIT equipped with a user-friendly GUI. We hope and expect that this software environment will be able to serve as a source to generate impedance measurements along with electric field data in aid of development and evaluation of thoracic EIT reconstruction algorithms.

Conflict of Interests

The authors have no conflict of interests to disclose.

Acknowledgment

The authors would like to thank the Minnesota Supercomputing Institute for providing the computation resources.

References

[1] R. P. Henderson and J. G. Webster, "Impedance camera for spatially specific measurements of thorax," *IEEE Transactions on Biomedical Engineering*, vol. 25, no. 3, pp. 250–254, 1978.

[2] F. Yang and R. P. Patterson, "The contribution of the lungs to thoracic impedance measurements: a simulation study based on a high resolution finite difference model," *Physiological Measurement*, vol. 28, no. 7, pp. S153–S161, 2007.

[3] E. L. V. Costa, C. N. Chaves, S. Gomes, M. A. Beraldo, M. S. Volpe, M. R. Tucci et al., "Real-time detection of pneumothorax using electrical impedance tomogyaphy," *Critical Care Medicine*, vol. 36, no. 4, pp. 1230–1238, 2008.

[4] H. J. Smit, A. V. Noordegraaf, J. T. Marcus, A. Boonstra, P. M. de Vries, and P. E. Postmus, "Determinants of pulmonary perfusion measured by electrical impedance tomography," *The European Journal of Applied Physiology*, vol. 92, no. 1-2, pp. 45–49, 2004.

[5] K. Boone, A. M. Lewis, and D. S. Holder, "Imaging of cortical spreading depression of EIT-implications for localization of elieptic foci," *Physiological Measurement*, supplement 2a, pp. A189–A198, 1994.

[6] C. T. Soulsby, M. Khela, E. Yazaki, D. F. Evans, E. Hennessy, and J. Powell-Tuck, "Measurements of gastric emptying during continuous nasogastric infusion of liquid feed: electric impedance tomography versus gamma scintigraphy," *Clinical Nutrition*, vol. 25, no. 4, pp. 671–680, 2006.

[7] M. J. Moskowitz, T. P. Ryan, K. D. Paulsen, and S. E. Mitchell, "Clinical implementation of electrical impedance tomography with hyperthermia," *International Journal of Hyperthermia*, vol. 11, no. 2, pp. 141–149, 1995.

[8] F. Yang and R. P. Patterson, "A novel impedance-based tomography approach for stenotic plaque detection: a simulation study," *International Journal of Cardiology*, vol. 144, no. 2, pp. 279–283, 2010.

[9] I. Frerichs, J. Scholz, and N. Weiler, "10: electrical impedance tomography and its perspectives in intensive care medicine," in *Yearbook of Intensive Care and Emergency Medicine*, pp. 437–447, Springer, Heidelberg, Germany.

[10] A. Adler, J. H. Arnold, R. Bayford et al., "GREIT: a unified approach to 2D linear EIT reconstruction of lung images," *Physiological Measurement*, vol. 30, no. 6, pp. S35–S55, 2009.

[11] F. Yang and R. Patterson, "Optimal transvenous coil position on active-can single-coil ICD defibrillation efficacy: a simulation study," *Annals of Biomedical Engineering*, vol. 36, no. 10, pp. 1659–1667, 2008.

[12] J. Malmivuo and R. Plonsey, "7: Volume source and volume conductor," in *Bioelectromagnetism: Principles and Applications of Bioelectric and Omagnetic Fields*, pp. 405–407, University Press, New York, NY, USA, 1995.

[13] W. R. B. Lionheart, "EIT reconstruction algorithms: pitfalls, challenges and recent developments," *Physiological Measurement*, vol. 25, no. 1, pp. 125–142, 2004.

[14] P. K. Kauppinen, J. A. Hyttinen, and J. A. Malmivuo, "Sensitivity distributions of impedance cardiography using band and spot electrodes analyzed by a three-dimensional computer model," *Annals of Biomedical Engineering*, vol. 26, no. 4, pp. 694–702, 1998.

[15] A. L. de Jongh, E. G. Entcheva, J. A. Replogle, R. S. Booker III, B. H. Kenknight, and F. J. Claydon, "Defibrillation efficacy of different electrode placements in a human thorax model," *Pacing and Clinical Electrophysiology*, vol. 22, no. 1, pp. 152–157, 1999.

[16] F. Aguel, J. C. Eason, N. A. Trayanova, G. Siekas, and M. G. Fishler, "Impact of transvenous lead position on active-can ICD defibrillation: a computer simulation study," *Pacing and Clinical Electrophysiology*, vol. 22, no. 1, pp. 158–164, 1999.

[17] J. Wtorek, "Relations between components of impedance cardiogram analyzed by means of finite element model and sensitivity theorem," *Annals of Biomedical Engineering*, vol. 28, no. 11, pp. 1352–1361, 2000.

[18] D. Mocanu, J. Kettenbach, M. O. Sweeney, R. Kikinis, B. H. Kenknight, and S. R. Eisenberg, "A comparison of biventricular and conventional transvenous defibrillation: a computational study using patient derived models," *Pacing and Clinical Electrophysiology*, vol. 27, no. 5, pp. 586–593, 2004.

19

Application of Principal Component Analysis in Automatic Localization of Optic Disc and Fovea in Retinal Images

Asloob Ahmad Mudassar[1] and Saira Butt[2]

[1] *Department of Physics and Applied Mathematics, Pakistan Institute of Engineering and Applied Sciences, Nilore, Islamabad 45650, Pakistan*
[2] *Isotope Application Division, Pakistan Institute of Nuclear Science and Technology, Nilore, Islamabad 45650, Pakistan*

Correspondence should be addressed to Asloob Ahmad Mudassar; asloob@yahoo.com

Academic Editor: Nicusor Iftimia

A retinal image has blood vessels, optic disc, fovea, and so forth as the main components of an image. Segmentation of these components has been investigated extensively. Principal component analysis (PCA) is one of the techniques that have been applied to segment the optic disc, but only a limited work has been reported. To our knowledge, fovea segmentation problem has not been reported in the literature using PCA. In this paper, we are presenting the segmentation of optic disc and fovea using PCA. The PCA was trained on optic discs and foveae using ten retinal images and then applied on seventy retinal images with a success rate of 97% in case of optic discs and 94.3% in case of fovea. Conventional algorithms feed one patch at a time from a test retinal image, and the next patch separated by one pixel part is fed. This process is continued till the full image area is covered. This is time consuming. We are suggesting techniques to cut down the processing time with the help of binary vessel tree of a given test image. Results are presented to validate our idea.

1. Introduction

This paper presents an extension of the application of principal component analysis (PCA) to retinal images. Localization cases of optic disc and fovea have been presented in the literature [1–16] using techniques other than PCA except the optic disc localization by PCA which is discussed in [2]. In this paper, application of PCA is presented for two different cases: (1) to automatically locate the position of the optic disc in a retinal image, and (2) to automatically locate the position of the fovea in a retinal image. To our knowledge, the latter application is novel and has not been reported in the literature. The former application has been discussed in the literature [2]. The information contained in [2] does not fully appreciate the scope of PCA in optic disc localization. This paper will elaborate the work of optic disc localization and will extend the scope of this work to the localization of fovea. The algorithm we have developed for the localization of optic disc and fovea works faster than the one reported in [2]. The application of PCA to determine the location of

the fovea is a relatively difficult problem as compared with locating the optic disc because the fovea usually has lower contrast compared with the optic disc in a retinal image.

Knowledge of the optic disc location and its diameter is important in the automatic analysis of retinal images. A variety of techniques to automatically determine the location of the optic disc in a retinal image have been described in the literature [1–10]. Most of the techniques do not give satisfactory results. One of the main reasons for the failure of such techniques is that the exudates in a retinal image are sometimes comparable or greater than the size of the optic disc. However, the techniques work very well in a normal retinal image that has no exudates or where the size of the exudates is small enough compared with the size of the optic disc.

The usefulness of finding the location of an optic disc is that once a point near the centre of an optic disc is determined, then some techniques like the active contour method or other techniques [4, 7] can be used to determine the exact diameter of the optic disc. An increase in the size of the optic disc is an indication of some sort of eye abnormality named peripapillary atrophy.

Principal component analysis (PCA) is a powerful statistical technique that can be used to identify patterns in data of high dimensions. The technique has applications in face compression and pattern recognition. For pattern recognition problems, some training images are stored in principal component form. Later, this data is used to find the similarity or dissimilarity between an unknown data set and the stored data. Identification of the optic disc and fovea is an example of pattern recognition problems. These two problems have been addressed in this paper using PCA. Application of PCA to retinal images cannot be appreciated without its detailed description which is given in the next section.

2. Mathematical Modeling of PCA with Reference to Retinal Images for Localization of Optic Disc and Fovea

In PCA a set of images is required for training purposes. Let N be the total number of images used in the training process with each training image having $n \times n$ dimensions. Training images can be designated by $I_1, I_2, I_3, \ldots, I_N$. These are the intensity images taken from the same part of N different retinal images. I_i may represent a patch centred at an optic disc in a retinal image i or may represent a patch centred at the fovea in a retinal image i. I_i for $i = 1, 2, 3, \ldots, N$ represents N patches centred at the optic disc or fovea in N different retinal images. Let us define mn_i and mx_i as the minimum and maximum values in the training images as follows:

$$mn_i = \text{Min}\left[I_i\right],$$
$$mx_i = \text{Max}\left[I_i - mn_i\right]. \tag{1}$$

Normalized training images are obtained using (2):

$$I_i^n = \frac{I_i - mn_i}{mx_i} 255, \tag{2}$$

where I_i^n is a normalised version of I_i with intensity in the range from 0 to 255 assuming that the images are grey scale 8-bit images. The mean image of the training images is calculated using the following equation:

$$M = \frac{1}{N} \sum_{i=1}^{N} I_i^n. \tag{3}$$

If we consider the optic discs as training images, then I_i^n would represent either the optic discs taken from the right-eye retinal images or the discs taken from the left-eye retinal images. The right-eye images have optic discs located on the right side in a retinal image, and such images are flipped to align them with the left-retinal images. The mean image M is subtracted from each normalised training image I_i^n, and the resultant image may be designated with a new symbol I_i^{nm} and is obtained from the following equation:

$$I_i^{nm} = I_i^n - M. \tag{4}$$

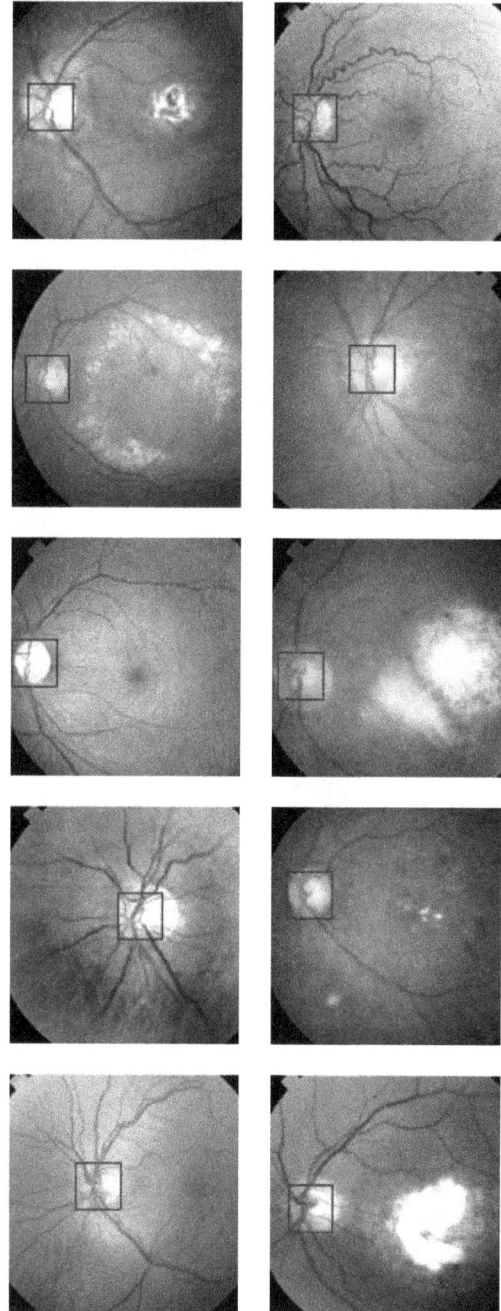

Figure 1: Ten retinal images used as training images in PCA. Areas marked within black squares were used in the training process of PCA.

The matrix form of I_i^{nm} can be written as given below:

$$I_i^{nm} = \begin{bmatrix} I_i^{nm}[1,1] & I_i^{nm}[1,2] & \cdots & \cdots & I_i^{nm}[1,n] \\ I_i^{nm}[2,1] & I_i^{nm}[2,2] & \cdots & \cdots & I_i^{nm}[2,n] \\ \cdots & \cdots & \cdots & \cdots & \cdots \\ \cdots & \cdots & \cdots & \cdots & \cdots \\ I_i^{nm}[n,1] & I_i^{nm}[n,2] & \cdots & \cdots & I_i^{nm}[n,n] \end{bmatrix}. \tag{5}$$

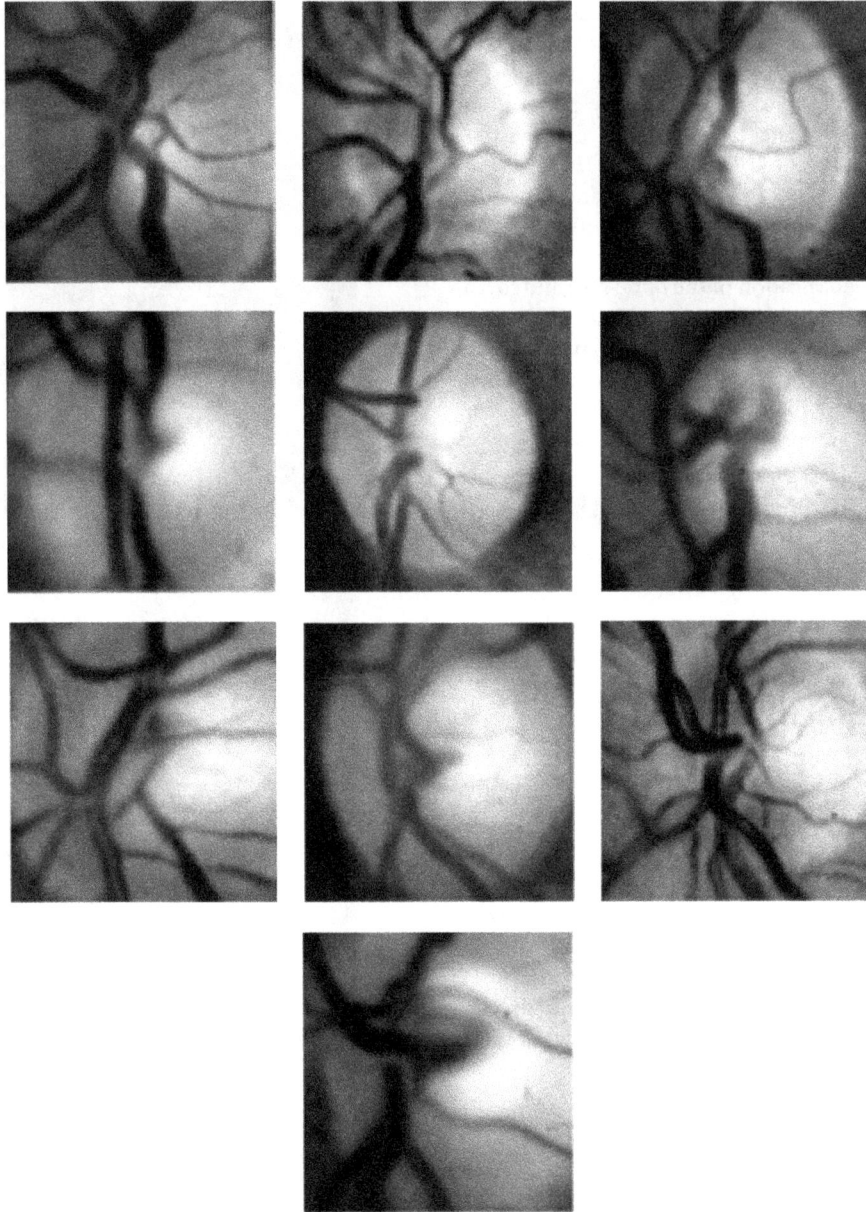

FIGURE 2: Optic discs indicated in Figure 1 used in the training process of PCA.

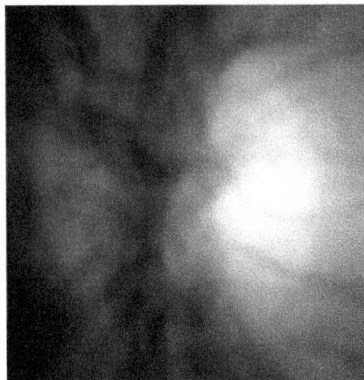

FIGURE 3: Average image of the ten optic disc images shown in Figure 2.

The matrix elements are flattened to form a set of n^2 elements, and the flattened list can be designated with the symbol $_f I_i^{nm}$:

$$_f I_i^{nm} = \left\{ \begin{array}{l} I_i^{nm}[1,1], I_i^{nm}[1,2], \ldots, I_i^{nm}[1,n], \\ I_i^{nm}[2,1], I_i^{nm}[2,2], \ldots, I_i^{nm}[2,n], \ldots, I_i^{nm}[n,n] \end{array} \right\}. \tag{6}$$

The required data, D, for PCA is formed by combining $_f I_i^{nm}$ for $i = 1, 2, 3, \ldots N$ elements as given below:

$$D = \left\{ \left\{ _f I_i^{nm}[k,l] \; \forall 1 \le i \le N \right\}, \forall 1 \le l \le n, \forall 1 \le k \le n \right\}. \tag{7}$$

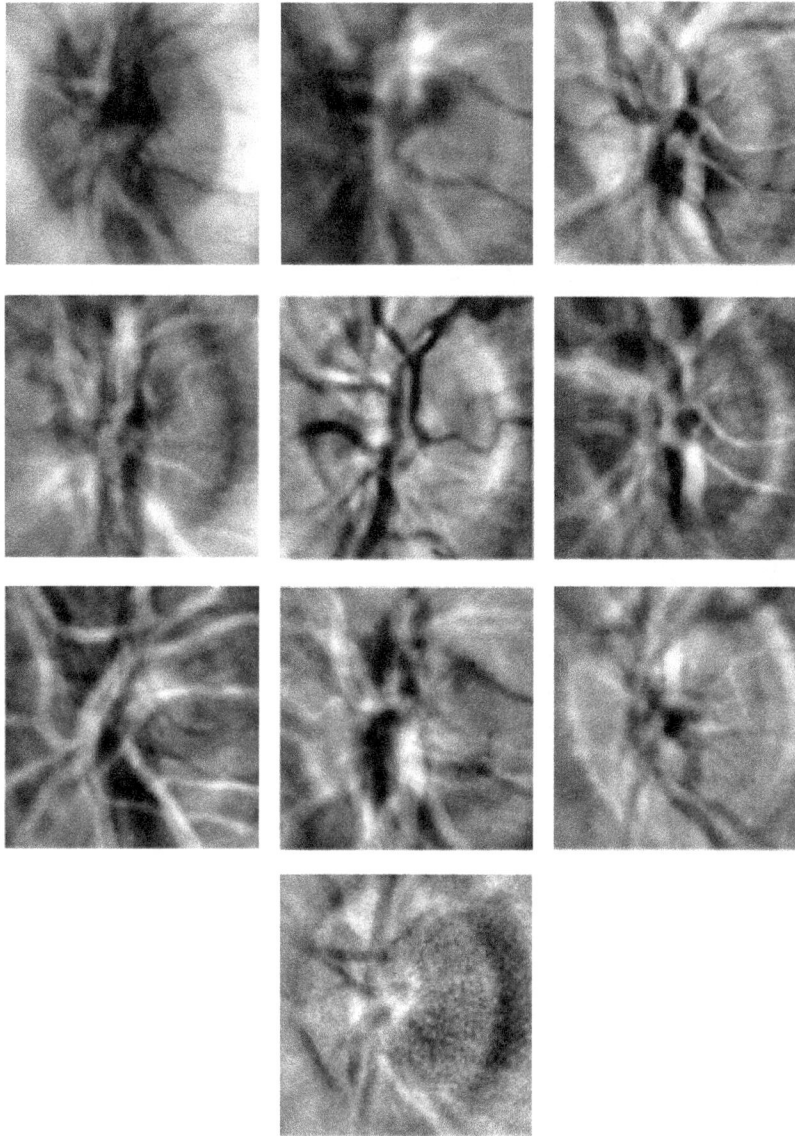

FIGURE 4: First ten eigendiscs obtained by applying PCA to images in Figure 2.

The data D has a dimension of (n^2, N), and the covariance matrix C of D has dimension (N, N) and is given by

$$C = \frac{D^T D}{N}. \tag{8}$$

A feature vector featvec having dimension (N, N) is formed by

$$\text{featvec} = \text{Eigen value} [C], \tag{9}$$

featvec is used to compute the finaldat_1 having dimension (N, n^2) and sum_1 as given by

$$\text{finatdat}_1 = \text{featvec} \cdot D^T, \tag{10}$$

$$\text{sum}_1 = \sum_{p=1}^{N} \sum_{q=1}^{n^2} \text{finaldat}_1^2 [p, q]. \tag{11}$$

In words, D^T has N elements each of length n^2, and each element of D^T represents one training image. For any number of unknown image patches from a retinal image each having dimension $(1, n^2)$ to resemble to one of the training images, the error as defined below must have a minimum value for the most similar patch:

$$\text{error} = \text{sum}_1 - \text{sum}_2, \tag{12}$$

where sum_2 has the following definition:

$$\text{sum}_2 = \sum_{p=1}^{N} \sum_{q=1}^{n^2} \text{finaldat}_2^2 [p, q], \tag{13}$$

finaldat_2 is found from the following equation:

$$\text{finatdat}_2 = (\text{featvec} [1])^T \cdot \{\text{patch}\}, \tag{14}$$

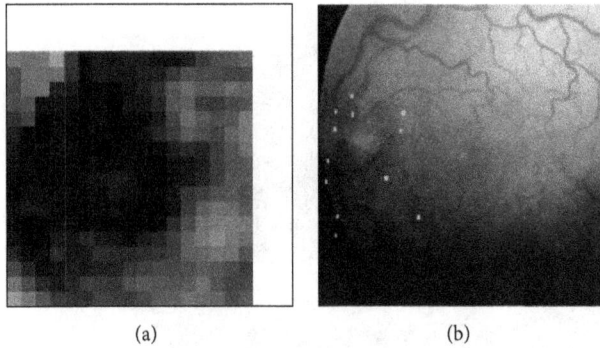

FIGURE 5: Euclidian distance map (a) of the input image (b). First few minima in the distance map are indicated in (b).

FIGURE 6: Indication of precise minimum position on the patch with minimum in the Euclidian distance map in Figure 5(a).

$(\text{featvec}[1])^T$ has a dimension of $(N, 1)$, and each {patch} has a dimension of $(1, n^2)$, and as a result of matrix multiplication finaldat_2 is of dimension (N, n^2).

The mathematical modeling given above explains clearly how the training images are used to determine the similarity between the unknown image and the images in the training set. The training images in our case may consist of either the region centered around the optic disc or the region centered around the fovea. The algorithm that we used in localization of optic disc and fovea follows the steps explained by the mathematical modeling.

3. PCA and Optic Disc Localization

The first step in applying PCA to localization of the optic disc is to train the PCA using a set of optic discs from a variety of different retinal images. The training process should cover most of the features of different types and forms of optic discs. For the training purpose, we chose ten retinal images with a variety of optic disc forms. The ten retinal images are shown in Figure 1. The retinal images shown in Figure 1

indicate the position of a square of size 100×100 pixels nearly centred at the optic disc covering most of the area surrounded by the optic disc. Some of the images, shown in Figure 1, were right-eye images and were converted to left-eye images using the mirror-image technique. The magnified version of the marked optic discs in Figure 1 is shown in Figure 2. The images shown in Figure 2 were used as training images for the PCA used in localising the optic disc in retinal images.

The training images shown in Figure 2 were normalised in the range 0 to 255 using (1) and (2). These normalised images were then used to compute the average image using (3). The average image so obtained is shown in Figure 3. The average image of Figure 3 was subtracted from each of the images shown in Figure 2 according to (4). The resulting images were then flattened in rows each of length 10000. Ten rows corresponding to ten training images were obtained, and using (6) they were combined to form a matrix of order (10,10000). The transpose of this matrix gives data of dimensions (10000,10). The covariance matrix of the data written in the form of (7) was then calculated using (8). The covariance matrix so obtained had dimensions (10,10). The eigenvalues of the covariance matrix were labelled as featvec having dimensions (10,10). It is the featvec that has the characteristics of all the training images. The featvec is used to compute finaldat_1 which has dimension (10,10000). The ten elements of finaldat_1 are called eigendiscs and can be represented as ten images, as shown in Figure 4. The first eigendisc (image at the top left in Figure 4) corresponds to the first principal component, and the tenth eigendisc (image at the bottom right in Figure 4) corresponds to the tenth principal component; sum_1 is then calculated using (11).

To explain how PCA is used to locate the position of an optic disc in a retinal image, we chose a retinal image as shown in Figure 5(b). The image was split into subimages each of size 100×100 called patches. Each patch was substituted in (14), and finaldat_2 was computed; sum_2 was then determined using (13), and the error was computed using (12). The error value is then assigned to an area patch in an image which gives the Euclidian distance map and is shown in Figure 5(a). Each patch of size 100×100 pixels in the Euclidian distance map represents the value of the error corresponding to the patch at that location in Figure 5(b). Once the Euclidian distance map is formed, the first few minima were computed from the map and are shown in Figure 5(b). A square of size 100×100 pixels with its left bottom corner at the minimum position and sides parallel to the image axes is then placed on the input image to mark the location of the optic disc in the input retinal image.

The patches in the Euclidian distance map in Figure 5 each has dimension (25×25) because the patches of size 100×100 pixels with a step size of 25 were extracted from Figure 5(b) and were presented to the PCA. So the exact location of the optic disc is uncertain in an area of 50×50 pixels about the minimum position found from the Euclidian distance map. To find an exact location for the optic disc, an area of size 50×50 pixels is selected on the input image where the Euclidian distance map gives the minimum, and patches of size 100×100 pixels with a step size of 1 are presented to the PCA, and the minimum error then gives the exact location

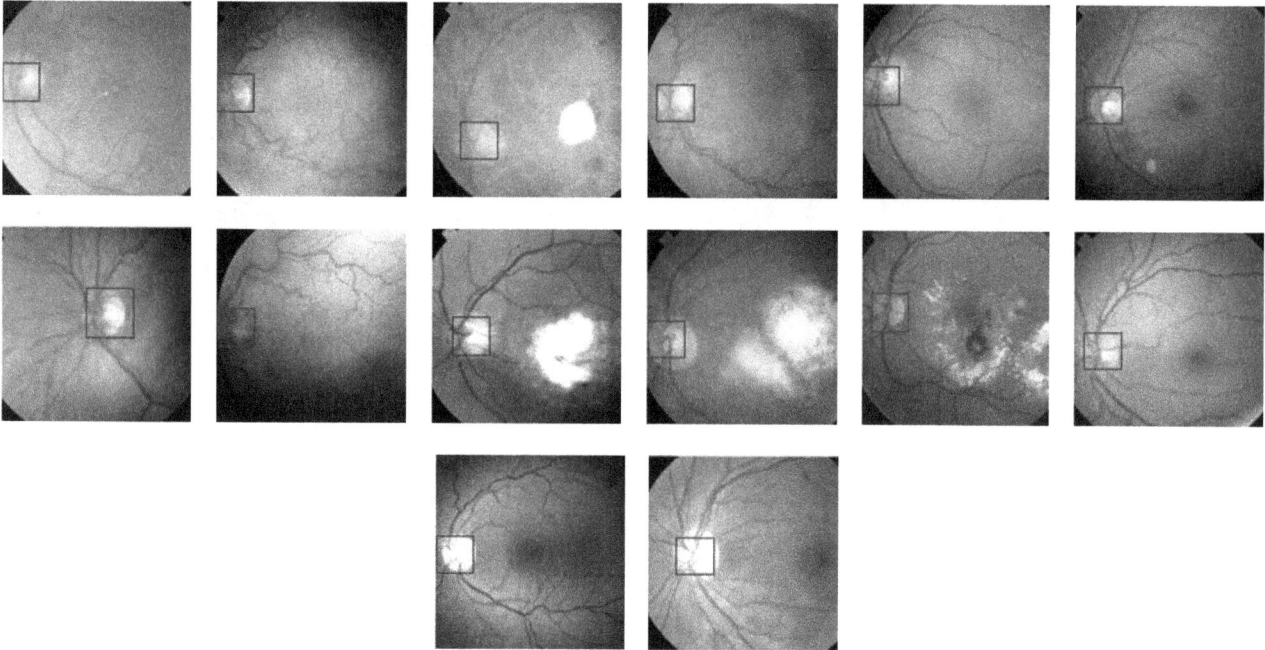

FIGURE 7: Location of optic discs on the retinal images as determined by PCA.

of the optic disc. Such an error image has been shown in Figure 6 where the grey spot precisely locates the position of the optic disc.

Application of the PCA to the input retinal images saves a lot of computational time when the patches of sizes 100×100 pixels with a step size of 25 pixels along both axes are chosen. The simulation work on this problem has revealed that in a very few cases the minimum at the desired location may be missed. To avoid this, the first three minima found from the Euclidian distance map are further explored to find the exact location of the minimum using the techniques described above and shown in Figure 6. It is possible to sort out the candidate regions by clustering the brightest pixels in the input retinal image as mentioned in [2]. To further cut down the computational time, we applied the PCA to the candidate regions only with the strategy described above. This process has significantly reduced the computational time.

We have also explored a novel idea to cut down the simulation time and to make the PCA effective for retinal images. First, a thick binary vessel tree is extracted from the input retinal image using the matched filtering technique [17]. The PCA was then applied to the input retinal image only at those points of the input retinal image where the binary vessel tree gave unit values. In a binary vessel tree, the vessels are represented by 1 s and the background by 0 s. In this way, only a limited area of the retinal image is explored in the investigation of optic disc localization.

4. Optic Disc Localization Results

First, the PCA was trained using 10 optic discs as training images. The training images are shown in Figure 2. PCA was then applied to 50 retinal images out of which only ten images have been shown in Figure 7. The retinal images shown in Figure 7 are the input images to PCA. For each input

image in Figure 7, a Euclidian distance map is constructed like the one shown in Figure 5(a). A square boundary of size 100×100 pixels is then fitted at the position in the input retinal image corresponding to the minimum position given by the Euclidian distance map. Square boundaries have been drawn on the input retinal images in Figure 7 as determined by PCA. Some input images in Figure 7 are normal retinal images, while others have brighter areas even bigger than the size of the optic disc. The results show that the localization of optic discs in retinal images can be determined sufficiently accurately using PCA. Some of the input retinal images in Figure 7 have faint optic discs with low contrasts where simple techniques, which rely on the brightest region localization as optic discs, cannot work and where the PCA has given good results. The optic discs shown in Figure 2 taken from the input retinal images shown in Figure 1 have got unequal diameters, and the optic discs with unequal diameter were used in training the PCA. Some of the optic discs in the input images shown in Figure 7 also have unequal diameters, but the PCA has worked well in each case. The more the variety in the optic discs used in the training process, the more successful the results obtained from the PCA, and the success rate can be much improved.

The only drawback with the PCA is that it is computationally very expensive in time as far as the decision on an input retinal image is concerned. The training process is easy and fast. The decision time for an input retinal image can be greatly reduced by taking measures as explained in the previous section.

5. PCA and Fovea Localization

The first step in the application of PCA to locate foveae in the input retinal images is to choose foveae regions from retinal images for training purposes. Figure 8 shows ten

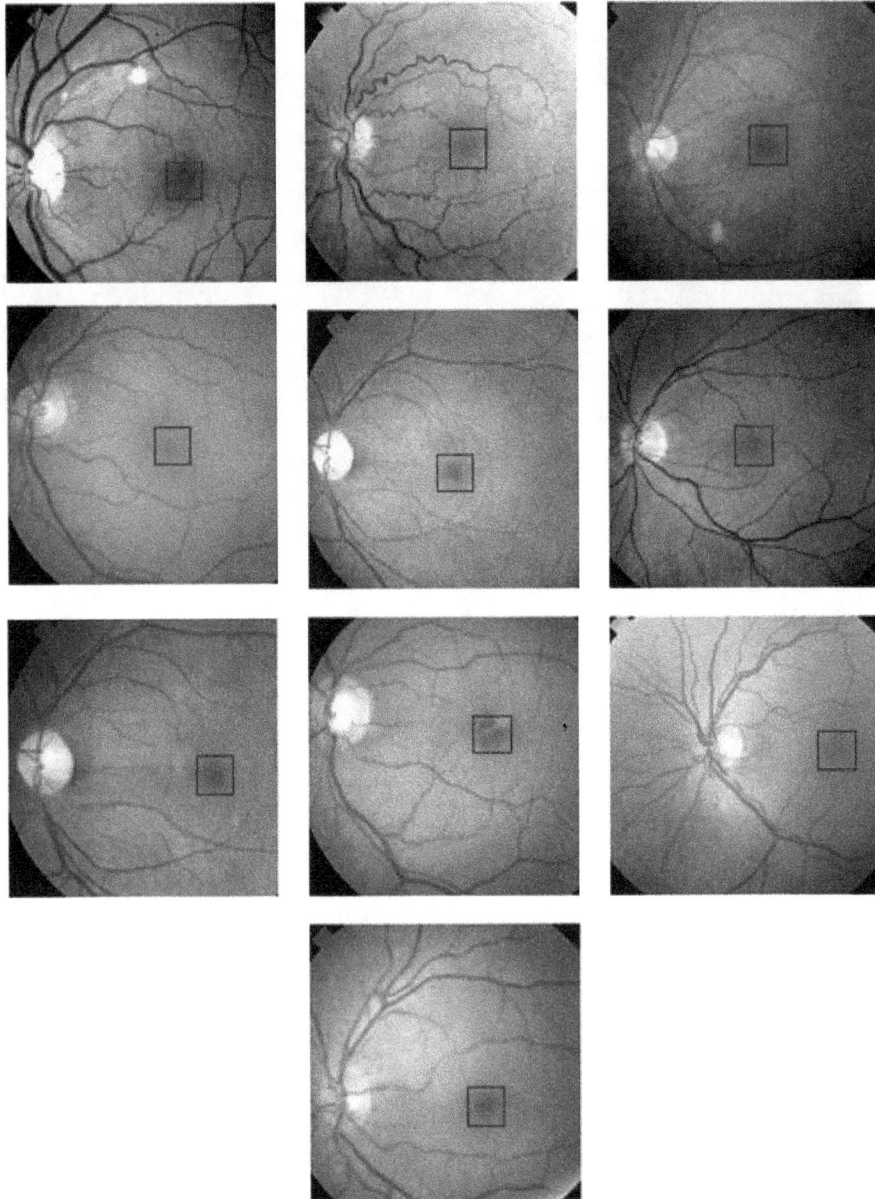

FIGURE 8: Ten retinal images used as training images in the PCA. Areas marked within black squares were used in the training process of the PCA.

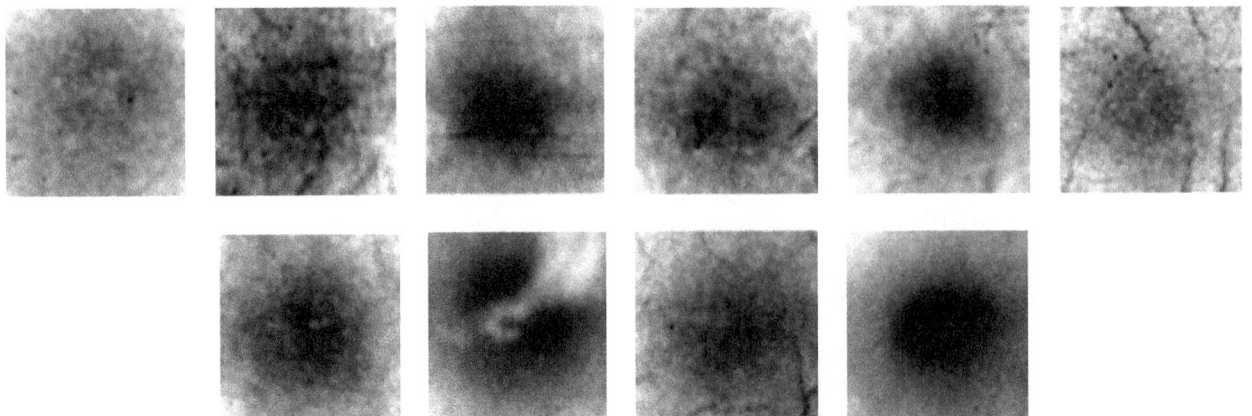

FIGURE 9: Foveae indicated in Figure 8 used in the training process of PCA.

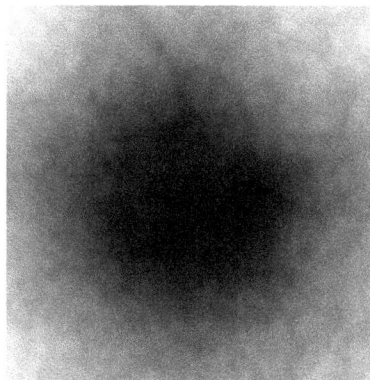

FIGURE 10: Average image of the ten foveae images shown in Figure 9.

retinal images chosen for training the PCA. Foveae regions have been indicated on the retinal images in Figure 8, and their magnified versions are shown in Figure 9. The foveae regions in Figure 9 are all different from one another in order to introduce versatility into the training process. The size of each fovea in Figure 9 is 70×70 pixels.

The fovea images of Figure 9 were normalised using (2), and the average image was then found using (3). The average image is shown in Figure 10. Following the steps given by (4) to (10), finaldat$_1$ was computed having dimensions (10,4900). Finaldat$_1$ has 10 data sets each of length 4900. These data sets, called eigenfoveae, have been displayed in Figure 11 as ten images each of size 70×70 pixels. The final step in the learning process was to compute sum$_1$ using (11).

In Figure 12(b), there is an input retinal image, on which one to four numbers are marked in the region around fovea. The numbers in the region of fovea indicate the first four minima as given by the Euclidian distance map in Figure 12(a). The Euclidian distance map is actually a representation of the error signal computed using (12).

Sum$_2$ varies from patch to patch when the PCA is applied to an input retinal image in Figure 12(b). The first four minima on the input image in Figure 12 represent four patches each of dimension 70×70 pixels on the Euclidian distance map. These patches can further be analysed by the PCA with a step size of one to precisely identify the location of the foveae as described earlier and presented in Figure 6.

To determine the effectiveness of the PCA in localization of the fovea in retinal images, 70 retinal input images were chosen, and the PCA was tested on them. The success rate was found to be more than 94.3%. A few of the input retinal images on which the PCA was tested have been given in Figure 13. Each of the 20 input retinal images in Figure 13 indicates the location of the fovea as determined by the PCA. Foveae in some of the images in Figure 13 have very poor contrast that even a human eye cannot discriminate properly from the surrounding areas, but the PCA has determined their locations to an outstanding precision. Localization of foveae using the PCA in an input retinal image is more time consuming computationally compared with localization of the optic disc. This is because the PCA can be applied to

specified regions when localising the optic disc by choosing candidate regions as clusters of the brightest pixels in a retinal image or by applying the PCA only at thick vessel points using the binary vessel tree of the retinal image.

The fovea is the darkest region in a retinal image with nearly the same intensity as the blood vessels. The centre of the fovea is usually located from the centre of the optic disc at about 2.5 times the diameter of the optic disc. Even if the location of the optic disc is unknown, the PCA can be applied to the candidate regions, which are clusters of the darkest pixels excluding the blood vessels (using information from a binary vessel tree). This technique then provides a way out to cut down the computation time in locating the fovea by the PCA. The retinal images in Figure 13 have their darkest areas at the edges, which are not part of the retina and can be filled with intensities from adjacent regions.

6. General Discussion on PCA and Related Issues

PCA is generally regarded as a computationally expensive technique which has extremely limited its domain of applications. The size of each retinal image used in fovea localization by PCA was 500 by 500 pixels. The size of the patches from the foveae regions used in training the PCA was 70 by 70 pixels. The same was the size of the patches taken from the retinal images which were presented to PCA for evaluation. The machine on which the computations were done had the following specifications: Pentium(R) D CPU 3.4 GHz, 2 GB of RAM, Microsoft Windows XP Professional Service Pack 2 Version 2002, and Mathematica 4.1 in a stand-alone mode. Foveae from ten retinal images were extracted as pointed out by a human grader and presented to PCA for training. For evaluation, sixteen images were chosen, and patches from one image at a time with a step of 1 pixel along each image axes were selected and fed to PCA, and an automatic square boundary was fit to fovea in each retinal image presented to PCA. The time for this activity excluding the time taken by the human grader in providing the training set of images was 80 hours, 14 minutes, 38 sec, and 109 milliseconds. When the above procedure was repeated with a step size of 25 pixels along each image dimension and the first four minima in the Euclidian distance map were further investigated by taking patches with a 1 pixel step in the region of 4 patches, the measured time was found to be 40 minutes, 13 sec, and 210 milliseconds. The computational time was further reduced by using the vessel tree of each retinal image presented for evaluation as a supporting image. Only those patches from each retinal image were taken for evaluation purposes in which the corresponding patches from the binary vessel trees did not contain any thick vessel. This limited the number of patches used in the evaluation procedure. The time for this activity with a step of 1 pixel was found to be 19 minutes, 9 sec, and 301 milliseconds. This time included the time used to obtain binary vessels tree for each retinal image. The binary vessel tree for each retinal image was obtained using the technique described in [17].

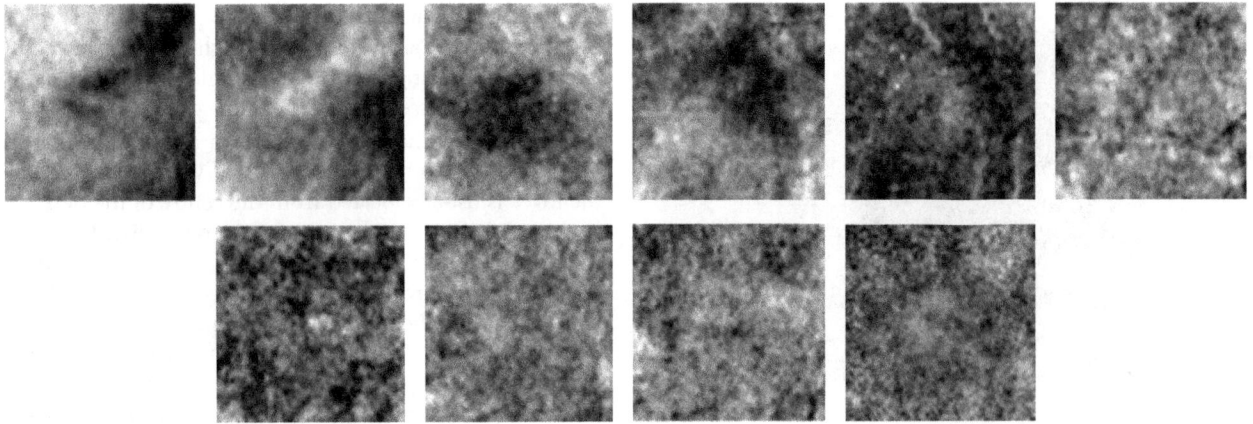

FIGURE 11: First ten eigenfoveae obtained by applying PCA to images shown in Figure 9.

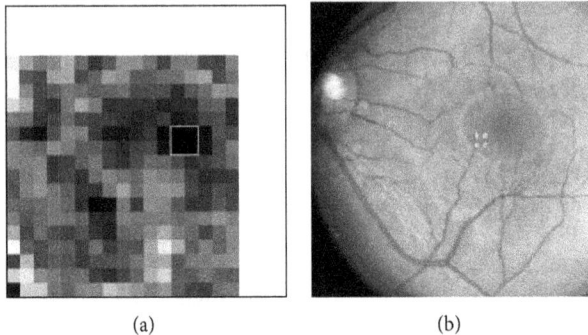

(a)   (b)

FIGURE 12: Euclidian distance map (a) of the input image (b). First four minima (contained in a square box with white boundary) in the distance map are indicated in (b).

The computational time for the application of PCA for the localization of optic nerve head is not available, but we expect a trend in the time measurements similar to that taken by different steps of PCA for fovea localization. The patch sizes were 100 by 100 pixels, and obviously the computational time would be greater in comparison with the time taken by PCA for localization of fovea, where the patch sizes were 70 by 70 pixels, but the size of the retinal images for both activities was 500 by 500 pixels. Binary vessel trees were also used to minimize the number of patches taken from a retinal image for the purpose of evaluation. Only those patches from each retinal image were taken for evaluation purposes in which the corresponding patches from the binary vessel trees contained thick vessel, keeping in view the fact the optic disc is located at the junction of thick vessels. Keeping in view the measured time data available for the application of PCA to fovea localization, we believe that with the help of binary vessels tree we have succeeded in reducing the candidate regions for evaluation purposes, and therefore we have succeeded in reducing the computational time to a reasonably good extent.

The accuracy of the algorithm was tested on 70 retinal images for foveae and optic discs localization. The size of the optic disc patches was 100 by 100 pixels, and the size

of the patches for fovea localization was 70 by 70 pixels. Three human graders were asked to give their best judgment about the centre of optic nerve head and the fovea for each of the images. The accumulated results showed that on the average the results of mean disc centre and the mean fovea centre varied by ±5 pixels leading to a human grader accuracy of 10% and 14.2% for the optic disc and fovea cases, respectively. When 70 retinal images were presented to PCA for evaluation, 68 of the images met the criterion of 10% accuracy leading to an efficiency of 97%, whereas in case of fovea localization, 66 images met the criterion of 14.2% accuracy leading to an efficiency of 94.3%. The 70 retinal images used for the above analysis were normal retinal images with respect to optic disc and fovea. A similar analysis for retinal images corrupted with respect to optic disc and fovea will be dealt in future as the analysis would comprise an extensive and perhaps more difficult work.

The performance of the algorithm was determined on retinal images which can be regarded as normal images with respect to optic disc and fovea. Some of the images in Figure 7 are abnormal images in terms of exudates present in them, but they may be regarded as normal images as far as optic discs are concerned. The two images which failed to fulfill the above-mentioned criterion as far as the optic disc localization is concerned were the images having large exudates present in them, and thick vessels passing through the exudates were prominent resembling much with the original optic discs. But when these two images were presented to PCA in the presence of binary vessel tree, then the problem was resolved. In the absence of assistance from binary tree, the optic disc was located outside the region of actual optic discs, but including the contribution from binary tree the optic discs were found in the designated region, but still the above-mentioned criterion was not met as the exudates peeped into the region close to optic disc covering some of the thick vessels in the candidate regions. These were examples of minor artifact at the optic discs.

The sixteen images which were presented to PCA for fovea localization are shown in Figure 13. All of the images are normal with respect to fovea. The four abnormal images consisting of large exudates in the candidate regions on which

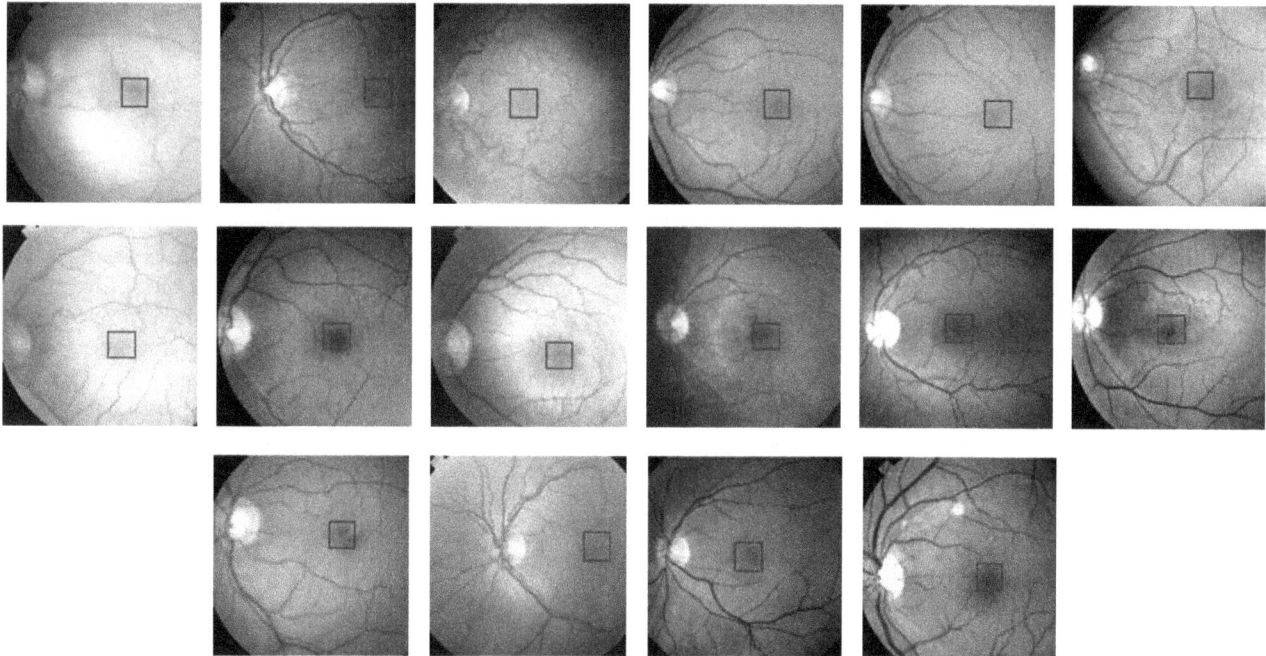

FIGURE 13: Location of fovea on the retinal images as determined by PCA.

PCA failed to locate foveae at the designated positions are shown in Figure 7. Even the assistance from corresponding binary tree images was useless. A limited number of abnormal images with respect to fovea have been dealt in this research; however, the possible reasons for the failure of PCA in localizing the fovea may be listed as follows: (a) foveal regions may be nonexistent in some retinal images as such regions may be occupied by some eye abnormality, for example, exudates may be covering the candidate regions, (b) the candidate region does exist but with a very poor contrast which may be harder even for a human grader to locate the centre with precision, (c) corneal reflection may mask the foveal region, (d) thin vessels in the vicinity of candidate regions may either be absent or have very poor visibility, in which case the assistance from the binary vessel tree will be useless resulting in the determination of the foveal centers somewhere else in the image, (e) there may be some image artifacts that could prevent the PCA localizing the foveal regions accurately, and the localization may occur far away from the actual region. Future work will address these problems in a greater detail.

The variability of features in the training set images matters a lot for the PCA to correctly identify the candidate regions with high accuracy. To establish that the variability is important, only two images (first one and the last one) from the second row in Figure 9 were chosen to train the PCA for fovea localization. The small number of training sets implies less variability of features in the training process of PCA. The sixteen images shown in Figure 13 were presented to PCA for fovea localization. It was found that only four images (4th image in the first row from the left, 2nd, 4th, and 6th images from the second row) satisfied the 14.2% accuracy leading to an efficiency of 25%. It may be concluded that variability

in the training set of images is an essential part of PCA for high performance in results. Variability in a training set may be improved by combining set members from a variety of different retinal images which at least appear different to a human grader.

Figure 12(b) has been labeled with first four minima which apparently do not correspond to the centre of the fovea as viewed by a grader. The first four minima were determined from the Euclidian distance map given in Figure 12(a). Some further details are necessary to understand the construction of the Euclidian distance map given in Figure 12 to point out why the first four minima indicated on the retinal image in Figure 12 do not lie close to the fovea centre. The input retinal image in Figure 12 was divided into patches of sizes 70 by 70 pixels with a step of 25 pixels along both the image axes. Each patch from the retinal image shown in Figure 12(b) was presented to PCA, and an error was calculated. This error was displayed as a single grey-level pixel in an empty image called Euclidian distance map at the coordinates corresponding to the lower-left-edge coordinate of the image patch. The number of pixels in the Euclidian distance map along an image axis is equal to the total number of pixels in the retinal image along the same image axes divided by the number of pixels in the step with which the image patches are extracted. The size of the Euclidian distance image is equal to the size of the retinal image divided by the dimensions of the step size. When the Euclidian distance image is displayed in comparative size to the retinal image, the pixeling effect becomes prominent as it is evident from the Euclidian distance map in Figure 12(a). To display the majority of the grey-level pixels in the Euclidian distance map in the lower range for determination of minima, the plot range was set such that the pixels with larger magnitude of errors are displayed white as evident on the right side and on the top side of the Euclidian distance map in Figure 12.

The four minima indicated on the retinal image in Figure 12 correspond to the first four minima obtained from the Euclidian distance map given in Figure 12. The region around each minimum is further investigated by taking image patches of sizes 70 by 70 pixels in the vicinity of each minimum, and a refined Euclidian distance map is obtained with an accuracy of a single pixel. A square of size 70 by 70 pixels is then drawn on the retinal image with its left-lower edge at the coordinates obtained from the refined Euclidian distance map. Each retinal image given in Figure 13 is, therefore, fitted with a square of size 70 by 70 pixels according to the description mentioned above.

The images used in training and in the evaluation for both the optic disc and fovea were selected randomly. Figure 1 shows retinal images used in the training of PCA for optic disc localization. Some of the images in Figure 1 are normal images, while others are abnormal or diseased images. The diseased images in Figure 1 show large-sized exudates, but they do not corrupt optic discs. As far as the optic discs are concerned, the images in Figure 1 may be regarded as normal images, and the images for evaluation as in Figure 7 may also be regarded as normal images. The images used in the training (Figure 8) and in the evaluation (Figure 13) of PCA for foveae localization may also be regarded as normal retinal images with respect to foveae. The application of PCA to the diseased retinal images with respect to optic discs and foveae requires separate investigation, and this problem will be dealt separately in future work.

7. Critical Analysis

Segmentation of optic disc in retinal image is well documented [1–16]. Limited work on this subject has been reported using PCA [2]. The segmentation of fovea using PCA is the main novelty of this paper. The segmentation of optic disc using PCA is a relatively simple problem in comparison with the segmentation of fovea using PCA. Fovea has the lowest contrast in a retinal image comparable to the contrast of weak vessels, whereas the contrast of optic disc is generally the highest in a retinal image being the brightest component of the retinal image. It is because the previous attempts for the localization or the segmentation of fovea have been unsuccessful. We have successfully demonstrated the segmentation capability of PCA for fovea.

There is one major drawback of the PCA, and that is it is a time consuming technique and cannot be applied for real time processing of retinal images for the localization of optic disc and fovea. We have proposed some measures that make PCA work faster especially for the processing of retinal images. In normal process for localization or segmentation of either optic disc or fovea using PCA, test patches from a test retinal image are extracted in a square window which is displaced by one pixel along the horizontal axis or one pixel along the vertical axis for the extraction of the next patch. These patches are fed to PCA for processing which is time consuming. The processing time can be cut down

by displacing the window to half of its dimension in either direction, and the patches which are now reduced in number are fed for PCA processing. The patch for which the Euclidian distance map gives the minimum is selected, and a normal procedure of displacing the window by one pixel is applied to refine the localization of optic disc or fovea around that particular patch.

In a normal retinal image optic disc is the brightest part of a retinal image, and due to this reason its segmentation is simpler and easier than the segmentation of fovea as far as the PCA is concerned. To cut down the processing time further, thick binary vessel tree from the test retinal image is extracted. Optic disc is at the junction of these vessels. Only those test patches are fed to PCA which contain vessels, and this information is obtained from the binary vessel tree of the test image. This reduces the number of test patches for the processing of optic disc localization. At the same time, patches that do not contain the blood vessels are used for the localization of fovea as no vessels pass through this region. This suggests that by selecting candidate regions for the optic disc and fovea, the processing time can be cut down drastically.

8. Conclusion

This paper has described the application of principal component analysis (PCA) to the localization of the optic disc and fovea in a retinal image. Ten optic discs and ten foveae from a variety of retinal images were chosen as training images. The size of the optic disc training images was 100×100 pixels and that of the fovea was 70×70 pixels. PCA was then applied to 70 retinal images for localization of optic discs and foveae. The two cases were treated separately. The success rate was 97% in the case of optic disc segmentation and 94.3% in case of fovea segmentation. PCA was found to be a promising technique, but it is a computationally time consuming approach. Methods to make the technique faster have been identified separately for both the optic disc and the fovea. The technique cannot be recommended for real time analysis, but it can be utilised in "batch processing" for the localization of optic discs and foveae in retinal images that have been captured previously.

The PCA technique gives correlation between two objects by comparing Euclidian distance and maximum correlation occuring when the error term is the minimum. The area where the error is the minimum is shown by a square patch. It is assumed that the point lying on the cross-section of two diagonals in that square patch is the most probable centre of the optic disc and the fovea.

Acknowledgments

The authors acknowledge the financial support for this work from the Higher Education Commission (HEC) of Pakistan and from Heriot-Watt University, Edinburgh, UK.

References

[1] C. Sinthanayothin, J. F. Boyce, H. L. Cook, and T. H. Williamson, "Automated localisation of the optic disc, fovea, and retinal blood vessels from digital colour fundus images," *British Journal of Ophthalmology*, vol. 83, no. 8, pp. 902–910, 1999.

[2] H. Li and O. Chutatape, "Automatic location of optic disk in retinal images," in *Proceedings of the International Conference on Image Processing*, vol. 2, pp. 837–840, 2001.

[3] H. Li and O. Chutatape, "Automated feature extraction in color retinal images by a model based approach," *IEEE Transactions on Biomedical Engineering*, vol. 51, no. 2, pp. 246–254, 2004.

[4] H. Li and O. Chutatape, "Boundary detection of optic disk by a modified ASM method," *Pattern Recognition*, vol. 36, no. 9, pp. 2093–2104, 2003.

[5] H. Li and O. Chutatape, "A model-based approach for automated feature extraction in fundus images," in *Proceedings of the 9th IEEE International Conference on Computer Vision*, vol. 1, pp. 394–399, October 2003.

[6] N. Patton, T. M. Aslam, T. MacGillivray et al., "Retinal image analysis: concepts, applications and potential," *Progress in Retinal and Eye Research*, vol. 25, no. 1, pp. 99–127, 2006.

[7] H. Li and O. Chutatape, "Automatic detection and boundary estimation of the optic disk in retinal images using a model-based approach," *Journal of Electronic Imaging*, vol. 12, no. 1, pp. 97–105, 2003.

[8] M. Niemeijer, M. D. Abràmoff, and B. van Ginneken, "Fast detection of the optic disc and fovea in color fundus photographs," *Medical Image Analysis*, vol. 13, no. 6, pp. 859–870, 2009.

[9] J. Gutiérrez, I. Epifanio, E. de Ves, and F. J. Ferri, "An active contour model for the automatic detection of the fovea in fluorescein angiographies," in *Proceedings of the 15th International Conference on Pattern Recognition (ICPR '00)*, vol. 4, 2000.

[10] S. Sekhar, W. Al-Nuaimy, and A. K. Nandi, "Automated localisation of optic disk and fovea in retinal fundus images," in *Proceedings of the 16th European Signal Processing Conference (EUSIPCO '08)*, Lausanne, Switzerland, August 2008.

[11] F. Zana, I. Meunier, and J. C. Klein, "A region merging algorithm using mathematical morphology: application to macula detection," in *Proceedings of the 4th International Symposium on Mathematical Morphology and Its Applications to Image and Signal Processing (ISMM '98)*, pp. 423–430, Norwell, Mass, USA, 1998.

[12] M. V. Ibañez and A. Simó, "Bayesian detection of the fovea in eye fundus angiographies," *Pattern Recognition Letters*, vol. 20, no. 2, pp. 229–240, 1999.

[13] O. Chutatape, "Fundus foveal localization based on vessel model," *Proceedings of the Annual International Conference of the IEEE Engineering in Medicine and Biology Society*, vol. 1, pp. 4440–4444, 2006.

[14] K. Estabridis and R. J. P. de Figueiredo, "Automatic detection and diagnosis of diabetic retinopathy," in *Proceedings of the 14th IEEE International Conference on Image Processing (ICIP '07)*, pp. II445–II448, September 2007.

[15] K. Estabridis and R. Defigueiredo, "Fovea and vessel detection via multi-resolution parameter transform," in *Medical Imaging*, Proceedings of SPIE, February 2007.

[16] A. Pinz, S. Bernögger, P. Datlinger, and A. Kruger, "Mapping the human retina," *IEEE Transactions on Medical Imaging*, vol. 17, no. 4, pp. 606–619, 1998.

[17] S. Butt and A. A. Mudasar, "Extraction of blood vessels in retinal images using line cross-section of image data," in *Proceedings of The International Bhurban Conference on Applied Sciences and Technology*, Islamabad, Pakistan, January 2010.

Permissions

The contributors of this book come from diverse backgrounds, making this book a truly international effort. This book will bring forth new frontiers with its revolutionizing research information and detailed analysis of the nascent developments around the world.

We would like to thank all the contributing authors for lending their expertise to make the book truly unique. They have played a crucial role in the development of this book. Without their invaluable contributions this book wouldn't have been possible. They have made vital efforts to compile up to date information on the varied aspects of this subject to make this book a valuable addition to the collection of many professionals and students.

This book was conceptualized with the vision of imparting up-to-date information and advanced data in this field. To ensure the same, a matchless editorial board was set up. Every individual on the board went through rigorous rounds of assessment to prove their worth. After which they invested a large part of their time researching and compiling the most relevant data for our readers. Conferences and sessions were held from time to time between the editorial board and the contributing authors to present the data in the most comprehensible form. The editorial team has worked tirelessly to provide valuable and valid information to help people across the globe.

Every chapter published in this book has been scrutinized by our experts. Their significance has been extensively debated. The topics covered herein carry significant findings which will fuel the growth of the discipline. They may even be implemented as practical applications or may be referred to as a beginning point for another development. Chapters in this book were first published by Hindawi Publishing Corporation; hereby published with permission under the Creative Commons Attribution License or equivalent.

The editorial board has been involved in producing this book since its inception. They have spent rigorous hours researching and exploring the diverse topics which have resulted in the successful publishing of this book. They have passed on their knowledge of decades through this book. To expedite this challenging task, the publisher supported the team at every step. A small team of assistant editors was also appointed to further simplify the editing procedure and attain best results for the readers.

Our editorial team has been hand-picked from every corner of the world. Their multi-ethnicity adds dynamic inputs to the discussions which result in innovative outcomes. These outcomes are then further discussed with the researchers and contributors who give their valuable feedback and opinion regarding the same. The feedback is then collaborated with the researches and they are edited in a comprehensive manner to aid the understanding of the subject.

Apart from the editorial board, the designing team has also invested a significant amount of their time in understanding the subject and creating the most relevant covers. They scrutinized every image to scout for the most suitable representation of the subject and create an appropriate cover for the book.

The publishing team has been involved in this book since its early stages. They were actively engaged in every process, be it collecting the data, connecting with the contributors or procuring relevant information. The team has been an ardent support to the editorial, designing and production team. Their endless efforts to recruit the best for this project, has resulted in the accomplishment of this book. They are a veteran in the field of academics and their pool of knowledge is as vast as their experience in printing. Their expertise and guidance has proved useful at every step. Their uncompromising quality standards have made this book an exceptional effort. Their encouragement from time to time has been an inspiration for everyone.

The publisher and the editorial board hope that this book will prove to be a valuable piece of knowledge for researchers, students, practitioners and scholars across the globe.

List of Contributors

Salim Lahmiri and Mounir Boukadoum
Department of Computer Science, University of Quebec at Montreal, 201 President-Kennedy, Local PK-4150, Montreal, QC, Canada H2X 3Y7

Mika Ruohonen and Jarmo Alander
Faculty of Technology, University of Vaasa, P.O. Box 700, 65101 Vaasa, Finland

Katri Palo
Dental Services of the City of Vaasa, Social and Health Administration, P.O. Box 241, 65101 Vaasa, Finland

J. Pan and J. Vossoughi
Fischell Department of Bioengineering, University of Maryland, College Park, MD 20742, USA
Engineering and Scientific Research Associates, Olney, MD 20832, USA

A. Johnson
Engineering and Scientific Research Associates, Olney, MD 20832, USA

A. Saltos and D. Smith
School of Medicine, University of Maryland, Baltimore, MD 21201, USA

Tushar Kanti Bera and J. Nagaraju
Department of Instrumentation and Applied Physics, Indian Institute of Science Bangalore, Bangalore, Karnataka 560012, India

Kyrin Liong and Heow Pueh Lee
Department of Mechanical Engineering, National University of Singapore, Singapore 117576

Shu Jin Lee
Division of Plastic, Reconstructive and Aesthetic Surgery, National University Hospital, Singapore 119074

Werner Winter and Daniel Klein
Department of Mechanical Engineering, University of Erlangen-Nuremberg, Egerlandstraße 5, 91058 Erlangen, Germany

Matthias Karl
Department of Prosthodontics, University of Erlangen-Nuremberg, Glueckstraße 11, 91054 Erlangen, Germany

K. Satyanarayana J. Sravan and M. Malini
Department of Biomedical Engineering, Osmania University, Hyderabad 500 007, India

A. D. Sarma
Research and Training Unit for Navigational Electronics, Osmania University, Hyderabad 500 007, India

G. Venkateswarlu
Department of ECE, Vasavi College of Engineering, Hyderabad 500 031, India

Zhenyu Zhang and Nejat Olgac
Department of Mechanical Engineering, ALARM Lab, University of Connecticut, Storrs, CT 06269, USA

Benjamin Leporq, Sorina Camarasu-Pop, Eduardo E. Davila-Serrano and Olivier Beuf
Universit´e de Lyon, CREATIS, CNRS UMR 5220, InsermU1044, INSA-Lyon, Universit´e Lyon 1, 69622 Villeurbanne Cedex, France

Frank Pilleul
Université de Lyon, CREATIS, CNRS UMR 5220, InsermU1044, INSA-Lyon, Université Lyon 1, 69622 Villeurbanne Cedex, France
Departement d'imagerie Digestive, Hospices Civils de Lyon, CHU Edouard Herriot, 69008 Lyon, France

Adriana Cajiao, Ezra Kwok and Bhushan Gopaluni
University of British Columbia, Chemical and Biological Engineering, 2360 East Mall, Vancouver, BC, Canada V6T 1Z3

Jayachandran N. Kizhakkedathu
University of British Columbia, Centre for Blood Research, Department of Pathology and Laboratory Medicine and Department of Chemistry, 2350 Health Sciences Mall, Vancouver, BC, Canada V6T 1Z3

Jameel Shaik
Institute for Micromanufacturing, Louisiana Tech University, Ruston, LA 71272, USA
Biomedical Engineering Program, Louisiana Tech University, Ruston, LA 71272, USA
School of Bio Sciences & Technology, VIT University, Vellore 632014, India

Javeed Shaikh Mohammed
Institute for Micromanufacturing, Louisiana Tech University, Ruston, LA 71272, USA
Biomedical Technology Department, King Saud University, Riyadh 11433, Saudi Arabia

Michael J. McShane
Institute for Micromanufacturing, Louisiana Tech University, Ruston, LA 71272, USA
Biomedical Engineering Program, Louisiana Tech University, Ruston, LA 71272, USA
Biomedical Engineering Program, Texas A&M University, College Station, TX 77843, USA

David K. Mills
Institute for Micromanufacturing, Louisiana Tech University, Ruston, LA 71272, USA
Biomedical Engineering Program, Louisiana Tech University, Ruston, LA 71272, USA
School of Biological Sciences, Louisiana Tech University, Ruston, LA 71272, USA

P. J. Ogrodnik, C. I. Moorcroft and P. Wardle
Staffordshire University, Stafford ST18 0AD, UK

Xiaoyong Zhang, Noriyasu Homma and Makoto Yoshizawa
Research Division on Advanced Information Technology, Cyberscience Center, Tohoku University, 6-6-05 Aoba, Aramaki, Aoba-ku, Sendai 980-8579, Japan

Shotaro Goto, Makoto Abe and Norihiro Sugita
Graduate School of Engineering, Tohoku University, 6-6-05 Aoba, Aramaki, Aoba-ku, Sendai 980-8579, Japan

Yosuke Kawasumi and Tadashi Ishibashi
Tohoku University Graduate School of Medicine, Tohoku University, 2-1 Seiryo-mashi, Aoba-ku, Sendai 980-8575, Japan

Bailin Zhang, Juan Manuel Tamez-Vela, Steven Solis, Gilbert Bustamante, Ralph Peterson, Shafiqur Rahman, Andres Morales, Liang Tang and Jing Yong Ye
University of Texas at San Antonio, One UTSA Circle, San Antonio, TX 78249, USA

Phuc Van Pham, Lam Tan Khuat and Ngoc Kim Phan
Laboratory of Stem Cell Research and Application, University of Science, Vietnam National University, Ho Chi Minh City, Vietnam

Khanh Hong-Thien Bui
University of Medical Center, Ho Chi Minh University of Medicine and Pharmacy, Ho Chi Minh City, Vietnam

Dat Quoc Ngo
Department of Pathology, University of Medicine and Pharmacy, Ho Chi Minh City, Vietnam

Marjan Bahraminasab and Mohd Roshdi Hassan
Department of Mechanical and Manufacturing Engineering, Universiti Putra Malaysia, 43400 Selangor, Malaysia

Ali Jahan
Faculty of Engineering, Semnan Branch, Islamic Azad University, Semnan, Iran

Barkawi Sahari
Department of Mechanical and Manufacturing Engineering, Universiti Putra Malaysia, 43400 Selangor, Malaysia
Institute of Advanced Technology, ITMA, Universiti Putra Malaysia, Selangor, Malaysia

Manohar Arumugam
Department of Orthopedic Surgery, Faculty of Medicine & Health Science, Universiti Putra Malaysia, Selangor, Malaysia

Mahmoud Shamsborhan
Department of Mechanical Engineering, K.N. Toosi University of Technology, Tehran, Iran

Joel A. Cort
Department of Kinesiology, University of Windsor, 401 Sunset Avenue, Windsor, ON, Canada N9B 3P4

James P. Dickey
School of Kinesiology, The University of Western Ontario, 1151 Richmond Street, London, ON, Canada N6A 3K7

Jim R. Potvin
Department of Kinesiology, McMaster University, 1280 Main Street West, Hamilton, ON, Canada L8S 4L8

Fei Yang
Washington University School of Medicine, Saint Louis, MO 63110, USA

Jie Zhang
Division of Radiological Medical Physics, University of Kentucky, Lexington, KY 40536, USA

Robert Patterson
Institute of Engineering in Medicine, University of Minnesota, Minneapolis, MN 55455, USA

Asloob Ahmad Mudassar
Department of Physics and Applied Mathematics, Pakistan Institute of Engineering and Applied Sciences, Nilore, Islamabad 45650, Pakistan

Saira Butt
Isotope Application Division, Pakistan Institute of Nuclear Science and Technology, Nilore, Islamabad 45650, Pakistan

www.ingramcontent.com/pod-product-compliance
Lightning Source LLC
Chambersburg PA
CBHW050457200326
41458CB00014B/5219

9 781632 402783